Lecture Notes in Mathematics 1473

Editors:
A. Dold, Heidelberg
B. Eckmann, Zürich
F. Takens, Groningen

Yoshiyuki Hino Satoru Murakami
Toshiki Naito

Functional Differential Equations with Infinite Delay

Springer-Verlag
Berlin Heidelberg New York
London Paris Tokyo
Hong Kong Barcelona
Budapest

Authors

Yoshiyuki Hino
Department of Mathematics
Chiba University
Yayoicho, Chiba 260, Japan

Satoru Murakami
Department of Applied Mathematics
Okayama University of Science
Ridaicho, Okayama 700, Japan

Toshiki Naito
Department of Mathematics
The University of Electro-Communications
Chofu, Tokyo 182, Japan

Mathematics Subject Classification (1980): 34Kxx, 34C25, 34C27, 34D05, 34D10, 34D20, 45J05, 45M10, 47D05, 26A42

ISBN 3-540-54084-9 Springer-Verlag Berlin Heidelberg New York
ISBN 0-387-54084-9 Springer-Verlag New York Berlin Heidelberg

Printed in Germany

Printing and binding: Druckhaus Beltz, Hemsbach/Bergstr.
2146/3140-543210 - Printed on acid-free paper

Dedicated to Professor Taro Yoshizawa on his 70th birthday

PREFACE

The aim of this text is to deal with functional differential equations with infinite delay on an abstract phase space characterized by several axioms which are satisfied by different kinds of function spaces. The standard spaces which we have in mind are the one of continuous functions on $(-\infty,0]$ that are endowed with some restriction on their asymptotic behavior at $-\infty$, and the one of measurable functions on $(-\infty,0]$ that are integrable with respect to some Borel measure equipped with mild conditions. These spaces are carefully adopted as phase spaces for equations with infinite delay so as to solve each problem which we encounter in applications, nevertheless many fundamental properties of these equations hold good independently of the choice of phase spaces. The axiomatic approach is not only advantageous to summarize these properties, but also supply fruitful ideas and methods to investigate the mathematical structure which reflects the effect of infinite delay. Hence, we now intend to develop a unified theory of this field in terms of functional analysis and dynamical systems. For the sake of clear understanding of our ground, many elementary facts and proofs are given as completely as possible. But, few attempts had been made to give examples of equations in applications, and only a limited number of references are presented at the end of this text.

This text consists of nine chapters. Chapter 1 contains the formulation of axioms of the phase space together with many examples. Chapter 2 is a presentation of basic theory of existence, uniqueness, continuous dependence, etc. of solutions. After a brief introduction to Stieltjes integrals in Chapter 3, the theory of linear equations is developed from Chapter 4 through Chapter 6. Chapter 7 is an introduction to fading memory spaces. In Chapter 8, the stability problem in functional differential equations on a fading memory space is studied in connection with limiting equations. In succession, the existence of periodic and almost periodic solutions of functional differential equations is discussed in Chapter 9.

The authors are very grateful to Professor Taro Yoshizawa for his various instructions in their studying of differential equations for a long time, and express their hearty thanks to Professor Jack K.

Hale and Professor Junji Kato for their helpful suggestions and comments on this field. They also thank to Professor Shigeru Haruki for his eager advice to prepare this text.

Yoshiyuki Hino
Satoru Murakami
Toshiki Naito

CONTENTS

CHAPTER 1

PHASE SPACES

1.1. Introduction.

Functional differential equations of retarded type are the differential equations which contains retarded arguments of the unknown function. We formulate these equations as

$$\dot{x}(t) = f(t, x_t), \tag{1.1}$$

where $\dot{x}(t)$ is the derivative of $x(t)$. The symbol x_t stands for the retarded arguments of the unknown function x at time t.

How to describe the retarded arguments generally? Several formulations are possible; each one has advantages and disadvantages. In this note, having in mind the application of the theory of dynamical systems, we take the formulation that, for every t, x_t stands for a function in a fixed function space, called the phase space, which is characterized by some axioms. The crucial assumption is that the motion of x_t in the phase space is continuous for t. Though many results are derived under our axioms for the phase space as will be shown, it is still important to consider more general formulations to cover the whole branches of delay equations in the axiomatic manner. We add the recent work by Kato [5] in this direction as an appendix at the end of this note.

Let $r(t)$ be the time delay such that all the retarded arguments of x at time t are on the interval $[t-r(t), t]$. Put $r = \sup r(t)$, $0 \le r \le \infty$. Then, for every t, such a situation is reflected in the term x_t if we define x_t to be a function on $[-r, 0]$ as $x_t(\theta) = x(t+\theta)$, $-r \le \theta \le 0$. If r is finite, the equation is called the one with finite delay; otherwise, the one with infinite delay. The latter is our main subject, so x_t is defined as

$$x_t(\theta) = x(t+\theta), \quad -\infty < \theta \le 0, \tag{1.2}$$

and called the t-segment of x, the t-section of x or the history of x up to t.

Consider the case of finite delay. If x is a solution satisfying Equation (1.1) on $[\sigma, \sigma+a)$, for some $a > r$, then x is continuous on this interval. This implies that for $t \geq \sigma+r$, x_t belongs to the space of continuous functions on $[-r,0]$. If this space is endowed with the topology induced by the supremum norm, the motion of x_t in this space is continuous at least for $t \geq \sigma+r$. Thus, we usually take this space as the phase space for equations with finite delay.

In the case of infinite delay, contrarily, we have several possibilities for the choice of the phase spaces. For example, consider the delay differential equation

$$\dot{x}(t) = a_0(t)x(t) + \sum_{n=1}^{\infty} \frac{a_n(t)}{n!} [x(t-n)]^n,$$

where $a_n(t)$ are continuous functions on $\mathbb{R} = (-\infty,\infty)$ for $n = 1, 2,\cdots$. This equation is written as (1.1) formally if we set

$$f(t,\varphi) = a_0(t)\varphi(0) + \sum_{n=1}^{\infty} \frac{a_n(t)}{n!}[\varphi(-n)]^n$$

for a function φ on $(-\infty,0]$. Suppose $a_n(t)$ are bounded uniformly for t and n. Then, $f(t,\varphi)$ is well defined for any function φ in the set BC, the family of continuous functions φ on $(-\infty,0]$ such that $|\varphi| := \sup\{|\varphi(\theta)| : -\infty < \theta \leq 0\}$ is finite. However, some kind of unbounded functions may be substituted. For a positive continuous function g on $(-\infty,0]$, let C_g be the space of continuous functions φ on $(-\infty,0]$ such that

$$|\varphi|_g := \sup\{|\varphi(\theta)|/g(\theta) : -\infty < \theta \leq 0\}$$

is finite. Take a function g whose graph is a polygon combining the points $(0,1)$, $(-n,[n!]^{1/n})$ for $n = 1,2,\cdots$ successively. Then, it is easy to see that $f(t,\varphi)$ is also well defined for t in $\mathbb{R}$ and for φ in C_g such that $\lim\sup|\varphi(\theta)|/g(\theta) < 1$ as $\theta \to -\infty$. Thus we have another candidate C_g for the phase space.

The requirement that the motion of x_t is continuous in the phase space is satisfied by putting some restriction on these spaces. For example, it suffices that BC and C_g are respectively replaced by BC^0 and C_g^0, where $BC^0 = \{\varphi \in BC : \lim_{\theta\to-\infty} \varphi(\theta) = 0\}$ and $C_g^0 = \{\varphi \in C_g : \lim_{\theta\to-\infty} \varphi(\theta)/g(\theta) = 0\}$. Note that C_g^0 contains every bounded continuous functions on $(-\infty,0]$ provided $g(\theta) \to \infty$ as $\theta \to -\infty$.

A Volterra integral differential equation

$$\dot{x}(t) = \int_0^t a(s,x(s))ds, \quad t \geq 0,$$

is also regarded as an equation with infinite delay. For this equation, we have an unbounded function $r(t) := -t$, $t \geq 0$. If we set formally

$$f(t,\varphi) = \int_{-t}^{0} a(t+\theta,\varphi(\theta))d\theta$$

for a function φ on $(-\infty,0]$, the above equation is written as (1.1). The choice of the proper phase space depends on the function $a(s,x)$. If $a(s,x)$ is a continuous function, the function f is well defined on $[0,\infty) \times BC^0$ and on $[0,\infty) \times C_g^0$ for any positive continuous function g. If $a(s,x) = b(s)x^2$ for a locally bounded and measurable function b on $[0,\infty)$, we can choose the space L^2, the set of functions on $(-\infty,0]$ whose square are integrable on $(-\infty,0]$.

It is important to deal with each equation on the phase space taken properly for the problem. However, it is also true that there are many facts which hold independently of each concrete phase space. These results will be expected to be summarized from the discussion of the equation on an abstract phase space defined by some axioms induced from many examples for the phase space. We will develop the theory from such a point of view.

The first axiomatic approach for equations with infinite delay was given by Coleman and Mizel [1-3]. The typical norm considered by them is $|\varphi| = \{|\varphi(0)|^p + \int_{-\infty}^{0} |\varphi(s)|^p k(s)ds\}^{1/p}$. It is easily seen that the above norm does not assure the continuity of the right hand side even in a simple differential difference equation. To clarify this situation, Hale [2] introduced some other axioms, and contributions to these axioms have been brought by Hino [1-4] and Naito [1-3]. Furthermore, Hale and Kato [1], and Schumacher [1,2] gave more systematic development of this subject, independently, and related papers have been published by several authors (see Hagemann and Naito [1], Hale [3,4], Hino [5-9], Hino and Murakami [1], Hino and Yoshizawa [1] Kaminogo [1], Kappel and Schappacher [1], Kato [2-5] Murakami [1,4], Murakami and Naito [1], Naito [4-6], Sawano [1]). In 1980, a survey paper on functional differential equations with infinite delay was published by Corduneanu and Lakshmikantham [1]. Included in their paper is a discussion of numerous topics and a list of 289 references.

Recently, some attention have been focused on the space C_g as a phase space for functional differential equations with infinite delay; especially, Volterra integro-differential equations. The space C_g frequently arises in a natural way for many types of integro-differential equations, and setting up integro-differential equations as functional differential equations on the space has the advantage in several points of view rather than doing as those on the space BC (as usual). In fact, Arino, Burton and Haddock [1] were the first who introduced the space C_g, and they have established the existence theorem of periodic solutions of some integro-differential equations by applying Horn's fixed point theorem (cf. Burton [3]). For more details on this topic, we refer the reader to Burton's book [2, Chapter 4] (also, see Haddock [1]). On the other topics, e.g., convergence, stabilities, invariance principle, comparison theorems, compactness of positive orbits etc., there are some works which show the effectiveness of the space C_g in the analysis of functional differential equations with infinite delay. For such topics, see Atkinson and Haddock [1], Haddock [1], Haddock and Hornor [1], Haddock, Krisztin and Terjéki [1,2], and Murakami and Naito [1].

<u>1.2. Axioms of the phase space.</u> The symbol $\mathbb{R}$ denotes the set of real numbers, $\mathbb{C}$ the set of complex numbers and $\mathbb{E}$ a real or complex Banach space of finite dimension. A norm or a seminorm on a linear space X is denoted by $|\cdot|_X$; the index X will be often omitted if the meaning is clear. On the other hand, for a function f mapping a topological space S into $\mathbb{E}$, and for a subset K of S, we sometimes set

$$|f|_K = \sup\{|f(x)| : x \in K\}.$$

If $|f|_K$ is finite for every compact set K of S, f is said to be locally bounded on S. If a sequence of functions f^n mapping S into $\mathbb{E}$ converges to a function f uniformly on every compact set in S, we say that $\{f^n\}$ converges to f compactly on S. For a function $x : (-\infty, a) \to \mathbb{E}$ and for $t < a$, we define a function $x_t : (-\infty, 0] \to \mathbb{E}$ by Relation (1.2).

The phase space $\mathcal{B}$ for equations with infinite delay is a linear space, with a seminorm $|\cdot|_{\mathcal{B}}$, consisting of functions mapping $(-\infty, 0]$ into $\mathbb{E}$. The fundamental axioms assumed on $\mathcal{B}$ are the followings.

(A) *If* x *is a function mapping* $(-\infty,\sigma+a)$ *into* $\mathbb{E}$, $a > 0$, *such that* x *is in* $\mathcal{B}$ *and* x *is continuous on* $[\sigma,\sigma+a)$, *then for every* t *in* $[\sigma,\sigma+a)$ *the following conditions hold;*

(i) x_t *is in* $\mathcal{B}$,

(ii) $|x(t)|_E \le H|x_t|_{\mathcal{B}}$,

(iii) $|x_t|_{\mathcal{B}} \le K(t-\sigma)\sup\{|x(s)|_E : \sigma\le s\le t\} + M(t-\sigma)|x_\sigma|_{\mathcal{B}}$,

where H *is a constant,* K, $M : [0,\infty) \to [0,\infty)$, K *is continuous*, M *is locally bounded, and they are independent of* x.

(A1) *For the function* x *in (A),* x_t *is a* $\mathcal{B}$*-valued continuous function for* t *in* $[\sigma,\sigma+a)$.

These two axioms are always assumed on $\mathcal{B}$ in this note. The function x arising in (A) is sometimes said to be admissible (with respect to $\mathcal{B}$ on the interval $[\sigma,\sigma+A)$). The space C_g^0 introduced in Section 1.1 satisfies (A) and (A1) if the function g is nonincreasing. The space BC satisfies (A) but not (A1). Examples of the space $\mathcal{B}$ satisfying (A) and (A1) will be shown in Section 1.3.

The elements in $\mathcal{B}$ will be usually denoted by φ, ψ,$\cdots$, and called histories. Since $|\cdot|_{\mathcal{B}}$ is a seminorm, $|\varphi - \psi|_{\mathcal{B}} = 0$ does not necessarily mean that $\varphi(\theta) = \psi(\theta)$ for all $\theta \le 0$. But from (ii) in (A) (in short, (A-ii)), it is sure that $\varphi(0) = \psi(0)$ if $|\varphi - \psi|_{\mathcal{B}} = 0$. Indeed, (A-ii) is rewritten as

$$\text{(A) (ii)}' \qquad |\varphi(0)|_E \le H|\varphi|_{\mathcal{B}} \quad \textit{for} \quad \varphi \in \mathcal{B}.$$

The value $\varphi(0)$ is called the present state of φ.

First of all, we observe the following fact induced from Axiom (A). Let C_{00} be the set of continuous functions on $(-\infty,0]$ into $\mathbb{E}$ with compact support, and $\operatorname{supp}\varphi$ the support of φ in C_{00}.

Proposition 2.1. Any function φ in C_{00} belongs to $\mathcal{B}$. If $\operatorname{supp}\varphi$ is contained in $[-r,-s]$, $0 \le s \le r < \infty$, then

$$|\varphi|_{\mathcal{B}} \le \min\{K(r),\ K(r-s)M(s)\}|\varphi|_{(-\infty,0]}.$$

Proof. If φ is as in the above, then the $-r$-segment φ_{-r} belongs to the linear space $\mathcal{B}$ being the zero vector of $\mathcal{B}$. Since $\varphi(t)$ is continuous on $[-r,0]$, Axiom (A) implies that the 0-segment of φ, φ itself, belongs to $\mathcal{B}$. Applying (A-iii) for this function, we have $|\varphi|_{\mathcal{B}} \le K(r)|\varphi|_{[-r,0]}$, since $\varphi_{-r} = 0$. However, in the similar manner, we have another estimate that $|\varphi|_{\mathcal{B}} \le M(s)|\varphi_{-s}|_{\mathcal{B}}$,

$|\varphi_{-s}|_{\mathscr{B}} \le K(r-s)|\varphi|_{[-r,-s]}$. Summarizing these results, we have the inequality in the proposition.

For a function x in Axiom (A), define functions y and z by

$$y(t) = \begin{cases} x(t) & t < \sigma \\ x(\sigma) & \sigma \le t < \sigma+a. \end{cases}$$

and

$$z(t) = \begin{cases} 0 & t < \sigma \\ x(t) - x(\sigma) & \sigma \le t \le \sigma+a. \end{cases}$$

Then, x is decomposed as $x = y + z$, which is called the fundamental decomposition of x (at $t = \sigma$), and the function y is called the static continuation of x (at $t = \sigma$). This decomposition yields the one of the t-segment: $x_t = y_t + z_t$. Axiom (A) says that both segments y_t and z_t belong to $\mathscr{B}$ for t in $[\sigma,\sigma+a)$. The segment z_t is in C_{00}; the segment y_t is represented in terms of a semigroup as follows. For each φ in $\mathscr{B}$, define a static continuation $s(\varphi)$ of φ as $s(\varphi)(t) = \varphi(t)$ for $t \le 0$ and $s(\varphi)(t) = \varphi(0)$ for $t > 0$. Denote by $s_t(\varphi)$ the t-segment of $s(\varphi)$, which is in $\mathscr{B}$ for $t \ge 0$ from Axiom (A). Thus, we can define operators $S(t)$, $t \ge 0$, on $\mathscr{B}$ by

$$[S(t)\varphi](\theta) = [s_t(\varphi)](\theta) = \begin{cases} \varphi(0) & -t \le \theta \le 0 \\ \varphi(t+\theta) & \theta < -t \end{cases}$$

for φ in $\mathscr{B}$. They make a family of linear operators on $\mathscr{B}$ having the semigroup property : $S(t+s) = S(t)S(s)$ for $t, s \ge 0$. Immediately, the following result holds from Axiom (A).

<u>Proposition 2.2</u>. For each fixed φ in $\mathscr{B}$, $S(t)\varphi$ is continuous in $t \ge 0$. For each $t \ge 0$, $S(t)$ is a continuous linear operator on $\mathscr{B}$, whose operator norm satisfies

$$|S(t)| \le HK(t) + M(t).$$

Therefore, this class of operators, denoted by $S(t)$ simply, is a strongly continuous semigroup of bounded linear operators on $\mathscr{B}$. Since $s(\varphi)(t)$ satisfies the trivial equation $\dot{x}(t) = 0$ for $t > 0$,

$S(t)$ is sometimes called the solution semigroup of the trivial equation. Using this semigroup, the above segment y_t is written as $y_t = S(t-\sigma)y_\sigma = S(t-\sigma)x_\sigma$ for $t \geq \sigma$; consequently, we have

$$x_t = S(t-\sigma)x_\sigma + z_t \quad \text{for} \quad t \geq \sigma.$$

If the initial function x_σ is given in $\mathscr{B}$, we have an equation for z from Equation (1.1) by the transform $x = y + z$. This equation for z is regarded as an equation with finite delay as far as we consider local problems. We can deal with the fundamental theorems in this manner under Axioms (A) and (A1).

The theory of linear equations will be developed along the same line, too. However, in Chapters 5 and 6 we need more axioms to know the detailed structure of solutions. One axiom is concerned with the Bochner integral on $\mathscr{B}$, and the other axiom is concerned with the realization of some elements in $\mathscr{B}$.

(B) <u>The space $\mathscr{B}$ is complete.</u>

(C1) <u>If $\{\varphi^n\}$ is a Cauchy sequence in $\mathscr{B}$ with respect to the seminorm, and if $\{\varphi^n(\theta)\}$ converges to a function $\varphi(\theta)$ compactly on $(-\infty,0]$, then φ is in $\mathscr{B}$ and $|\varphi^n - \varphi|_{\mathscr{B}} \to 0$ as $n \to \infty$.</u>

For φ in $\mathscr{B}$, the symbol $\hat{\varphi}$ denotes the equivalence class $\{\psi : |\psi - \varphi|_{\mathscr{B}} = 0\}$, and $\hat{\mathscr{B}}$ denotes the quotient space $\{\hat{\varphi} : \varphi \in \mathscr{B}\}$, which becomes a normed linear space with the norm $|\hat{\varphi}|_{\hat{\mathscr{B}}} = |\varphi|_{\mathscr{B}}$. Axiom (B) is equivalent to saying that $\hat{\mathscr{B}}$ is a Banach space. The theory of linear equations will be mainly developed on this Banach space.

In Chapter 7, we introduce the concept of fading memory to discuss the stability of solutions and the existence of periodic solutions and almost periodic solutions. Roughly speaking, this means that $|x_t|_{\mathscr{B}} \to 0$ as $t \to \infty$ if $|x(t)|_E \to 0$ as $t \to \infty$. This property is related to the boundedness of $K(t)$ and the decay of $M(t)$ as $t \to \infty$. However, instead of assuming these properties on K and M, we will start from the following axiom which is similar to (C1). We say that a sequence of functions φ^n on $(-\infty,0]$ is uniformly bounded if $\sup\{|\varphi^n|_{(-\infty,0]} : n = 1,2,\cdots\}$ is finite.

(C2) <u>If a uniformly bounded sequence $\{\varphi^n(\theta)\}$ in C_{00} converges to a function $\varphi(\theta)$ compactly on $(-\infty,0]$, then φ is in $\mathscr{B}$ and $|\varphi^n - \varphi|_{\mathscr{B}} \to 0$ as $n \to \infty$.</u>

This axiom immediately implies that bounded continuous functions are in $\mathscr{B}$, $K(t)$ can be assumed to be bounded for $t \geq 0$ and that $|x_t|_{\mathscr{B}} \to 0$ as $t \to \infty$ as long as $x(t)$ is continuous and bounded on $\mathbb{R}$ and $|x(t)|_E \to 0$ as $t \to \infty$ (see Chapter 7). More detailed definition of fading memory spaces will be given in terms of the operators $S_0(t)$, the restriction of $S(t)$ on the closed subspace

$$\mathscr{B}_0 = \{\varphi \in \mathscr{B} : \varphi(0) = 0\}.$$

Clearly, the family $S_0(t)$, $t \geq 0$, is also a strongly continuous semigroup of bounded linear operators on $\mathscr{B}_0$; it is given explicitly as

$$[S_0(t)\varphi](\theta) = \begin{cases} 0 & -t \leq \theta \leq 0 \\ \varphi(t+\theta) & \theta < -t \end{cases}$$

for φ in $\mathscr{B}_0$.

Finally, one more axiom on $\mathscr{B}$ will be assumed in Chapters 8 and 9 to discuss the limiting equations. Main interest is in the equation defined by an almost periodic function ; that is, the function $f(t,\varphi)$ in Equation (1.1) is almost periodic in t uniformly for φ in $\mathscr{B}$. Let $T(f)$ be the family of functions f^τ, τ in $\mathbb{R}$, where f^τ is defined as

$$f^\tau(t,\varphi) = f(\tau+t,\varphi).$$

The limiting equation is an equation which is defined by a limit function of a sequence $\{f^{\tau(k)}\}$ in $T(f)$ for some sequence $\{\tau(k)\}$ with $\lim \tau(k) = \infty$ as $k \to \infty$. We assume the following axiom to have the normality of $T(f)$ with respect to the compact open topology:

(D) $\mathscr{B}$ <u>is a separable space.</u>

<u>1.3. Examples</u>. We introduce some standard examples for the phase space $\mathscr{B}$, and examine which axioms they satisfy. To do so, it suffices to consider the case that $\sigma = 0$ in Axiom (A) and (A-1) (cf. Kato [5] in Appendix 1). We say that a function $x : (-\infty,\infty) \to \mathbb{E}$ is admissible with respect to $\mathscr{B}$ if x_0 is in $\mathscr{B}$ and x is continuous on $[0,\infty)$. Denote by $C(S,E)$ the space of continuous functions mapping a topological space S into $\mathbb{E}$.

(a) The spaces BC, BU, C^{∞} and C^{0}. Define

$$BC = \{\varphi \in C((-\infty,0],E) : |\varphi(\theta)| \text{ is bounded on } (-\infty,0]\}$$
$$BU = \{\varphi \in BC : \varphi \text{ is uniformly continuous on } (-\infty,0]\}$$
$$C^{\infty} = \{\varphi \in BC : \lim_{\theta\to-\infty} \varphi(\theta) \text{ exists in } E\}$$
$$C^{0} = \{\varphi \in BC : \lim_{\theta\to-\infty} \varphi(\theta) = 0\},$$

and set

$$|\varphi|_{\mathcal{B}} = \sup\{|\varphi(\theta)| : -\infty < \theta \le 0\}$$

for $\mathcal{B}$ = BC, BU, C^{∞} and C^{0}.

Theorem 3.1. We have the following table:

	A	A1	B	C1	C2	D
BC	o	×	o	o	×	×
BU	o	o	o	o	×	×
C^{∞}	o	o	o	o	×	o
C^{0}	o	o	o	o	×	o

For example, read that BC satisfies Axiom (A), and that BC does not satisfy Axiom (A1), etc. For these spaces, we can take H, K(t) and M(t) in Axiom (A) as

$$H = 1,\ K(t) = M(t) = 1 \quad \text{for} \quad t \ge 0.$$

Proof. We check only Axiom (D), the separability of these spaces. The other part is almost obvious. The space C^{∞} is separable since it is isometric to the separable Banach space $C([-1,0],E)$ with the supremun norm. To see that BU is not separable, consider the set $2^{\mathbb{N}^-}$, the collection of the all subsets of the set $\mathbb{N}^- = \{-1,-2,\cdots\}$. For n in $\mathbb{N}^-$, we define a function χ_n on $(-\infty,0]$ by

$$\chi_n(s) = \begin{cases} 2(s-n+1/2) & n-1/2 \le s \le n \\ -2(s-n-1/2) & n \le s \le n+1/2 \\ 0 & \text{otherwise.} \end{cases}$$

and for A in $2^{\mathbb{N}^-}$, we set

$$\chi_A(s) = \sum_{n\in A} \chi_n(s).$$

Then, χ_A is a function in BU for every A in $2^{\mathbb{N}^-}$, and the correspondence of A with χ_A is one to one. Suppose that BU has a dense subset $\{p_m\}$, $m = 1,2,\cdots$. Then, for every A in $2^{\mathbb{N}^-}$, there exists at least one m such that $|p_m - \chi_A|_{BU} \le 1/4$. Take such an m, and set $m = m(A)$. It is clear that $m(A) \ne m(B)$ for $A \ne B$ in $2^{\mathbb{N}^-}$. This is a contradiction since $\{\chi_A : A \in 2^{\mathbb{N}^-}\}$ is an uncountable set.

(b) The space C_g and its subspaces. Let g be a positive continuous function on $(-\infty,0]$. We define

$$C_g = \{\varphi \in C((-\infty,0],\mathbb{E}) : |\varphi(\theta)|/g(\theta) \text{ is bounded on } (-\infty,0]\},$$
$$UC_g = \{\varphi \in C_g : \varphi(\theta)/g(\theta) \text{ is uniformly continuous on } (-\infty,0]\},$$
$$C_g^0 = \{\varphi \in C_g : \lim_{\theta\to-\infty} \varphi(\theta)/g(\theta) = 0\},$$

and set

$$|\varphi|_g = \sup\{|\varphi(\theta)|/g(\theta) : -\infty < \theta \le 0\}$$

for φ in these spaces. If we set $\psi = \varphi/g$ for φ in C_g, then ψ is in BC: C_g is isometric to BC. Also, UC_g and C_g^0 are isometric to BU and C^0, respectively. Consider the following conditions on g.

(g-1) $G(t) := \sup\{g(\theta+t)/g(\theta) : -\infty < \theta \le -t\}$ is locally bounded for $t \ge 0$.

(g-2) $g(\theta) \to \infty$ as $\theta \to -\infty$.

Then, we have the following table.

Theorem 3.2.

	A	B	C1	C2	D
C_g	(g-1)	o	o	(g-2)	×
C_g^0	(g-1)	o	o	(g-2)	o

For example, read that C_g satisfies Axiom (A) if (g-1) holds and that C_g satisfies Axiom (C2) if (g-2) holds, etc.

Proof. Obvious.

As already seen in Theorem 3.1, there exists some spaces which satisfy Axiom (C1) but not Axiom (C2). In turn, we shall give an example of the space $\mathcal{B}$ which satisfy Axiom (C2) but not Axiom (C1). We define

$$C_g^\infty = \{\varphi : (-\infty, 0] \to \mathbb{E} : \lim_{\theta \to -\infty} \frac{\varphi(\theta)}{g(\theta)} = \tilde{\varphi}(-\infty) \text{ exists}\}$$

and set

$$|\varphi|_{g;\infty} = |\varphi(0)| + |\tilde{\varphi}(-\infty)|$$

for φ in C_g^∞. Clearly, $|\cdot|_{g;\infty}$ is a seminorm of C_g^∞.

Theorem 3.3. Let g be a positive continuous function on $(-\infty, 0]$ satisfying the condition:

(g-3) $\lim_{s \to -\infty} \frac{g(t+s)}{g(s)} =: a_t$ exists for each t, and a_t is continuous in t.

Then we have the following table:

	A	B	C1	C2	D
C_g^∞	o	o	×	(g-2)	o

Proof. We shall only check that Axiom (C1) does not hold. The other parts are almost obvious. Now, for each positive integer n, we define a continuous function ψ^n on $(-\infty, 0]$ by

$$\psi^n(s) = \begin{cases} 1 & \text{if } s \le -n \\ 0 & \text{if } -n+1 \le s \le 0. \\ \text{linear} & \text{if } -n < s < -n+1 \end{cases}$$

and set

$$\varphi^n(s) = \psi^n(s) g(s), \quad s \le 0.$$

Clearly, $\varphi^n \in C_g^\infty$, $\{\varphi^n\}$ is a Cauchy sequence with respect to the seminorm $|\cdot|_{g;\infty}$ and $\{\varphi^n(\theta)\}$ converges to the zero compactly on $(-\infty,0]$. On the other hand, we have $|\varphi^n - 0|_{g;\infty} = |\varphi^n(0)| + |(\varphi^n)^\sim(-\infty)| = 1$ for all n. Thus, Axiom (C1) does not hold.

To discuss Axiom (A1), we use functions which look like subadditive functions.

<u>Lemma 3.4</u>. Let f be a nonnegative function on $(-\infty,a]$, $a \in \mathbb{R}$. Suppose that there exist positive constants r and A such that

$$(3.1) \qquad f(s+\theta) \le Af(\theta) \quad \text{for} \quad -r \le s \le 0 \quad \text{and for} \quad \theta \le a.$$

Then, we have

$$(3.2) \qquad f(s+\theta) \le ce^{\alpha s}f(\theta) \quad \text{for} \quad s \le 0 \quad \text{and} \quad \theta \le a,$$

where $c = \max(1,A)$ and $\alpha = -r^{-1}\log A$.

<u>Proof</u>. If $-nr < s \le -(n-1)r$ for some $n = 1,2,\cdots$, it follows from (3.1) that $f(s+\theta) \le Af(r+s+\theta) \le \cdots \le A^{n-1}f((n-1)r+s+\theta) \le A^n f(\theta)$ for $\theta \le a$, which implies (3.2).

<u>Lemma 3.5</u>. Suppose that g is a positive function on $(-\infty,0]$. Then,

(g-4) $\quad \log g(\theta)$ is uniformly continuous on $(-\infty,0]$

if and only if, for every $\varepsilon > 0$, there exists a $\gamma(\varepsilon) < 0$ such that

$$(3.3) \qquad e^{-\varepsilon}e^{-\gamma(\varepsilon)s} \le g(s+\theta)/g(\theta) \le e^{\varepsilon}e^{\gamma(\varepsilon)s} \quad \text{for} \quad s,\ \theta \le 0.$$

<u>Proof</u>. If (g-4) holds, for every $\varepsilon > 0$ there exists a $\delta(\varepsilon) > 0$ such that $|\log g(s+\theta) - \log g(\theta)| < \varepsilon$ for $-\delta(\varepsilon) \le s \le 0$ and for $\theta \le 0$. This is written as $g(s+\theta)/g(\theta)$, $g(\theta)/g(s+\theta) < e^{\varepsilon}$ for $-\delta(\varepsilon) \le s \le 0$, $\theta \le 0$. Applying Lemma 3.4 to $g(\theta)$ and $g(\theta)^{-1}$, we have Inequality (3.3) with $\gamma(\varepsilon) = -\varepsilon\delta(\varepsilon)^{-1}$. The converse is obvious.

<u>Theorem 3.6 (Haddock [1])</u>. The space C_g^0 satisfies Axiom (A1) under Condition (g-1). For the space UC_g, we have the following table:

	A	A1	B	C1	C2	D
UC_g	(g-4)	(g-4)	o	o	(g-2)	×

For these spaces, H, K(t) and M(t) in Axiom (A) can be taken as $H = g(0)$, $K(t) = \sup\{1/g(\theta) : -t \le \theta \le 0\}$ and $M(t) = G(t) \le e^{\varepsilon}e^{-\gamma(\varepsilon)t}$ for $t \ge 0$.

Proof. The statement for C_g^0 is obvious. Suppose that Condition (g-4) holds and that $x : (-\infty,\infty) \to E$ is admissible with respect to UC_g. Though the space UC_g is isometic to BU, it is not trivial that $x_t(\theta)/g(\theta)$ is in BU for $t > 0$. To see this property, set $x_0 = \varphi$, and set

$$y^t(\theta) = x(t+\theta)/g(\theta) \quad \text{for} \quad \theta \le 0, \ t \ge 0.$$

Let $t \ge 0$ be fixed. Suppose that $\theta \le -t-1$ and $h \le 1$. Then $y^t(\theta) = \varphi(t+\theta)/g(\theta)$ and $y^{t+h}(\theta) = \varphi(t+h+\theta)/g(\theta)$. Hence we have that $|y^t(\theta+h) - y^t(\theta)| \le \eta^1(t,\theta,h) + \eta^2(t,\theta,h)$, where

$$\eta^1 = \left|\frac{\varphi(t+\theta+h)}{g(\theta+h)} - \frac{\varphi(t+\theta+h)}{g(\theta)}\right|$$

$$\eta^2 = \left|\frac{\varphi(t+\theta+h)}{g(\theta)} - \frac{\varphi(t+\theta)}{g(\theta)}\right|.$$

Rewrite η^1 and η^2 as

$$\eta^1 = \frac{g(t+\theta+h)}{g(\theta+h)} \frac{|\varphi(t+\theta+h)|}{g(t+\theta+h)} \left|1 - \frac{g(\theta+h)}{g(\theta)}\right|$$

$$\eta^2 = \frac{g(t+\theta)}{g(\theta)} \frac{1}{g(t+\theta)} |\varphi(t+\theta+h) - \varphi(t+\theta)|.$$

If we replace θ by $t+\theta+h$ and s by $-t$ in Lemma 3.5, we have $e^{-\varepsilon}e^{\gamma(\varepsilon)t} \le g(\theta+h)/g(t+\theta+h) \le e^{\varepsilon}e^{-\gamma(\varepsilon)t}$ for $\theta \le -t-1$, $h \le 1$. Hence, we have $g(t+\theta+h)/g(\theta+h) \le \varepsilon^{\varepsilon}e^{-\gamma(\varepsilon)t}$, which implies that

$$\eta^1 \le e^{\varepsilon}e^{-\gamma(\varepsilon)t}|\varphi|_g \left|1 - \frac{g(\theta+h)}{g(\theta)}\right|. \tag{3.4}$$

Similarly, we have that

$$\eta^2 \le e^{\varepsilon}e^{-\gamma(\varepsilon)t} \frac{1}{g(s)} |\varphi(s+h) - \varphi(s)|, \tag{3.5}$$

where $s = t+\theta \le -1$. Furthermore, we have

$$(3.6)\qquad \frac{1}{g(s)}\,|\varphi(s+h)-\varphi(s)| \le |\varphi(s+h)[\frac{1}{g(s)}-\frac{1}{g(s+h)}]| + |\frac{\varphi(s+h)}{g(s+h)}-\frac{\varphi(s)}{g(s)}|$$
$$\le |\varphi|_g\,|\frac{g(s+h)}{g(s)}-1| + |\frac{\varphi(s+h)}{g(s+h)}-\frac{\varphi(s)}{g(s)}|$$

for $s \le -1$ and $h \le 1$.

From (3.4) and Condition (g-4) we have immediately that $\eta^1(t,\theta,h) \to 0$ as $h \to 0$ uniformly for $\theta \le -t-1$. From (3.5) and (3.6) we have similarly that $\eta^2(t,\theta,h) \to 0$ as $h \to 0$ uniformly for $\theta \le -t-1$. Therefore, $y^t(\theta)$ is uniformly continuous for θ in $(-\infty,-t-1]$, which, of course, implies that x_t is in UC_g.

The estimate for $|x_{t+h} - x_t|_g$ is reduced to the estimate for $\eta^2(t,\theta,h)$; hence, we see that (A1) holds for UC_g.

(c) <u>The space C_γ.</u> For any real constant γ, we set

$$C_\gamma = \{\varphi \in C((-\infty,0],E) : \lim_{\theta\to-\infty} e^{\gamma\theta}\varphi(\theta) \text{ exists in } E\},$$

and set

$$|\varphi|_\gamma = \sup\{e^{\gamma\theta}|\varphi(\theta)| : -\infty < \theta \le 0\} \quad \text{for } \varphi \text{ in } C_\gamma.$$

For a function φ in C_γ, we define a function u by

$$u(s) = \begin{cases} e^{\gamma s/1+s}\varphi(s/1+s) & -1 < s \le 0 \\ \tilde{\varphi}(-\infty) = \lim_{\theta\to-\infty} e^{\gamma\theta}\varphi(\theta) & s = -1. \end{cases}$$

Then, u is a function in $C([-1,0],E)$; the transform

$$\iota : C_\gamma \to C([-1,0],E)$$

defined by

$$(3.7)\qquad \iota(\varphi) = u \quad \text{for } \varphi \text{ in } C_\gamma$$

is an isometric isomorphism from C_γ onto $C([-1,0],E)$.

Theorem 3.7.

	A	A1	B	C1	C2	D
C_γ	o	o	o	o	$\gamma>0$	o

$H = 1$, $K(t) = \max(1, e^{-\gamma t})$ and $M(t) = e^{-\gamma t}$.

(d) The space $C \times L^p(g)$. Suppose that $1 \le p < \infty$, $0 \le r < \infty$ and g is a nonnegative Borel measurable function on $(-\infty,-r)$. Then, a positive Borel measure λ on $(-\infty,-r)$ is defined by

$$\lambda(E) = \int_E g(\theta)d\theta \quad \text{for } E \text{ in } \mathcal{B},$$

where $\mathcal{B}$ is the family of Borel sets of $(-\infty,-r)$. We denote this equation as $d\lambda = g(\theta)d\theta$. Let $\mathcal{B}_g$ be the completion of $\mathcal{B}$ with respect to this measure.

Let $C \times L^p(g)$ be the product space of $C = C([-r,0],E)$ and $L^p(g)$, which is a class of functions $\varphi : (-\infty,0] \to E$ such that φ is continuous on $[-r,0]$ and $|\varphi|^p$ is integrable with respect to λ on $(-\infty,-r)$. A seminorm in $C \times L^p(g)$ is defined by

$$|\varphi| = \sup\{|\varphi(\theta)| : -r \le \theta \le 0\} + \{\int_{-\infty}^{-r} |\varphi(\theta)|^p g(\theta)d\theta\}^{1/p}.$$

It is well known that $C \times L^p(g)$ satisfies Axiom (B), and that it satisfies Axiom (D) if g is locally integrable on $(-\infty,-r)$ (cf. Rudin [1, Theorem 3.11 and Theorem 3.14]). We consider the following conditions on g:

(g-5) $\int_u^{-r} g(\theta)d\theta < \infty$ for every u in $(-\infty,-r)$;

(g-6) $g(u+\theta) \le G(u)g(\theta)$ for $u \le 0$ and θ in $(-\infty,-r)\setminus N_u$ for a set $N_u \subset (-\infty,-r)$ with Lebesgue measure 0 and for a nonnegative function G which is locally bounded on $(-\infty,0]$;

(g-7) $\int_{-\infty}^{-r} g(\theta)d\theta < \infty$.

Then we have the following table.

Theorem 3.8.

	A	A1	B	C1	C2	D
$C \times L^p(g)$	(g-5,6)	(g-5,6)	(g-5)	(g-5)	(g-7)	(g-5)

H, K(t) and M(t) in (A) can be taken as follows:

$$H = 1$$

$$K(t) = \begin{cases} 1 & \text{for } 0 \le t \le r \\ 1 + [\int_{-t}^{-r} g(\theta)d\theta]^{1/p} & \text{for } r < t \end{cases}$$

$$M(t) = \begin{cases} \max\{1+[\int_{-r-t}^{-r} g(\theta)d\theta]^{1/p},\ G(-t)^{1/p}\} & 0 \le t \le r \\ \max\{[\int_{-r-t}^{-t} g(\theta)d\theta]^{1/p},\ G(-t)^{1/p}\} & r < t. \end{cases}$$

Furthermore, Condition (g-6) implies that

$$\int_{-r-t}^{-t} g(\theta)d\theta \le G(-t+r)\int_{-2r}^{-r} g(\theta)d\theta \quad \text{for } r < t.$$

Proof. As in the case (b), the point is to prove Axioms (A) and (A1) under Conditions (g-5,6). To do so, we first observe that

$$\lambda(E-t) \le G(-t)\lambda(E) \quad \text{for } t \ge 0$$

and for every set E in $\mathbb{B}_g$, where $E-t = \{\theta-t : \theta \in E\}$. Indeed, if E is a Borel set of $(-\infty,-r)$, then

$$\lambda(E-t) = \int_{E-t} g(\theta)d\theta = \int_E g(s-t)ds \le G(-t)\int_E g(s)ds = G(-t)\lambda(E)$$

by Condition (g-6). Hence, if $\lambda(E) = 0$, then $\lambda(E-t) = 0$. Suppose that E is in $\mathbb{B}_g$. Then, there exist Borel sets A and B of $(-\infty,-r)$ such that $A \subset E \subset B$ and $\lambda(B\setminus A) = 0$. It then follows that $A-t \subset E-t \subset B-t$, and that $\lambda((B-t)\setminus(A-t)) = \lambda((B\setminus A)-t) = 0$. This implies that $E-t$ is in $\mathbb{B}_g$, and that $\lambda(E-t) = \lambda(A-t) \le G(-t)\lambda(E)$.

Therefore, if $\varphi : (-\infty,-r) \to \mathbb{E}$ is λ-measurable, then $\varphi(t+\cdot)$, $t \ge 0$, is also λ-measurable on $(-\infty,-r-t)$, and

$$(3.8)\qquad \int_{-\infty}^{-r-t} |\varphi(t+\theta)|^p g(\theta)d\theta \le G(-t)\int_{-\infty}^{-r} |\varphi(\theta)|^p g(\theta)d\theta.$$

This implies that Axiom (A) holds for $C \times L^p(g)$. More precisely, suppose that $x : (-\infty,\infty) \to \mathbb{E}$ is admissible with respect to $C \times L^p(g)$. We have formally that

$$|x_t| = |x|_{[t-r,t]} + \{\int_{-\infty}^{-r} |x(t+\theta)|^p g(\theta)d\theta\}^{1/p}.$$

Set $x_0 = \varphi$. Suppose that $0 \le t \le r$. Then, we have that

$$|x|_{[t-r,t]} = \max\{|\varphi|_{[t-r,0]}, |x|_{[0,t]}\}$$

and

$$\int_{-\infty}^{-r} |x(t+\theta)|^p g(\theta)d\theta = \{\int_{-r-t}^{-r} + \int_{-\infty}^{-r-t}\}|\varphi(t+\theta)|^p g(\theta)d\theta$$

$$\le (|\varphi|_{[-r,t-r]})^p\int_{-r-t}^{-r} g(\theta)d\theta + G(-t)\int_{-\infty}^{-r} |\varphi(\theta)|^p g(\theta)d\theta.$$

Suppose that $r < t$. Then, we have that

$$\int_{-\infty}^{-r} |x(t+\theta)|^p g(\theta)d\theta$$

$$= \int_{-t}^{-r} |x(t+\theta)|^p g(\theta)d\theta + \{\int_{-t-r}^{-t} + \int_{-\infty}^{-t-r}\}|\varphi(t+\theta)|^p g(\theta)d\theta$$

$$\le (|x|_{[0,t-r]})^p\int_{-t}^{-r} g(\theta)d\theta + (|\varphi|_{[-r,0]})^p\int_{-t-r}^{-t} g(\theta)d\theta$$

$$+ G(-t)\int_{-\infty}^{-r} |\varphi(\theta)|^p g(\theta)d\theta.$$

Therefore, we have Axiom (A) for $C \times L^p(g)$ by setting H, $K(t)$ and $M(t)$ as in Theorem 3.8. The last inequality in the theorem is obtained by the substitution $\theta = -(t-r)+s$ and by condition (g-6). This inequality assures that $M(t)$ is locally bounded on $[0,\infty)$.

To discuss Axiom (A1), observe the seminorm $|x_{t+h} - x_t|$ for $t \ge 0$ and $h > 0$. It follows immediately that $|x_{t+h} - x_t|_{[-r,0]} \to 0$ as $h \to 0$ for $t > 0$, and as $h \to 0+$ for $t = 0$. To consider the integral part of $x_{t+h} - x_t$, we set

$$y_i(t,h) = \int_{I_i} |x(t+h+\theta)-x(t+\theta)|^p g(\theta)d\theta \quad i = 1,2,3,$$

where $I_1 = [-r-t,-r)$, $I_2 = [-r-t-h,-r-t)$ and $I_3 = (-\infty,-r-t-h)$. From the continuity of x on $[-r,\infty)$ and the condition (g-5), it follows at once that $y_1(t,h) \to 0$ as $h \to 0+$. Since $|x(t+h+\theta)|^p$ is continuous and bounded for θ in I_2, Condition (g-5) implies again that

$$\int_{-r-t-h}^{-r-t} |x(t+h+\theta)|^p g(\theta)d\theta \to 0 \quad \text{as} \quad h \to 0+.$$

Since $|x(t+\theta)|^p = |\varphi(t+\theta)|^p$ for θ in I_2, and since $|\varphi(t+\theta)|^p g(\theta)$ is integrable on $(-\infty,-r-t)$, we have that

$$\int_{-r-t-h}^{-r-t} |x(t+\theta)|^p g(\theta)d\theta \to 0 \quad \text{as} \quad h \to 0+.$$

Thus it follows that $y_2(t,h) \to 0$ as $h \to 0+$. From the definition of φ, we can write

$$y_3(t,h) = \int_{-\infty}^{-r-t-h} |\varphi(t+h+\theta) - \varphi(t+\theta)|^p g(\theta)d\theta.$$

Since λ is a regular Borel measure due to Condition (g-5), (cf. Rudin [1, Theorem 2.18]), for every $\varepsilon > 0$ there exists a continuous function φ^ε, with compact support in $(-\infty,-r)$, such that

$$\int_{-\infty}^{-r} |\varphi(\theta) - \varphi^\varepsilon(\theta)|^p g(\theta)d\theta < \varepsilon,$$

(cf. Rudin [1, Theorem 3.14]). Then, from Inequality (3.8) it follows that

$$\int_{-\infty}^{-r-t-h} |\varphi(t+h+\theta) - \varphi^\varepsilon(t+h+\theta)|^p g(\theta)d\theta \le G(-t-h)\varepsilon,$$

and that

$$\int_{-\infty}^{-r-t-h} |\varphi(t+\theta) - \varphi^\varepsilon(t+\theta)|^p g(\theta)d\theta \le \int_{-\infty}^{-r-t} |\varphi(t+\theta) - \varphi^\varepsilon(t+\theta)|^p g(\theta)d\theta$$

$$\le G(-t)\varepsilon.$$

Hence, we have that

$$\limsup_{h\to 0+} y_3(t,h) \le \varepsilon\{G(-t) + \limsup_{h\to 0+} G(-t-h)\}$$

$$+ \limsup_{h\to 0+}\int_{-\infty}^{-r-t-h} |\varphi^\varepsilon(t+h+\theta) - \varphi^\varepsilon(t+\theta)\}^p g(\theta)d\theta.$$

The last term in this inequality vanishes since φ^ε is uniformly continuous and g satisfies Condition (g-5). Since $\varepsilon > 0$ is arbitrary, we can conclude that $y_3(t,h) \to 0$ as $h \to 0+$.

Therefore, we have proved that $|x_{t+h} - x_t| \to 0$ as $h \to 0+$ for each $t \ge 0$. Similarly, one can prove that $|x_{t+h} - x_t| \to 0$ as $h \to 0-$ for each $t > 0$.

(e) The space $\mathbb{E}\times M$. Let $\mathbb{E}\times M$ be the space of all functions φ such that

$$\sup\{\int_{-n-1}^{-n} |\varphi(\theta)|d\theta : n = 1,2,\cdots\} < \infty$$

supplied with the seminorm

$$|\varphi| = |\varphi(0)| + \sup_n\int_{-n-1}^{-n} |\varphi(\theta)|d\theta.$$

Kappel and Schappacher [1] claimed that this space does not satisfy Axiom (A-1).

Theorem 3.9.

	A	A1	B	C1	C2	D
$\mathbb{E}\times M$	o	×	o	o	×	o

Proof. We only give the proof that (A1) does not hold for $\mathbb{E}\times M$. Indeed, define the function φ by

$$\varphi(\theta) = \begin{cases} 0 & \text{for } \theta \in [-2k,-2k+1] \\ 2k-1 & \text{for } \theta \in (-2k+1,-2k+1+\frac{1}{2k-1}] \\ 0 & \text{for } \theta \in (-2k+1+\frac{1}{2k-1},-2k+2] \end{cases}$$

and for $k = 1,2,\cdots$. Then, φ is in $\mathbb{E}\times M$, and $|S(t)\varphi - \varphi| = 1$ for any $t > 0$. For the detail of the space $\mathbb{E}\times M$, see (Massera-Schäffer [1], Corduneanu[1]).

<u>(f) Supplementary remarks on the space BC.</u> Seifert [3] presented a simple example:

$$x(t) = \sin t^2 \quad \text{for} \quad t \le 0 \quad \text{and} \quad x(t) = t \quad \text{for} \quad 0 \le t < 1.$$

The t-segment x_t is in BC for every $t < 1$, but $\limsup_{h \to 0} |x_{t+h} - x_t| = 2$. Let $f : BC \to \mathbb{R}$ be a linear operator defined by

$$f(\varphi) = \sup\{\varphi(-\sqrt{2\pi n}) : n = 1, 2, \cdots\}.$$

Then, the norm of f is given by $|f| = 1$. Set $\alpha = t/\sqrt{2\pi}$ for t in $[0,1)$. If α is rational, then $f(x_t) < 1$; if α is irrational, then $f(x_t) = 1$. Hence, $f(x_t)$ is not continuous, but it is measurable.

However, the function x of this example is at least a Caratheodory type solution of the equation $\dot{x} = f(x_t)$. Sawano [3] investigated the fundamental theorems for equations on the space BC along this line. We only refer to the following result in relation with the Seifert's example.

<u>Theorem 3.10.</u> Assume that $f : \mathbb{R} \times BC \to \mathbb{R}^n$ satisfies the following hypotheses:

(H1) If $x : (-\infty, a) \to \mathbb{R}^n$ is bounded and continuous, then $f(t, x_t)$ is measurable in $t \in (-\infty, a)$.

(H2) For any bounded set V in BC, there exists a function $m(t) = m_V(t)$, locally integrable on $\mathbb{R}$, such that $|f(t,\varphi)| \le m(t)$ for any φ in V.

(H3) $f(t,\varphi)$ is continuous in φ for each t in $\mathbb{R}$.

Then, for any (σ, φ) in $\mathbb{R} \times BC$, there exists a Caratheodory type solution of Equation (1.1) through (σ, φ). Moreover, if for any bounded set $V \subset BC$, there is a locally integrable function $n_V(t)$ such that

$$|f(t,\varphi) - f(t,\psi)| \le n_V(t)|\varphi - \psi| \quad \text{on} \quad \mathbb{R} \times V,$$

then the solution of Equation (1.1) is unique for a given initial condition.

Can we omit the hypothesis (H1) in the above theorem? Seifert gives a negative answer to this question.

Theorem 3.11 (Seifert [5]). There exist an $\mathbb{R}^2$-valued function $x(t)$, bounded and continuous on $\mathbb{R}$, such that, given any $\mathbb{R}^2$-valued function $g(t)$ on $\mathbb{R}$ satisfying $|g(t)| \le 1$ for t in $\mathbb{R}$, there exists a continuous linear operator $f : BC \to \mathbb{R}^2$ such that $f(x_t) = g(t)$ or $f(x_t) = g(-t)$ for t in $\mathbb{R}$.

Theorem 3.12 (Seifert [5]). There exists a continuous linear functional F on BC to $\mathbb{R}$, and an initial function φ in BC such that the initial value problem

$$\dot{y}(t) = F(y_t) \qquad y_0 = \varphi$$

has no solution of Caratheodory type.

1.4. Discussions on the weight function g. We assumed Conditions (g-1) and (g-6) in the previous section. They look different; g appears in the denominator in the norm of C_g, and in the numerator in the norm of $C \times L^p(g)$. So there is a confusion in the definition of G, but we leave untouched following the usage in several past papers. Observe that G is defined on $[0,\infty)$ in (g-1), and on $(-\infty,0]$ in (g-6). In this section, we consider the necessity of these conditions in relation with Axiom (A) and (A-1).

(a) The case of C_g.

Proposition 4.1. Let g be a positive continuous function on $(-\infty,0]$. Then, the space C_g satisfies Axiom (A) if and only if Condition (g-1) holds.

Proof. To see the only if part, let e be a unit vector in $\mathbb{E}$, i.e., $|e| = 1$, and define continuous functions $x^n(t)$, $n = 1,2,\cdots$, by

$$x^n(t) = \begin{cases} g(t)e & -\infty < t \le 0 \\ (1-nt)g(0)e & 0 < t < 1/n \\ 0 & 1/n \le t. \end{cases}$$

Suppose that C_g satisfies Axiom (A). Since x^n is admissible with respect to C_g, and since $|x_0| = 1$, it follows that $|x^n{}_t| \le K(t)|g(0)| + M(t)$ for $n = 1,2,\cdots$. Taking the limit as $n \to \infty$, we then have that

$$G(t) \le K(t)g(0) + M(t)$$

for $t \ge 0$. Therefore, $G(t)$ is locally bounded on $[0,\infty)$ since $K(t)$ is continuous and $M(t)$ is locally bounded.

Suppose that $\mathscr{B} = C_g$, UC_g or C_g^0. In any case, $S_0(t)$ is a linear operator on $\mathscr{B}_0$.

Theorem 4.2. Suppose that Condition (g-1) holds. Then, for each case $\mathscr{B} = C_g$, UC_g or C_g^0, the norm of $S_0(t)$ is given by $|S_0(t)| = G(t)$ for $t \ge 0$.

Proof. For simplicity, denote by $|S_0(t)|_g$, $|S_0(t)|_U$ and $|S_0(t)|_0$ the operator norm of $S_0(t)$ as an operator on C_g, UC_g and C_g^0, respectively. Then $|S_0(t)|_0 \le |S_0(t)|_U \le |S_0(t)|_g$ since $C_g^0 \subset UC_g \subset C_g$.

Let $t \ge 0$ be fixed, and let φ be a function in C_g^0 such that $\varphi(0) = 0$ and that $|\varphi|_g \le 1$ or $|\varphi(\theta)| \le g(\theta)$ for $\theta \le 0$. Then, it follows immediately that $|S_0(t)\varphi|_g \le G(t)$; hence, $|S_0(t)|_g \le G(t)$.

Next, let $\{s_k\}$, $s_k < 0$, be a sequence such that $g(s_k)/g(s_k-t) \to G(t)$ as $k \to \infty$, and for each k define a function φ^k in C_g^0 by

$$\varphi^k(\theta) = \begin{cases} \theta g(\theta)/s_k & s_k \le \theta \le 0 \\ (\theta - s_k + 1)g(\theta) & s_k - 1 < \theta < s_k \\ 0 & \theta \le s_k - 1. \end{cases}$$

Clearly, $\varphi^k(0) = 0$ and $|\varphi^k|_g = 1$. Thus

$$|S_0(t)|_0 \ge |S_0(t)\varphi^k| = \sup\{|\varphi^k(t+\theta)|/g(\theta) : \theta \le -t\}$$

$$\ge |\varphi^k(s_k)|/g(s_k - t) = g(s_k)/g(s_k - t).$$

Hence, we have $|S_0(t)|_0 \ge G(t)$; consequently, $G(t) \le |S_0(t)|_0 \le |S_0(t)|_U \le |S_0(t)|_g \le G(t)$, which shows Theorem 4.2.

Example 4.3 (Shin [4]). For the space C_γ, we have that $|S_0(t)| = e^{-\gamma t}$, $t \ge 0$, and that

$$|S(t)| = \begin{cases} 1 & \text{if } \gamma \ge 0 \\ e^{-\gamma t} & \text{if } \gamma < 0. \end{cases}$$

(b) The case of $C \times L^p(g)$. At first, we introduce the example given in (Hille-Phillips [1]). Let $g(\theta) = \exp \theta^2$ for $\theta \le 0$, and $\chi(t) = (1+t^2)^{-1}\exp(-t^2)$ for t in $\mathbb{R}$. Then, χ_0 is in $C \times L^1(g)$, but the t-segment χ_t is not in this space even if the positive t is taken as small as we will. In this case, we obtain $G(-t) =$ ess $\sup\{g(-t+\theta)/g(\theta) : \theta \le -t\} = \infty$ for every $t > 0$.

We have been considering the measure $g(\theta)d\theta$, that is, the measure which is locally absolutely continuous on $(-\infty,0]$ with respect to the Lebesgue measure. Why do we take such a measure? Is $\mathcal{B}$ not chosen from the space $L^p(\mu)$ defined by a more general positive Borel measure on $(-\infty,0]$? The answer is that such a space $L^p(\mu)$ is actually isomorphic to the space $\mathbb{E} \times L^p(g)$ for some function g with the property (g-6) as long as Axioms (A) and (A-1) hold for $L^p(\mu)$. We show this result by following the works of Coleman-Mizel [1,2] and the one by Lima [1].

Let μ be a positive Borel measure on $(-\infty,0]$, and $L^p(\mu)$, $1 \le p < \infty$, the space of μ-measurable functions $\varphi : (-\infty,0] \to \mathbb{E}$ such that

$$|\varphi| := [\int_{-\infty}^{0} |\varphi(\theta)|^p d\mu(\theta)]^{1/p}$$

is finite. Coleman and Mizel put the following postulates on $L^p(\mu)$:

Postulate A. The measure μ is non-trivial and finite on every compact set of $(-\infty,0]$.

Postulate B. If φ is in $L^p(\mu)$, then $S(t)\varphi$ is in $L^p(\mu)$ for $t \ge 0$. Furthermore, if $|\varphi| = 0$, then $|S(t)\varphi| = 0$ for $t \ge 0$.

Postulate C. If φ is in $L^p(\mu)$, then every s-segment φ_s for $s \le 0$ is in $L^p(\mu)$.

It is easy to see that the couple of Postulates A and B is equivalent to Axiom (A) for the space $L^p(\mu)$, cf. Schumacher [1] and Kappel-Schappacher [1]. Postulate C does not appear in our system of Axioms. Coleman-Mizel set a value on Postulate C : in other words, it requires that the Radon-Nikodim derivative $k(\theta)$ of measure μ must not decay to zero too fast as $\theta \to -\infty$. However, we can develop the fundamental theory of equations with infinite delay without this postulate.

Before starting, we recall some basic facts and definitions of measure theory. Refer to the book of Rudin [1] for those results. A Borel measure μ defined on a locally compact Hausdorff space X is said to be regular if

$$\mu(E) = \inf\{\mu(V) : E \subset V, V \text{ open}\}$$
$$= \sup\{\mu(K) : K \subset E, K \text{ compact}\}$$

for every Borel set E of X. Assume that every open set of X is a countable union of compact sets. From Rudin [1, Theorems 2.14, 2.17 and 2.18], we can then assume that any Borel measure μ on X is regular and has an extension to the measure, denoted by μ again, defined on a σ-algebra $\mathbb{M}$ in X with the following properties :

(a) If E is in $\mathbb{M}$ and $\varepsilon > 0$, there are a closed set F and an open set V such that $F \subset E \subset V$ and $\mu(V \setminus F) < \varepsilon$.

(b) If E is in $\mathbb{M}$, then there are sets A and B such that A is in F_σ, B is in G_δ, $A \subset E \subset B$ and $\mu(B-A) = 0$.

(c) If E is in $\mathbb{M}$, $A \subset E$, and $\mu(E) = 0$, then A is in $\mathbb{M}$.

As a corollary of (b) and (c), every set E in $\mathbb{M}$ is a union of a set in F_σ and a set of measure 0.

Let $L^p(\mu)$ be the space defined previously. Since the set $(-\infty,0]$ satisfies the above conditions on X, μ has the properties cited above. If $L^p(\mu)$ is a nontrivial space and satisfies Axiom (A), μ has at least the following properties :

(μ-1) μ is nontrivial, $\mu((-\infty,0]) > 0$, and $\mu(K)$ is finite for every compact set K of $(-\infty,0]$.

(μ-2) If φ is in $L^p(\mu)$, then $S(t)\varphi$ is in $L^p(\mu)$ for $t \geq 0$, and if $|\varphi - \psi| = 0$ for φ, ψ in $L^p(\mu)$, then $|S(t)\varphi - S(t)\psi| = 0$ for $t \geq 0$.

From now on, we extend our problem to the one of studying the space $L^p(\mu)$ defined by the measure μ with the properties (μ-1,2).

<u>Lemma 4.4.</u> If (μ-1,2) hold, we have the followings:

(a) $\mu(\{0\}) > 0$.

(b) If $\mu(A) = 0$, then $\mu(A-t) = 0$ for $t \geq 0$.

<u>Proof</u>. Let h be a function such that $h(\theta) = 0$ for $\theta < 0$ and $h(0)$ is a unit vector of $\mathbb{E}$. Then, $|h|^p = \mu(\{0\})$, and $|S(t)h|^p = \mu([-t,0])$ for $t \geq 0$. If $\mu(\{0\}) = 0$, then (μ-2) implies that $\mu([-t,0]) = 0$ for all $t \geq 0$, which contradicts (μ-1).

Suppose that $\mu(A) = 0$. Then, the result (a) implies that $A \cap \{0\}$

$= \phi$; in other words, $\chi_A(0) = 0$, where χ_A is the indicator function of A. Set $\varphi(\theta) = \chi_A(\theta)e$ for $\theta \le 0$ and for some unit vector e of E. The condition $\mu(A) = 0$ implies that φ is in $L^p(\mu)$ and $|\varphi| = 0$. Observe that $[S(t)\varphi](\theta) = \chi_{A-t}(\theta)e$ for $\theta \le 0$. Hence, from (μ-2) we have that $\mu(A-t) = |S(t)\varphi|^p = 0$.

<u>Theorem 4.5.</u> For every $t \ge 0$, $S(t)$ is a bounded linear operator on $L^p(\mu)$.

<u>Proof</u>. Since $S(t)$ is a linear operator defined everywhere in the space $L^p(\mu)$, we may apply the closed graph theorem to prove the boundedness of $S(t)$. Hence, it suffices to show that $S(t)$ is a closed operator. Suppose that a sequence $\{\varphi^n\}$ in $L^p(\mu)$ converges to φ in $L^p(\mu)$, and that $\{S(t)\varphi^n\}$ converges to ψ in $L^p(\mu)$. By taking a subsequence, we can assume that $\varphi^n(\theta) \to \varphi(\theta)$ and $[S^n(t)\varphi](\theta) \to \psi(\theta)$ as $n \to \infty$ for θ in $(-\infty,0]\setminus N$, where $\mu(N) = 0$. Since $\mu(\{0\}) > 0$, the origin is not in N ; hence, $\varphi^n(0) \to \varphi(0)$ as $n \to \infty$. Since $[S(t)\varphi^n](\theta) = \varphi^n(0)$ for θ in $[-t,0]$, $\psi(\theta) = \varphi(0)$ for θ in $[-t,0]\setminus N$. On the other hand, $[S(t)\varphi^n](\theta) = \varphi^n(t+\theta)$ for $\theta < -t$, and $\varphi^n(t+\theta) \to \varphi(t+\theta)$ for θ in $(-\infty,-t]\setminus(N-t)$. Thus, we have that $\psi(\theta) = \varphi(t+\theta)$, or $\psi(\theta) = [S(t)\varphi](\theta)$, for θ in $(-\infty,0]\setminus[N \cup (N-t)]$. Since $\mu(N-t) = 0$ by Lemma 4.4, it follows that $\psi(\theta) = [S(t)\varphi](\theta)$ μ-a.e. in $(-\infty,0]$. Thus, $S(t)\varphi = \psi$ in $L^p(\mu)$, which shows that $S(t)$ is a closed operator.

We need the following well known lemma (Hille-Phillips [1, Theorem 7.4.1, p. 241]) to show that the norm

$$|S(t)| := \inf\{A : |S(t)\varphi| \le A|\varphi| \text{ for every } \varphi \text{ in } L^p(\mu)\}$$

is locally bounded on $(0,\infty)$.

<u>Lemma 4.6.</u> If $f(t)$ is measurable, subadditive and different from $+\infty$ in $(0,\infty)$, then $f(t)$ is bounded above in any compact subset of $(0,\infty)$.

<u>Proposition 4.7</u>. $|S(t)|$ is lower semicontinuous in $(0,\infty)$, and bounded in any compact subset of $(0,\infty)$.

<u>Proof</u>. Since the subset C_{00} (see Section 1.2) is dense in $L^p(\mu)$ from the regularity of μ (Rudin [1, Theorem 3.14, p. 71]), we have that $|S(t)| = |S(t)|_0$, where

$$|S(t)|_0 = \inf\{B : |S(t)\varphi| \le B|\varphi| \text{ for every } \varphi \text{ in } C_{00}\}$$

$$= \sup\{|S(t)\varphi| : |\varphi| \le 1, \ \varphi \in C_{00}\}.$$

Since $\mu(K) < \infty$ for every compact set K of $(-\infty,0]$, $|S(t)\varphi|$ is continuous for t in $[0,\infty)$ for each fixed φ in C_{00}. Hence, $|S(t)|_0$ is lower semicontinuous for t in $[0,\infty)$, which implies that $\log|S(t)|$ is lower semicontinuous. Hence, $\log|S(t)|$ is measurable, subadditive and different from $+\infty$ in $[0,\infty)$; Proposition 4.7 follows from Lemma 4.6.

Theorem 4.8. For each fixed φ in $L^p(\mu)$, $S(t)\varphi$ is continuous for t in $(0,\infty)$.

Proof. Let t be in $(0,\infty)$, and take constants a and b such that $0 < a < t < b < \infty$. Then, there is a constant C such that $|S(t)| \le C$ for $a \le t \le b$. Let φ be in $L^p(\mu)$. Then, for every $\varepsilon > 0$ there exists a φ^ε in C_{00} such that $|\varphi - \varphi^\varepsilon| < \varepsilon$. If $t+h$ is in $[a,b]$, we then have that

$$|S(t+h)\varphi - S(t)\varphi| \le |S(t+h)\varphi - S(t+h)\varphi^\varepsilon| + |S(t+h)\varphi^\varepsilon - S(t)\varphi^\varepsilon|$$

$$+ |S(t)\varphi^\varepsilon - S(t)\varphi|$$

$$\le 2C\varepsilon + |S(t+h)\varphi^\varepsilon - S(t)\varphi^\varepsilon|.$$

Since $|S(t+h)\varphi^\varepsilon - S(t)\varphi^\varepsilon| \to 0$ as $h \to 0$, it follows that $|S(t+h)\varphi - S(t)\varphi| \to 0$ as $h \to 0$.

Let $x : (-\infty,\infty) \to \mathbb{E}$ be admissible with respect to $L^p(\mu)$. Then, we have that, for each $t \ge 0$,

$$\int_{-\infty}^{0} |x(t+\theta)|^p d\mu(\theta) \le \left\{\int_{[-t,0]} + \int_{(-\infty,-t)}\right\} |x(t+\theta)|^p d\mu(\theta)$$

$$\le \mu([-t,0])(|x|_{[0,t]})^p + |S(t)x_0|^p,$$

or

$$|x_t| \le \mu([-t,0])^{1/p}|x|_{[0,t]} + |S(t)||x_0|.$$

Hence, Conditions (μ-1,2) imply a condition similar to Axiom (A). It is Axiom (A) itself if $|S(t)|$ is bounded on intervals containing the origin. Furthermore, from Theorem 4.8 we have that x_t is continuous for t in $(0,\infty)$ due to the decomposition $x_t = z_t + S(t)x_0$ appearing in Section 1.2. In view of the proof of Theorem 4.8 we know that x_t is right continuous at $t = 0$ if $|S(t)|$ is bounded on some intervals containing the origin. Hence, we proceed to the estimate of $|S(t)|$, which leads naturally to the discovery of more detailed structures of the measure μ.

We start from the following definition. Set

$$(4.1) \qquad -R = \sup\{\theta : \mu((-\infty,\theta)) = 0\}.$$

If $\mu((-\infty,\theta)) > 0$ for every $\theta \le 0$, we understand that $R = \infty$. Let m denote the Lebesgue measure on $(-\infty,\infty)$ and $\mathbb{L}$ the σ-algebra of the Lebesgue measurable sets of $(-\infty,0]$. Repeating the argument in the proof of Saks [1, Theorem 11.1, p. 91], we have the following result. The first part is due to Coleman-Mizel [1].

<u>Theorem 4.9.</u>

(a) $\mathbb{L} \subset \mathbb{M}$, and μ is absolutely continuous on $(-\infty,0)$ with respect to the Lebesgue measure m.

(b) $\mathbb{M} \subset \mathbb{L}$ in $[-R,0]$, and if $\mu(Y) = 0$ for Y in $\mathbb{M}$, then $m(Y \cap [-R,0]) = 0$, where $[-R,0]$ stands for $(-\infty,0]$ in case that $R = \infty$. Namely, m is absolutely continuous on $[-R,0]$ with respect to μ.

<u>Proof</u>. Let X be a G_δ set contained in (a,b) for some a, b with $-\infty < a < b < 0$, and assume that $m(X) = 0$. Set

$$X^* = \{(\theta,t) : \theta+t \in X,\ \theta \le 0,\ t \ge b\}$$

and set

$$X^*_t = \{\theta \le 0 : (\theta,t) \in X^*\} \quad \text{for } t \ge b$$

and

$$X^*_\theta = \{t \ge b : (\theta,t) \in X^*\} \quad \text{for } \theta \le 0.$$

Then, X^* is also a G_δ set in $(-\infty,0] \times [b,\infty)$, $X^*_t = X-t$ and $X^*_\theta = (X-\theta) \cap [b,\infty)$. Hence, Fubini's theorem implies that

$$\int_b^\infty \mu(X-t)dm(t) = \int_{-\infty}^0 m(X^*_\theta)d\mu(\theta).$$

Since $m(X^*_\theta) \le m(X-\theta) = m(X) - \theta$ for all $\theta \le 0$, we conclude that $\mu(X-t) = 0$ m-a.e. in $[b,\infty)$. Hence, there exists a negative t_0 in $[b,0]$ such that $\mu(X-t_0) = 0$. Then, we have $\mu(X) = \mu(X-t_0+t_0) = 0$ due to Lemma 4.4.

If X is a G_δ set in $(-\infty,0)$ with $m(X) = 0$, then $X_n = (-n,-\frac{1}{n}) \cap X$, $n = 1,2,\cdots$, is a G_δ set in $(-n,-\frac{1}{n})$ with $m(X_n) = 0$. Therefore, $\mu(X) = \lim_{n\to\infty} \mu(X_n) = 0$. Let X be a Lebesgue measurable set in $(-\infty,0)$ with $m(X) = 0$. Then, there is a G_δ set Y in $(-\infty,0)$ such that $X \subset Y$ and $m(Y) = 0$. Since $\mu(Y) = 0$ as verified in the above, we have that $X \in \mathbb{M}$ and $\mu(X) = 0$. Thus, μ is absolutely continuous on $(-\infty,0)$ with respect to m, and $\mathbb{L} \subset \mathbb{M}$ since every set in $\mathbb{L}$ is a union of a set in F_σ and a set of Lebesgue measure 0.

Suppose Y is a G_δ set in $(-\infty,0]$ such that $\mu(Y) = 0$. In the same manner as in the above, we have that

$$\int_0^\infty \mu(Y-t)dm(t) = \int_{-\infty}^0 m([Y-\theta] \cap [0,\infty))d\mu(\theta).$$

Since $\mu(Y-t) = 0$ for all $t \ge 0$ due to Lemma 4.4, the left side of this equation vanishes. Thus, there is a μ-null set N in $(-\infty,0]$ such that $m([Y-\theta]\cap[0,\infty)) = 0$ for θ in $(-\infty,0]\setminus N$. Since $[[Y-\theta]\cap[0,\infty)] + \theta = Y\cap[\theta,0]$, it follows that

$$m(Y\cap[\theta,0]) = 0 \quad \text{for} \quad \theta \quad \text{in} \quad (-\infty,0]\setminus N.$$

Since $\mu([-R,-R+\varepsilon)) > 0$ for any $\varepsilon > 0$ by the definition of R, the set $[-R,-R+\varepsilon)\cap N^c$ is a nonempty set for any $\varepsilon > 0$. Thus, there exists a sequence $\{\theta_n\}$ such that $\theta_n \ge -R$, $m(Y\cap[\theta_n,0]) = 0$ for $n = 1$ $2, \cdots$ and $\theta_n \to -R$ as $n \to \infty$, which implies that $m(Y\cap[-R,0]) = m(Y\cap(-R,0]) = 0$. The proof of (b) is complete.

The result (a) in Theorem 4.9 implies that there is a measurable function $g(\theta)$, locally integrable on $(-\infty,0)$, such that

$$\mu(A) = \int_A g(\theta)d\theta \tag{4.2}$$

for every Lebesgue measurable set A in $(-\infty,0)$: that is, g is the Radon-Nikodim derivative of μ on $(-\infty,0)$ with respect to m. The couple of (a) and (b) in Theorem 4.9 implies that $\mathbb{L} = \mathbb{M}$ on $[-R,0]$. Since $\mu((-\infty,-R)) = 0$ by Definition of R, it then follows that a subset A of $(-\infty,0]$ is μ-measurable if and only if $A \cap [-R,0]$ is

Lebesgue measurable, and that

$$\mu(A) = \mu(A \cap [-R,0]) - \mu(A\cap\{0\}) + \int_{A\cap[-R,0)} g(\theta)d\theta.$$

This implies that $g(\theta) = 0$ m-a.e. in $(-\infty,-R)$. Hence it follows that a function $\varphi : (-\infty,0] \to E$ is μ-measurable if and only if φ is Lebesgue measurable in $[-R,0]$, and that, for such a function φ,

$$\int_{-\infty}^{0} |\varphi(\theta)|^p d\mu(\theta) = |\varphi(0)|^p \mu_0 + \int_{-R}^{0} |\varphi(\theta)|^p g(\theta)d\theta.$$

where $\mu_0 = \mu(\{0\})$.

The following observation is due to Lima [1].

Theorem 4.10. $g(\theta) > 0$ m-a.e. in $[-R,0]$, and $g(\theta) = 0$ m-a.e. in $(-\infty,-R)$.

Proof. Since $N_g = \{\theta < 0 : g(\theta) = 0\}$ is a Lebesgue measurable set, we can apply Formula (4.2) for $A = N_g$: that is, $\mu(N_g) = 0$. Then, $m(N_g \cap [-R,0]) = 0$ from (b) in Theorem 4.9. This means that $g(\theta) > 0$ m-a.e. in $[-R,0]$. The second property of g is already proved directly from the definition of R.

This theorem suggests us to define a function $G(u)$ by

(4.3) $\quad G(u) = \text{ess sup}\{g(\theta+u)/g(\theta) : \theta \in [-R,0)\setminus N_g\}$ for $u < 0$,

where N_g is the above set, and ess sup is taken with respect to the Lebesgue measure. Then, $G(u)$ has the desired property that we have been having in mind. Set $L^p{}_0(\mu) = \{\varphi \in L^p(\mu) : \varphi(0) = 0\}$, which is a closed subspace of $L^p(\mu)$ since $\mu(\{0\}) > 0$. Let $S_0(t)$ be the restriction of $S(t)$ on $L^p{}_0(\mu)$. Then, $S_0(t)$ is a linear operator on $L^p{}_0(\mu)$ into itself. The following proof is due to Lima [1].

Theorem 4.11. $|S_0(t)| = G(-t)^{1/p}$ for $t > 0$.

Proof. It is obvious that $G(u) = 0$ for $u < -R$, and that $|S_0(t)| = 0$ for $t > R$. Let t be in $[0,R]$, and φ be in $L^p{}_0(\mu)$. Then, we have that

$$\int_{-\infty}^{0} |[S_0(t)\varphi](\theta)|^p d\mu(\theta) = \int_{-R}^{-t} |\varphi(t+\theta)|^p g(\theta)d\theta = \int_{-R}^{0} |\varphi(s)|^p g(s-t)ds$$

since $g(\theta) = 0$ m-a.e. in $(-\infty,-R)$. From the definition of $G(u)$, we thus obtain that $|S_0(t)\varphi| \le G(-t)^{1/p}|\varphi|$; hence, $|S_0(t)| \le G(-t)^{1/p}$. On the other hand, define $Q(\theta,u)$ for $\theta, u < 0$ by

$$Q(\theta,u) = \begin{cases} g(\theta+u)/g(\theta) & \text{if } g(\theta) > 0 \\ 0 & \text{if } g(\theta) = 0. \end{cases}$$

Since $m(N_g \cap [-R,0]) = 0$, we can write

$$\int_{-R}^{0} |\varphi(s)|^p g(s-t)ds = \int_{-R}^{0} |\varphi(\theta)|^p Q(\theta,-t)g(\theta)d\theta.$$

Therefore, if φ is in $L^p{}_0(\mu)$, then

$$\int_{-R}^{0} |\varphi(\theta)|^p Q(\theta,-t)g(\theta)d\theta = |S_0(t)\varphi|^p. \tag{4.4}$$

Let e be a unit vector of $\mathbb{E}$, and $f(\theta)$ an integrable function on $[-R,0]$ with respect to μ, and set $\varphi(\theta) = |f(\theta)|^{1/p}e$ for $\theta < 0$ and $\varphi(0) = 0$. Then, φ is in $L^p{}_0(\mu)$. Since $|S_0(t)\varphi| \le |S_0(t)||\varphi|$, it follows from (4.4) that

$$\int_{-R}^{0} |f(\theta)|Q(\theta,-t)g(\theta)d\theta \le |S_0(t)|^p \int_{-R}^{0} |f(\theta)|g(\theta)d\theta.$$

Since f is an arbitrary μ-integrable function on $[-R,0]$, it follows that

$$\mu\text{-ess sup}\{Q(\theta,-t) : -R \le \theta < 0\} \le |S_0(t)|^p$$

for each $t > 0$. However, this essential supremum with respect to μ agrees with $G(-t)$ since μ and m are mutually absolutely continuous on $[-R,0)$. Hence, $G(-t) \le |S_0(t)|^p$ for $t > 0$. The proof is complete.

The results up to this point are summarized as follows.

<u>Theorem 4.12.</u> Assume that μ is a positive Borel measure on $(-\infty,0]$ having the properties (μ-1) and (μ-2). Then, the following conditions hold:

(i) For $t > 0$, $S(t)$ is a bounded linear operator on $L^p(\mu)$. Moreover, for each fixed φ in $L^p(\mu)$, $S(t)\varphi$ is continuous in $t > 0$.

(ii) $\mu_0 := \mu(\{0\}) > 0$, and μ is absolutely continuous on $(-\infty,0)$ with respect to the Lebesgue measure.

(iii) Define R as in (4.1), and let $g(\theta)$ be the Radon-Nikodim derivative of μ on $(-\infty,0)$ with respect to the Lebesgue measure. Then, $g(\theta) = 0$ a.e. in $(-\infty,-R)$ and $g(\theta) > 0$ a.e. in $[-R,0]$. The Lebesgue measure is absolutely continuous on $[-R,0)$ with respect to μ.

(iv) A function $\varphi : (-\infty,0] \to E$ is in $L^p(\mu)$ if and only if $\varphi(\theta)$ is Lebesgue measurable on $[-R,0)$ and $|\varphi(\theta)|^p$ is Lebesgue integrable on $[-R,0)$: the norm in $L^p(\mu)$ is given by

$$|\varphi| = \{|\varphi(0)|^p\mu_0 + \int_{-R}^{0} |\varphi(\theta)|^p g(\theta)d\theta\}^{1/p}.$$

(v) Define $G(u)$, $u < 0$, as in (4.2). Then,

$$|S_0(t)| = G(-t)^{1/p} \quad \text{for} \quad t > 0.$$

Before the conclusion, we take a simple observation.

Proposition 4.13.

$$|S_0(t)| \le |S(t)| \le \max\{[1+\mu_0^{-1}\int_{-t}^{0} g(\theta)d\theta]^{1/p},\ |S_0(t)|\}$$

for $t > 0$.

Proof. The first inequality is obvious. Every function φ in $L^p(\mu)$ is decomposed as $\varphi = \psi + \xi$, where $\psi(0) = \varphi(0)$, $\psi(\theta) = 0$ for $\theta < 0$. Then, $S(t)\varphi = S(t)\psi + S(t)\xi = S(t)\psi + S_0(t)\xi$ for $t \ge 0$, from which the second inequality is derived easily.

Example 4.14. Let $p = 1$, $\mu_0 = 1$ and $g(\theta) = e^{\gamma\theta}$ for $\theta \le 0$. Then $|S_0(t)| = e^{-\gamma t}$ for $t \ge 0$. On the one hand, we have the following relation on $|S(t)|$ for $t \ge 0$:

$$|S(t)| = \begin{cases} 1+t & \text{if } \gamma = 0, \\ e^{-\gamma t} & \text{if } \gamma \le -1, \\ 1 + (1-e^{-\gamma t})/\gamma & \text{otherwise}. \end{cases}$$

Corollary 4.15. Let μ be a positive Borel measure on $(-\infty,0]$.

Then, $L^p(\mu)$ satisfies Axiom (A) if and only if μ has the properties (μ-1) and (μ-2) and the function $G(u)$ defined by (4.3) is locally bounded on $(-\infty,0]$. Furthermore, if $L^p(\mu)$ satisfies Axiom(A), it also satisfies Axiom (A-1).

<u>Proof</u>. If $L^p(\mu)$ satisfies Axiom(A), then μ has the properties (μ-1) and (μ-2). Since $|S_0(t)| \le M(t)$ for $t \ge 0$, $G(u)$ is locally bounded on $(-\infty,0]$ by (v) in Theorem 4.12. The inverse result follows from the remark after Theorem 4.8, Theorem 4.11 and Proposition 4.13. The same remark and Theorems also imply the second result.

Taking the advantage that we are now familiar with the space $L^p(\mu)$, we add the study of the asymptotic behavior of $g(\theta)$ in relation with the property of fading memory, cf. Coleman & Mizel [1,2].

The condition that $L^p(\mu)$ contains BC as a subset is equivalent that

$$(\mu\text{-}3) \qquad \mu_\infty := \mu((-\infty,0]) < \infty$$

This implies that, for every $c > 0$,

$$(4.5) \qquad m(\{\theta < 0 : g(\theta) > c\}) < \mu_\infty/c.$$

Consider the property of fading memory (see Section 7.1):

$$(4.6) \qquad |S_0(t)\varphi| \to 0 \quad \text{as} \quad t \to \infty \quad \text{for each} \quad \varphi \quad \text{in} \quad L^p{}_0(\mu).$$

We take a simple observation to examine this property. If f is integrable on $[a,b]$, then

$$\lim_{h\to 0} \frac{1}{h} \int_x^{x+h} f(t)dt = f(x)$$

a.e. in $[a,b]$. Hence, if A is a subset of $[a,b]$ with a positive measure, it contains at least one point x for which the above limit relation is valid.

<u>Theorem 4.16.</u> Suppose that Conditions (μ-1), (μ-2), (μ-3) and (4.6) hold. Then $G(u)$ is bounded on $(-\infty,d]$ for every $d < 0$, and

(4.7) $$\lim_{d\to-\infty} \operatorname{ess\,sup}\{g(\theta) : \theta < d\} = 0.$$

Proof. The first result follows from Condition (4.6), Theorem 4.12 and the Banach-Steinhauss theorem.

For any $c > 0$, we set $X_c = \{\theta < 0 : g(\theta) > c\}$. Suppose that

$$a := \lim_{d\to-\infty} \operatorname{ess\,sup}\{g(\theta) : \theta < d\} > 0.$$

Then, we have $m(X_{a/2} \cap (-\infty, d]) > 0$ for every $d < 0$.

From the observation before the theorem, for any $n = 1, 2, \cdots$ there is a point θ_n in $X_{a/2} \cap (-\infty, -2n]$ such that

$$\lim_{h\to 0} \frac{1}{h} \int_{\theta_n}^{\theta_n+h} g(\theta)d\theta = g(\theta_n).$$

On the other hand, since $m(X_{\mu_\infty/n}) < n$ by (4.5), it follows that $m([\theta_n+1, \theta_n+n+2]\setminus X_{\mu_\infty/n}) > 1$. Hence, there is a σ_n in $[\theta_n+1, \theta_n+n+2]\setminus X_{\mu_\infty/n}$ such that

$$\lim_{h\to 0} \int_{\sigma_n}^{\sigma_n+h} g(\theta)d\theta = g(\sigma_n).$$

Let $\varphi^{n,h}$ and $\psi^{n,h}$ be the indicator function of $[\sigma_n, \sigma_n+h]$ and $[\theta_n, \theta_n+h]$, $h > 0$, respectively, and set $t_n = \sigma_n - \theta_n$. Since $S_0(t_n)\varphi^{n,h} = \psi^{n,h}$, we have that $|\psi^{n,h}| \le |S_0(t_n)||\varphi^{n,h}|$, which is written as

$$\int_{\theta_n}^{\theta_n+h} g(\theta)d\theta \le |S_0(t_n)| \int_{\sigma_n}^{\sigma_n+h} g(\theta)d\theta.$$

Dividing both sides by $h > 0$, and taking the limit as $h \to 0+$, we have that $g(\theta_n) \le |S_0(t_n)|g(\sigma_n)$. Since θ_n is in $X_{a/2}$ and σ_n is in $(X_{\mu_\infty/n})^c$, it follows that $a/2 \le |S_0(t_n)|^p\mu_\infty/n$, or $na/2\mu_\infty \le |S_0(t_n)|^p$. Since $t_n \ge 1$ and n is an arbitrary positive integer, this contradicts that $|S_0(t)|$ is bounded for $t \ge 1$.

Theorem 4.17. Assume (μ-1) and (μ-2) hold, $G(u)$ is bounded on $(-\infty, d]$ for some $d < 0$ and Condition (4.7) holds. Then Condition (4.6) holds.

Proof. Obviously, Condition (4.6) is written as

$$(4.8) \qquad \int_{-\infty}^{0} |\varphi(\theta)|^p g(\theta-t)d\theta \to 0 \quad \text{as} \quad t \to \infty$$

for φ in $L^p(\mu)$ with $\varphi(0) = 0$. If $R < \infty$, this follows from Theorem 4.10. Suppose $R = \infty$. Let n be a positive integer. From Condition (4.7), there is a $d_n < 0$ such that

$$m(\{\theta < d_n : g(\theta) > 1/n\}) = 0.$$

Since the integral in (4.8) is not affected by the replacement of $g(\theta)$ in the set of measure 0, we can assume that $g(\theta) \le 1/n$ for all $\theta < d_n$ and $n = 1,2,\cdots$. This means that

$$g(\theta-t) \to 0 \quad \text{as} \quad t \to \infty \quad \text{for each} \quad \theta < 0.$$

If we set $A = \sup\{G(u) : u \le d\}$, then we have

$$|\varphi(\theta)|^p g(\theta-t) \le A|\varphi(\theta)|^p g(\theta)$$

for $t > -d$ and for θ in $(-\infty,0)\setminus N_g$. Since $m(N_g) = 0$ by Theorem 4.10, the dominated convergence theorem implies Relation (4.8).

Example 4.18. Suppose that $L^p(\mu)$ is defined by the norm

$$|\varphi| = |\varphi(0)| + \{\int_{-\infty}^{0} |\varphi(\theta)|^p \frac{1}{\sqrt{|\theta|}(1+|\theta|^2)} d\theta\}^{1/p}.$$

Then, Conditions (μ-1) and (μ-2) hold and $G(u) = 1$ for $u < 1$, and $g(\theta) = |\theta|^{-1/2}(1+\theta^2)^{-1} \to 0$ as $\theta \to -\infty$. Hence, $L^p(\mu)$ satisfies Axioms (A) and (A-1), $|S_0(t)| = 1$ for all $t > 0$, and $|S_0(t)\varphi| \to 0$ as $t \to \infty$ for each fixed φ in $L^p(\mu)$ with $\varphi(0) = 0$.

The function $g(\theta)$ in this example has the property; $\sup\{g(\theta) : \theta \le d\}$ is finite for every d, but $g(\theta) \to \infty$ as $\theta \to 0-$. In general, we can prove the following result in the same way as in the proof of Theorem 4.16. See also Coleman-Mizel [1, Theorem 2].

Theorem 4.19. Assume that Conditions (μ-1), (μ-2) and (μ-3) hold. Then, ess sup$\{g(\theta) : \theta \le d\}$ is finite for every $d < 0$.

CHAPTER 2

FUNDAMENTAL THEOREMS

2.1. Existence and uniqueness of solutions. Let f be a function mapping a subset Ω of $\mathbb{R} \times \mathscr{B}$ into $\mathbb{E}$, where $\mathbb{E} = \mathbb{R}^n$ or $\mathbb{C}^n$. We consider a functional differential equation with infinite delay

$$(1.1) \qquad \dot{x}(t) = f(t, x_t).$$

A solution of Equation (1.1) on an interval I of $\mathbb{R}$ is a function $x : \bigcup_{t\in I} (-\infty, t] \to \mathbb{E}$ such that (t, x_t) is in Ω for t in I, and that $x(t)$ is locally absolutely continuous on I and satisfies (1.1) almost everywhere on I. Suppose (σ, φ) is a given point in Ω. If a solution x of Equation (1.1) is defined on an interval $[\sigma, \sigma+a]$, $a > 0$, and satisfies

$$(1.2) \qquad x_\sigma = \varphi,$$

then x is called a solution of Equation (1.1) with the initial value φ at σ or simply a solution through (σ, φ). The symbol $x(\sigma, \varphi)$ denotes such a solution, $x(t, \sigma, \varphi)$ the $\mathbb{E}$-value of $x(\sigma, \varphi)$ at t, and $x_t(\sigma, \varphi)$ the t-segment of $x(\sigma, \varphi)$. To emphasize f as well, we denote $x(\sigma, \varphi, f)$, $x(t, \sigma, \varphi, f)$ and $x_t(\sigma, \varphi, f)$.

Consider the trivial equation

$$\dot{x} = 0.$$

The solution $x(t, \sigma, \varphi, 0)$ of this equation is given as follows;

$$x(t, \sigma, \varphi, 0) = \begin{cases} \varphi(0) & \text{for } t > \sigma \\ \varphi(\sigma - t) & \text{for } t \le \sigma, \end{cases}$$

or $x_t(\sigma, \varphi, 0) = S(t-\sigma)\varphi$ for $t \ge \sigma$. If we set

$$(1.3) \qquad g(t, \psi) = f(t, \psi + S(t-\sigma)\varphi)$$

for (t, ψ) in $\mathbb{R} \times \mathscr{B}$ as long as the right hand side is well defined, the function $y(t) := x(t, \sigma, \varphi, f) - x(t, \sigma, \varphi, 0)$ is a solution of the equation

$$\dot{y}(t) = g(t, y_t) \tag{1.4}$$

on the interval $I = [\sigma, \sigma+a]$ with the initial value 0 at σ. Since the converse transform is valid as well, the solution $x(t,\sigma,\varphi,f)$ exists on I if and only if the solution $y(t,\sigma,0,g)$ exists on I.

Suppose Ω is an open set of $\mathbb{R} \times \mathcal{B}$. A function $f : \Omega \to \mathbb{E}$ is said to satisfy the Caratheodory condition if $f(t,\varphi)$ is measurable in t for each fixed φ, continuous in φ for each fixed t, a.e., and for every (σ,φ) in Ω, there is a neighborhood $V(\sigma,\varphi)$ and a Lebesgue integrable function m such that

$$|f(t,\psi)| \le m(t), \quad (t,\psi) \in V(\sigma,\varphi).$$

Since Ω is an open set and $\{S(t)\}$ is a strongly continuous semigroup on $\mathcal{B}$, we can find a set

$$Q = \{(t,\psi) \in \mathbb{R}\times\mathcal{B} : \sigma \le t \le \sigma+a,\ |\psi| \le B\}, \tag{1.5}$$

$a, B > 0$, such that $g(t,\psi)$ is well defined and satisfies the Caratheodory condition on Q and that

$$|g(t,\psi)| \le m(t), \quad (t,\psi) \in Q. \tag{1.6}$$

A partition P of an interval $[a,b]$ is a finite sequence $\{t_j\}$, $j = 0,1,2,\dots,k$, such that $a = t_0 < t_1 < \cdots < t_k = b$. The mesh $m(P)$ is defined as $m(P) = \max\{|t_j - t_{j-1}| : j = 1,2,\cdots,k\}$. For such a partition, we define a function τ_P on $[a,b]$ by

$$\tau_P(t) = \begin{cases} t_{j-1} & \text{for } t_{j-1} < t \le t_j, \quad j = 1,2,\cdots,k, \\ t_0 & \text{for } t = t_0. \end{cases}$$

It is clear that $|\tau_P(t) - t| \le m(P)$ for $t \in [a,b]$.

<u>Theorem 1.1.</u> Suppose f satisfies the Caratheodory condition on an open set Ω of $\mathbb{R}\times\mathcal{B}$. Then for any (σ,φ) in Ω there is a solution $x(\sigma,\varphi,f)$ on an interval $I = [\sigma,\sigma+\alpha]$, $0 < \alpha \le a$, as long as

$$(1.7)\qquad K(s)\int_{\sigma}^{\sigma+s} m(u)du \le B \quad \text{for} \quad 0 \le s \le \alpha,$$

where a, B are the constants in (1.5), m the function in (1.6) and $K(s)$ is the function in Axiom (A).

Proof. It suffices to show that the solution $y(t,\sigma,0,g)$ of Equation (1.4) exists on $I = [\sigma,\sigma+\alpha]$. For a partition $P : \sigma = t_0 < t_1 < \cdots < t_k = \sigma+\alpha$, we can define an approximation solution $y(t)$ on $(-\infty,\sigma+\alpha]$ by

$$(1.8)\qquad y(t) = \begin{cases} 0 & t \le \sigma \\ y(t_{j-1}) + \int_{t_{j-1}}^{t} g(s,y_{t_{j-1}})ds & t \in (t_{j-1},t_j], \end{cases}$$

$j = 1,2,\cdots,k$. Indeed, suppose $y(t)$ is well defined and continuous on $(-\infty,t_{j-1}]$ and satisfies

$$(1.9)\qquad |y(t)| \le \int_{\sigma}^{t} m(s)ds$$

for t in $[\sigma,t_{j-1}]$. In view of Axiom (A-iii) and the condition $y_\sigma = 0$ we have

$$|y_t| \le K(t-\sigma)\int_{\sigma}^{t} m(s)ds \qquad \sigma \le t \le t_{j-1},$$

which implies $|y_t| \le B$ on $\sigma \le t \le t_{j-1}$ due to (1.7). Thus $y(t)$ is defined by (1.8) on $(t_{j-1},t_j]$ and satisfies (1.9) there.

Using the function $\tau = \tau_P$, we can write

$$y(t) = \int_{\sigma}^{t} g(s,y_{\tau(s)})ds \qquad t \in I.$$

Let $\{P_n\}$ be a sequence of partitions of I such that $m(P_n) \to 0$ as $n \to \infty$. Denote by y^n and τ_n the approximation solution and the "τ-function", respectively, corresponding to P_n. Then the above argument shows that the sequence $\{y^n(t)\}$ is uniformly bounded and equicontinuous on I. Hence, a subsequence, denoted by $\{y^n\}$ again, converges to a continuous function y uniformly on I. If we redefine $y(t) = 0$ for $t < \sigma$, we see that $\{y^n_t\}$ converges to y_t uniformly on I. Furthermore, since $|\tau_n(t) - t| \le m(P_n)$ for t in I, the sequence $\{y^n_{\tau_n(t)}\}$ also converges to y_t uniformly on I.

The sequence $\{g(t, y^n_{\tau_n(t)})\}$ is then a sequence of measurable function on I, dominated by $m(t)$ and converging to $g(t, y_t)$ at each point t in I. Therefore, from the dominated convergence theorem we have that

$$y(t) = \int_\sigma^t g(s, y_s)ds \quad \sigma \le t \le \sigma+\alpha,$$

which is the desired result.

In the case that $\mathbb{E}$ is a Banach space of infinite dimension, the approximation solution y^n is defined in the same manner, and the sequence $\{y^n\}$ is uniformly bounded and equicontinuous on I. Such a sequence is relatively compact in $C(I, \mathbb{E})$ if and only if the set $\{y^n(t) : n = 1, 2, \ldots\}$ is relatively compact in $\mathbb{E}$ for each $t \in I$. To derive this property, Shin [6] proposed some compactness type condition on f in terms of measure of noncompactness (see Section 4.3), and obtained the existence theorem of the solution of the initial value problem for Equation (1.1), where $f : \Omega \to \mathbb{E}$ is assumed to be uniformly continuous on some neighborhood of (σ, φ).

The existence theorem is also obtained by reducing the problem to the fixed point theorem for an operator defined from the integral equation which is equivalent to Equation (1.1). In this direction, Shin [3] has proved another existence theorem under some different condition of the compactness of f, where $\dim \mathbb{E} = \infty$ and f is assumed to be uniformly continuous as in the above.

Next, we shall discuss the uniqueness of solutions of Equation (1.1). A function $f : \Omega \to \mathbb{E}$ is said to satisfy the local Lipshitz condition on Ω, if for any $(\sigma, \varphi) \in \Omega$, there exist a neighborhood $V(\sigma, \varphi)$ of (σ, φ) and a Lebesgue integrable function $n(t)$ such that

$$|f(t, \psi_1) - f(t, \psi_2)| \le n(t)|\psi_1 - \psi_2|_{\mathcal{X}}$$

for $(t, \psi_1), (t, \psi_2) \in V(\sigma, \varphi)$.

<u>Theorem 1.2.</u> Suppose f satisfies the local Lipschitz condition on an open set Ω. Then, for any $(\sigma, \varphi) \in \Omega$ there exists at most one solution of Equation (1.1) through (σ, φ).

In particular, if there exists a constant L such that

$$|f(t,\varphi) - f(t,\psi)| \le L|\varphi - \psi|_{\mathcal{B}} \quad \text{on} \quad \Omega,$$

then there exists a continuous function $L(t)$ which satisfies

$$|x_t(\sigma,\varphi,f) - x_t(\sigma,\psi,f)|_{\mathcal{B}} \le L(t-\sigma)|\varphi - \psi|_{\mathcal{B}}, \qquad t \ge \sigma,$$

whenever (σ,φ), $(\sigma,\psi) \in \Omega$, and the solutions $x(\sigma,\varphi,f)$ and $x(\sigma,\psi,f)$ exist on $[\sigma,t]$.

Proof. Let x^1 and x^2 be solutions of Equation (1.1) through (σ,φ), and J^i the interval of existence of x^i, $i = 1,2$. To establish the first part of the theorem we must show that

$$(1.10) \qquad |x^1_t - x^2_t|_{\mathcal{B}} = 0 \quad \text{for all } t \in J^1 \cap J^2.$$

Suppose (1.10) does not hold. Set $\tau = \sup\{\mu \ge \sigma : \text{(1.10) holds on } [\sigma,\mu]\}$. Then $\tau \ge \sigma$ and

$$(1.11) \qquad |x^1_\tau - x^2_\tau|_{\mathcal{B}} = 0.$$

Let $V(\tau,x^1_\tau)$ and $n(t)$ be the ones for the local Lipschitz condition of $f(t,\varphi)$. Then there exists a $b > 0$ such that

$$(t,x^1_t),\ (t,x^2_t) \in V(\tau,x^1_\tau)$$

for all $t \in [\tau,\tau+b]$. Since $x^1(\tau) = x^2(\tau)$, we have, for $t \in [\tau,\tau+b]$,

$$(1.12) \quad |x^1(t)-x^2(t)| \le |x^1(\tau)-x^2(\tau)| + \int_\tau^t |f(s,x^1_s)-f(s,x^2_s)|ds$$

$$\le \int_\tau^t n(s)|x^1_s-x^2_s|_{\mathcal{B}}ds,$$

and

$$|x^1_t - x^2_t|_{\mathcal{B}} \le K(t-\tau) \sup_{\tau\le s\le t} |x^1(s)-x^2(s)| + M(t-\tau)|x^1_\tau-x^2_\tau|_{\mathcal{B}}$$

$$\le K(t-\tau)\int_\tau^t n(s)|x^1_s - x^2_s|_{\mathcal{B}}ds$$

by Axiom (A) and (1.11). Thus, by Gronwall's inequality, we have

$$|x^1_t - x^2_t|_{\mathcal{B}} \le \varepsilon(t-\tau)\int_\tau^t n(s)\exp[\int_s^t n(u)K(u-\tau)du]ds$$

for all $\varepsilon > 0$ and all $t \in [\tau,\tau+b]$. Hence $|x^1_t - x^2_t|_{\mathcal{B}} = 0$ on $[\tau,\tau+b]$, which contradicts the definition of τ. Thus, (1.10) holds. The second part of the theorem can easily be proved by applying Gronwall's inequality, again. Indeed, we may set

$$L(t) = HK(t) + M(t) + LK(t)\int_0^t \{HK(s)+M(s)\}\exp(L\int_s^t K(u)du)ds.$$

For the detail, see the proof of Theorem 4.1.2 below.

Shin has introduced several conditions for the uniqueness of solutions of the initial value problem (Shin [2,4,5]). For simplicity, suppose that there exists a function $\omega(t,r)$ such that

$$(1.13)\qquad |f(t,\varphi) - f(t,\psi)| \le \omega(t,|\varphi-\psi|) \quad \text{on } \Omega,$$

and let x^1, x^2, τ and b be as in the above proof. To estimate the functions $u(t) := |x^1_t - x^2_t|$ and $w(t) := \sup\{|x^1(s)-x^2(s)| : \tau \le s \le t\}$, we begin with the inequality that

$$(1.14)\qquad |x^1(t) - x^2(t)| \le \int_\tau^t \omega(s,|x^1_s - x^2_s|)ds \qquad \tau \le t \le \tau+b.$$

From Axiom (A) it follows at once that

$$u(t) \le K(t-\tau)\int_\tau^t \omega(s,u(s))ds.$$

Thus, the problem is reduced to the uniqueness of the zero solution of this integral inequality.

If $\omega(t,r)$ is nondecreasing with respect to r, it follows that $\omega(s,|x^1_s - x^2_s|) \le \omega(s,K(s-\tau)w(s))$. Hence, Inequality (1.14) implies that

$$w(t) \le \int_\tau^t \omega(s,K(s-\tau)w(s))ds.$$

The functions $u(t)$ and $w(t)$ also satisfy differential inequalities. To see this, denote by D^+x the right hand derivative of a function x, and $\overline{D}^+g$ the upper right hand derivative of a real function g.

Proposition 1.3 (Shin [4]). If $x : (-\infty, t+A) \to \mathbb{E}$, $A > 0$, is admissible on $[t, t+A)$, and if $[D^+x](t)$ exists, then

$$\overline{D}^+|x_t| \le K(0)|D^+x(t)| + \beta_\nu|x_t|,$$

where β_ν is defined as

$$\beta_\nu := \overline{D}^+|S|(0) = \limsup_{h \to 0+} \frac{|S(h)| - 1}{h}.$$

Proof. Taking the fundamental decomposition $x = y + z$ at t (see Section 1.2), we can write $x_{t+h} = y_{t+h} + S(h)x_t$, which implies that

$$|x_{t+h}| - |x_t| \le |y_{t+h}| + (|S(h)| - 1)|x_t|$$

$$\le K(h)\sup\{|x(t+s) - x(t)| : 0 \le s \le h\}$$

$$+ (|S(h)| - 1)|x_t|.$$

Since $[D^+x](t)$ exists, there exists a function $\rho(h) \ge 0$, $\rho(h) \to 0$ as $h \to 0+$, such that

$$\sup\{|x(t+s) - x(t)| : 0 \le s \le h\} \le |D^+x(t)|h + \rho(h)h.$$

These inequalities imply the desired result.

Applying this proposition to the function $u(t)$, we obtain the differential inequality that

$$\overline{D}^+u(t) \le K(0)\omega(t, u(t)) + \beta_\nu u(t)$$

under the assumption (1.13). It follows at once from this inequality that the function $L(t)$ in Theorem 1.2 can be taken as

$$L(t) = \exp(K(0)L + \beta_\nu)t.$$

The number β_ν is nonnegative for several spaces (see Shin [4]); it is an interesting problem to find a space having negative β_ν.

To deal with the function $w(t)$, we prepare the following Proposition.

Proposition 1.4 (Shin [4]). If $p(t)$ is a real, continuous

function on $[\sigma,\infty)$, then

$$\bar{D}^+\sup\{p(s) : \sigma \le s \le t\} \le |\bar{D}^+p(t)| \quad \text{for} \quad t \ge \sigma.$$

Proof. Set $q(t) = \sup\{p(s) : \sigma \le s \le t\}$. Since $q(t)$ is nondecreasing, the result is trivial in case that $q(t+h) = q(t)$ for some $h > 0$. Suppose that $q(t+h) > q(t)$ for all $h > 0$. If $p(t) < q(t)$, then $q(t+h) = q(t)$ for sufficiently small $h > 0$. Thus, we have that $p(t) = q(t)$, which implies that $q(t+h_2) = \sup\{p(t+s) : 0 \le s \le h_2\}$ for $h_2 > 0$. Since $q(t+h_2) > q(t) = p(t)$, there exists an h_1 in $(0,h_2]$ such that $q(t+h_2) = p(t+h_1)$. Hence, if $0 < h_2 \le h$, then

$$\frac{1}{h_2}\{q(t+h_2) - q(t)\} = \frac{1}{h_2}\{p(t+h_1) - p(t)\}$$

$$\le \frac{1}{h_1}\{p(t+h_1) - p(t)\}.$$

Consequently, we have that

$$\sup\{\frac{1}{h_2}[q(t+h_2) - q(t)] : 0<h_2<h\} \le \sup\{\frac{1}{\delta}[p(t+\delta) - p(t)] : 0<\delta\le h\},$$

which shows that $(\bar{D}^+q)(t) \le (\bar{D}^+p)(t)$.

Set $v(t) = |x^1(t) - x^2(t)|$. Since x^1 and x^2 are differentiable on $[\tau,\tau+b]$, it follows that $D^+v(t)$ exists and $D^+v(t) \le |\dot{x}^1(t) - \dot{x}^2(t)|$. Hence, if Inequality (1.13) holds and $\omega(t,r)$ is nondecreasing with respect to r, we have that

$$D^+w(t) \le \omega(t,K(t-\tau)w(t))$$

for $\tau \le t \le \tau+b$.

Thus the uniqueness of solutions is reduced to the uniqueness of the zero solution of integral inequalities and differential ones in the above. See Shin [2,4,5] for the details.

For ordinary differential equations, it is well known that the uniqueness of solutions is characterized by the existence of some kind of Liapunov function, see Yoshizawa [2]. Similar results are presented in a restricted sense for functional differential equations with finite delay in Furumochi [1].

<u>2.2. Noncontinuable solutions and continuous dependence.</u> For ordinary differential equations, continuous dependence theorems are well known (cf. Hartman [1, p.14, Theorem 3.2]). Similar results hold for functional differential equations with finite delay (cf. Rybakowskii [1,2]), and for functional differential equation with infinite delay (cf. Kato [4]). Following the idea of Rybakowskii, we now present the proof of the continuous dependence theorem for infinite delay equations by using Zorn's lemma. In this section, we set

$$|x|_S = \sup\{|x(t)| : t \in S\}$$

for a function x mapping a set S into a semi-normed space.

Suppose $x(t)$ and $y(t)$ are solutions with the same initial condition (1.2) and satisfies Equation (1.1) respectively on the intervals I and J whose left end points are σ. If I is contained in J and $x(t) = y(t)$ for t in I, we say that y is a continuation of x. If x has no continuation, it is called a noncontinuable solution, or a maximal solution. employing Zorn's lemma, one can show that every solution $x(\sigma,\varphi,f)$ has a maximal continuation.

Consider a family of functional differential equations with infinite delay

$$\dot{x} = f^k(t,x_t), \tag{2.1}$$

$k = 1,2,\cdots$. Let x^k be a noncontinuable solution of (2.1) on $[\sigma(k),\tau(k))$, with the initial condition

$$x^k_{\sigma(k)} = \varphi^k. \tag{2.2}$$

Let $\mathbb{N}$ be the set of all positive integers.

<u>Theorem 2.1.</u> Let f and f^k, $k \in \mathbb{N}$, be continuous functions mapping an open set Ω of $\mathbb{R} \times \mathscr{B}$ into $\mathbb{E} := \mathbb{R}^n$ or $\mathbb{C}^n$ such that the sequence $\{f^k\}$ converges to f compactly on Ω. Let x^k be as cited above, and assume that

$$(\sigma(k),\varphi^k) \to (\sigma,\varphi) \in \Omega \quad \text{as} \quad k \to \infty. \tag{2.3}$$

Then there are a maximal solution x of the equation

$$\dot{x} = f(t, x_t), \quad x_\sigma = \varphi$$

defined on $[\sigma, \tau)$, and a subsequence $\{k(i)\}$ of positive integers such that the following conditions hold:

(i) $\tau \le \liminf_{i\to\infty} \tau(k(i))$

(ii) $|x^{k(i)} - x|_{[s(k(i)),t]} \to 0$ as $i \to \infty$ for every t in (σ, τ), where

$$s(k) = \max\{\sigma, \sigma(k)\}. \tag{2.4}$$

The proof will be divided into several parts as follows. For an interval I with $\sigma = \inf I$ in I, and for a function $\varphi : (-\infty, 0] \to E$, we denote by $C_I(\varphi)$ the family of functions, with $x_\sigma = \varphi$, and which is continuous on I. Also, $C_I(\mathcal{B})$ denotes the union of $C_I(\varphi)$ for all φ in $\mathcal{B}$.

Lemma 2.2. Assume $\mathcal{Q} = \{\varphi^k : k \in \mathbb{N}\}$ is a relatively compact subset of $\mathcal{B}$, x^k is a function in $C_{I(k)}(\varphi^k)$ for some interval $I(k) = [\sigma(k), \tau(k)]$, $\{\tau(k) - \sigma(k) : k \in \mathbb{N}\}$ is bounded, and that $\{x^k|_{I(k)} : k \in \mathbb{N}\}$ equicontinuous. Then the family of $\mathcal{B}$-valued functions x^k_t, as a function of t in $I(k)$, $k \in \mathbb{N}$, is also equicontinuous.

Proof. It follows from Axiom (A) that

$$|x^k_{t+h} - x^k_t| \le K(t-\sigma(k))\sup\{|x^k(s+h) - x^k(s)| : \sigma(k) \le s \le t\}$$
$$+ M(t-\sigma(k))|x^k_{\sigma(k)+h} - x^k_{\sigma(k)}|$$

provided $\sigma(k) \le t \le t+h \le \tau(k)$, and that

$$|x^k_{\sigma(k)+h} - x^k_{\sigma(k)}| \le |x^k_{\sigma(k)+h} - S(h)x^k_{\sigma(k)}|$$
$$+ |S(h)x^k_{\sigma(k)} - x^k_{\sigma(k)}|$$
$$\le K(h)\sup\{|x^k(\sigma(k)+s) - x^k(\sigma(k))| : 0 \le s \le h\}$$
$$+ |S(h)\varphi^k - \varphi^k|.$$

The conclusion follows from these estimates, the assumptions in Lemma and the fact that $S(h)\varphi$ is uniformly continuous for (h, φ) in the relatively compact set $[0,1] \times \mathcal{Q}$.

Proposition 2.3. Suppose the conditions of Theorem 2.1 are satisfied, and set $B_* = \sup\{|f^k(\sigma(k),\varphi^k)| : k \in \mathbb{N}\}$. Then B_* is finite, and for every $B > B_*$, there is an $r > 0$ such that $[\sigma(k),\sigma(k)+r] \subset [\sigma(k),\tau(k))$ and

$$(2.5) \qquad |\dot{x}^k(t)| \le B \quad \text{for} \quad t \in [\sigma(k),\sigma(k)+r],$$

and for $k \in \mathbb{N}$.

Proof. Since $\{f^k(\sigma(k),\varphi^k)\}$ converges to $f(\sigma,\varphi)$ due to the assumptions for $\{f^k\}$ and $\{(\sigma(k),\varphi^k)\}$, the quantity B_* is finite. Let $I(k) = [\sigma(k),\tau(k))$, $B(k) = \sup\{|\dot{x}^k(t)| : t \in I(k)\}$, $K_1 = \{k : B(k) \le B\}$ and $K_2 = \{k : B(k) > B\}$.

Set $a_1 = \inf\{\tau(k)-\sigma(k) : k \in K_1\}$. If K_1 is a finite set, then $a_1 > 0$. Suppose $a_1 = 0$. Then K_1 is an infinite set; choosing a subsequence in K_1 if necessary, one can assume that $\tau(k)-\sigma(k) \to 0$ as $k \to \infty$. The function x^k, k in K_1, is extended to the interval $(-\infty,\tau(k)]$ so that $|\dot{x}^k(t)| \le B$ for t in $[\sigma(k),\tau(k)]$. It follows from Lemma 2.2 that the family of functions x^k_t, where t is in $[\sigma(k),\tau(k)]$, is uniformly equicontinuous. Thus, if we set $\psi^k = x^k_{\tau(k)}$, the inequality $|\psi^k - \varphi| \le |x^k_{\tau(k)} - x^k_{\sigma(k)}| + |\varphi^k - \varphi|$ implies $(\tau(k),\psi^k) \to (\sigma,\varphi)$ as $k \to \infty$. Since (σ,φ) is a point in the open set Ω, one has that $(\tau(k),\psi^k)$ lies in Ω for sufficiently large k. This contradicts the assumption that x^k is a noncontinuable solution on $I(k)$. Therefore, if we set $r_1 = a_1/2$, then r_1 is positive, and Condition (2.5) holds for $r = r_1$ and for k in K_1.

Let k be in K_2. Since $|\dot{x}^k(\sigma(k))| = |f^k(\sigma(k),\varphi^k)| \le B_* < B$, there is a $t(k)$ in $(\sigma(k),\tau(k))$ such that

$$|\dot{x}^k(t)| \le B \quad \text{for} \quad t \in [\sigma(k),t(k)) \quad \text{and} \quad |\dot{x}^k(t(k))| = B.$$

Set $a_2 = \inf\{t(k)-\sigma(k) : k \in K_2\}$. If K_2 is a finite set, then $a_2 > 0$. Suppose $a_2 = 0$. In the same manner as before, one can assume $(t(k),x^k_{t(k)}) \to (\sigma,\varphi)$ as $k \to \infty$, hence $f^k(t(k),x^k_{t(k)}) \to f(\sigma,\varphi)$ as $k \to \infty$. This is a contradiction since $|f^k(t(k),x^k_{t(k)})| = |\dot{x}^k(t(k))| = B > B_* \ge \lim_{k\to\infty} |f^k(\sigma(k),\varphi^k)| = |f(\sigma,\varphi)|$. Therefore, Condition (2.5) holds for $r = a_2 > 0$ and for k in K_2.

From Proposition 2.3, it follows immediately that $s(k)$ defined by (2.4) is in $[\sigma,\tau(k))$ for sufficiently large k, and

(2.6) $(s(k), x^k_{s(k)}) \to (\sigma, \varphi)$ as $k \to \infty$.

Proposition 2.4. Suppose that the conditions in Theorem 2.1 are satisfied, and that there is a sequence $r(k)$ having the following properties :

(i) $s(k) \le r(k) \le \tau(k)$, $r(k) \to r-0$ as $k \to \infty$ for some $r > \sigma$,

(ii) the family $\{x^k|_{I(k)}\}$, where $I(k) = [s(k), r(k)]$, is uniformly bounded and equicontinuous. Then there are a function x in $C_I(\varphi)$, where $I = [\sigma, r]$, and an increasing sequence of integers $\{k(i)\}$ such that

(2.7) $\sup\{|x^{k(i)}_t - x_t| : t \in I(k(i))\} \to 0$ as $i \to \infty$.

Furthermore, the limit function x is a solution of Equation (1.1) through (σ, φ) on any interval $[\sigma, \rho]$ with $\sigma < \rho \le r$ whenever (t, x_t) lies in Ω for t in $[\sigma, \rho]$.

Proof. From the assumption on the family $\{x^k|_{I(k)}\}$, it is easy to see that $|x^{k(i)} - x|_{I(k(i))} \to 0$ as $i \to \infty$ for some sequence $\{k(i)\}$ and for some continuous function $x : I \to E$. Since $x(\sigma) = \lim_{k\to\infty} x^k(s(k)) = \varphi(0)$ from (2.6), the function x lies in $C_I(\varphi)$ by setting $x(t) = \varphi(t-\sigma)$ for $t < \sigma$, which implies that

$$(s(k(i)), x^{k(i)}_{s(k(i))}) \to (\sigma, x_\sigma) \quad \text{as} \quad i \to \infty.$$

Condition (2.7) then follows immediately, and the set

$$\{(t, x_t) : t \in I\} \cup [\bigcup_{i=1}^{\infty} \{(t, x^{k(i)}_t) : t \in I(k(i))\}]$$

is a compact set of $\mathbb{R} \times \mathcal{B}$. Furthermore, the members in the bracket are of course contained in Ω, but the first member may touch the boundary of Ω. This situation can be avoided by replacing r with some ρ in (σ, r) so that $(t, x_t) \in \Omega$ for t in $[\sigma, \rho]$, which is possible since $(\sigma, x_\sigma) = (\sigma, \varphi)$ is in the open set Ω. In addition, it can be assumed that $s(k(i)) \le \rho \le r(k(i))$ for $i = 1, 2, \cdots$. As a result, the set

$$\mathcal{Q} = \{(t, x_t) : t \in J\} \cup [\bigcup_{i=1}^{\infty} \{(t, x^{k(i)}_t) : t \in J(k(i))\}]$$

is a compact set of Ω, where $J = [\sigma, \rho]$ and $J(k(i)) = [s(k(i)), \rho]$.

From the assumption on $\{f^k\}$ and f, the sequence $\{f^k\}$ converges to f uniformly on $\mathcal{Q}$, and f is uniformly continuous on $\mathcal{Q}$. Hence, it follows from (2.7) that

$$\sup\{|f^{k(i)}(s, x^{k(i)}_s) - f(s, x_s)| : s \in J(k(i))\} \to 0 \quad \text{as} \quad i \to \infty.$$

Taking the limit in the relation

$$x^k(t) = x^k(s(k)) + \int_{s(k)}^{t} f^k(s, x^k_s)ds, \quad t \in J(k)$$

for $k = k(i)$, $i = 1, 2, \cdots$, we have finally that

$$x(t) = \varphi(0) + \int_{\sigma}^{t} f(s, x_s)ds, \quad t \in J.$$

<u>Proof of Theorem 2.1.</u> Let X be a collection of all solutions $x = x(\sigma, \varphi, f)$ defined on $I(x) = [\sigma, \rho(x))$ such that Conditions (i) and (ii) in Theorem 2.1 hold for some subsequence $\{k(i)\}$, where τ is read as $\rho(x)$. Then X is a nonempty set from Propositions 2.3 and 2.4. Observe that a solution $x = x(\sigma, \varphi, f)$ on $I(x)$ belongs to X if and only if there exist a subsequence $\{k(i)\}$ and a sequence $\{t(i)\}$ such that $s(k(i)) < t(i) < \tau(k(i))$, $t(i) \to \rho(x)$-0 as $i \to \infty$ and

$$|x^{k(i)} - x|_{[s(k(i)), t(i)]} \to 0 \quad \text{as} \quad i \to \infty. \tag{2.8}$$

The set X becomes an inductive ordered set in the standard order relation: $x \leq y$ for x and y in X if y is a continuation of x. To see X is an inductive set, take a totally ordered subset Y of X. Then the union of $[\sigma, \rho(y))$ for all y in Y is an interval $[\sigma, \rho_0)$, and the function $z : (-\infty, \rho_0) \to \mathbb{E}$ is well defined by

$$z(t) = y(t) \quad \text{for} \quad t \quad \text{in} \quad I(y) \quad \text{and for some} \quad y \in Y.$$

This function z is clearly a solution of Equation (1.1) on $[\sigma, \rho_0)$ with the initial condition $z_\sigma = \varphi$. To show that z is in X, take a sequence $\{t(i)\}$ in (σ, ρ_0) converging to ρ_0. Since $\rho_0 = \sup\{\rho(y) : y \in Y\}$, there is a y^i in Y such that $t(i) < \rho(y^i)$ for $i = 1, 2, \cdots$. From the second assumption for elements in X, for each i there exist infinite number of k such that

$$|x^k - y^i|_{[s(k),t(i)]} < 1/i.$$

Thus, there is an increasing sequence $\{k(i)\}$ such that the above inequality holds for $k = k(i)$ and for $i = 1,2,\cdots$, in other words, Relation (2.8) holds for $x = z$, where $\rho(z) = \rho_0$.

Let x be a maximal element of X. Since any element of X satisfies Conditions (i) and (ii) in Theorem, it remains to show that x is a noncontinuable solution. Suppose this is not the case. Then x is bounded and uniformly continuous on $I(x) = [\sigma,\rho(x))$. Let $\{k(i)\}$ and $\{t(i)\}$ be the sequences which arise from the condition for x to belong to X. Then the family $\{x^{k(i)}|_{[s(k(i)),t(i)]}\}$ is uniformly bounded and uniformly equicontinuous. Furthermore, the existence of a continuation of x guarantees that (t,x_t) converges to a point $(\rho(x),\psi)$ as $t \to \rho(x)-0$, and that $(\rho(x),\psi)$ lies in the open set Ω. Combining this fact with Lemma 2.2, we have $(t(i),\psi^i) \to (\rho(x),\psi)$ as $i \to \infty$, where $\psi^i = x^{k(i)}_{t(i)}$, $i = 1,2,\cdots$. Also, since $x^{k(i)}$ is a noncontinuable solution through $(t(i),\psi^i)$, from Proposition 2.3 there are $r_0 > 0$ and $B_0 > 0$ such that $x^{k(i)}$ is defined on $[t(i),t(i)+r_0]$ and $|x^{k(i)}(t)| \le B_0$ for t in this interval and for $i = 1,2,\cdots$. Consequently the family $\{x^{k(i)}|_{J(i)}\}$, where $J(i) = [s(k(i)),t(i)+r_0]$, is again uniformly bounded and uniformly equicontinuous. Hence, there exist a subsequence $\{i(j)\}$, $j = 1,2,\cdots$, and a function y in $C_J(\varphi)$, where $J = [\sigma,\rho(x)+r_0]$ such that

$$(2.9) \qquad |y^j_t - y_t|_{J(i(j))} \to 0 \quad \text{as} \quad j \to \infty,$$

where $y^j = x^{k(i(j))}$ for $j = 1,2,\cdots$. It is clear that $y(t) = x(t)$ for $t < \rho(x)$, which implies $(t,y_t) = (\rho(x),\psi)$ at $t = \rho(x)$. Since $(\rho(x),\psi)$ is in the open set Ω, there is an r_1 in $(0,r_0)$ such that (t,y_t) still lies in Ω for t in $[\sigma,\rho(x)+r_1]$. Thus, from Proposition 2.4 $y(t)$ is a solution of Equation (1.1) through (σ,φ) on the interval $[\sigma,\rho(x)+r_1]$. It belongs to X due to Relation (2.9); hence, $x \le y$, $x \ne y$ and $y \in X$. This is a contradiction.

If the solution $x(\sigma,\varphi,f)$ is unique, it follows immediately from Theorem 2.1 that $x_t(\sigma(k),\varphi^k,f) \to x_t(\sigma,\varphi,f)$ as $k \to \infty$ compactly on the interval of the existence of $x(\sigma,\varphi,f)$, cf. Hino [2].

Theorem 2.1 is also extended to the case that $\mathbb{E}$ is a Banach space of infinite dimension (Shin [6]). As in the existence theorem for such a case, some compactness conditions for the functions f and $f - f^k$, $k = 1,2,\cdots$, are assumed to show that the set $\{x^k(t) : k \in \mathbb{N}\}$ and the set $\{x^k_t : k \in \mathbb{N}\}$ are relatively compact in $\mathbb{E}$ and in $\mathcal{B}$, respectively, for each fixed t. See the above paper for the details.

From Proposition 2.3 we can also derive the limiting behavior of a noncontinuable solution near the boundary of its domain.

<u>Theorem 2.5.</u> Assume Ω is an open set of $\mathbb{R} \times \mathcal{B}$, and $\mathcal{F}$ a relatively compact subset of $C(\Omega,\mathbb{E})$ in the compact open topology. Then, for any compact subset $\mathcal{Q}$ of Ω, there is an $r > 0$ such that, for any (σ,φ) in $\mathcal{Q}$ and for any f in $\mathcal{F}$, the interval $[\sigma,\sigma+r]$ is contained in the interval of existence of every noncontinuable solution $x(\sigma,\varphi,f)$.

<u>Proof.</u> Suppose a compact subset $\mathcal{Q}$ of Ω has not this property. Then, there is a sequence of noncontinuable solutions $x^k = x(\sigma(k),\varphi^k,f^k)$ defined on $[\sigma(k),\tau(k))$, $k \in \mathbb{N}$, such that $(\sigma(k),\varphi^k) \in \mathcal{Q}$ and $f^k \in \mathcal{F}$ for $k \in \mathbb{N}$ and that $\tau(k) - \sigma(k) \to 0$ as $k \to \infty$. From the compactness condition on $\mathcal{Q}$ and on $\mathcal{F}$, we can assume that $\{(\sigma_k,\varphi_k)\}$ converges to a point (σ,φ) in $\mathcal{Q}$, and that $\{f^k\}$ converges to a function f in $C(\Omega,\mathbb{E})$ compactly. Then the conclusion of Proposition 2.3 contradicts that $\tau(k) - \sigma(k) \to 0$ as $k \to \infty$.

<u>Theorem 2.6</u>. Assume Ω is an open set of $\mathbb{R} \times \mathcal{B}$, $f : \Omega \to \mathbb{E}$ is continuous, and x is a noncontinuable solution of Equation (1.1) defined on $I = [\sigma,\tau)$. Then, for every compact subset $\mathcal{Q}$ of Ω, there is a $t_{\mathcal{Q}}$ in I such that $(t,x_t) \bar{\in} \mathcal{Q}$ for t in $(t_{\mathcal{Q}},\tau)$.

<u>Proof</u>. It suffices to consider the case τ is finite. Let r be a positive number having the property in Theorem 2.5 for the singleton family $\mathcal{F} = \{f\}$ and the given $\mathcal{Q}$. If we set $t_{\mathcal{Q}} = \max\{\tau-r, \sigma\}$, then the conclusion immediately follows from Theorem 2.5.

<u>Theorem 2.7 (Hale & Kato [1])</u>. Let Ω, f and x be as in Theorem 2.6, and assume that f takes closed bounded subsets of Ω into bounded sets. Then, for every closed bounded subset W of Ω there is a sequence $t(k) \to \tau-0$ such that $(t(k),x_{t(k)}) \bar{\in} W$. Moreover, if $\Omega = \mathbb{R} \times \mathcal{B}$ or if $\mathcal{B}$ is bestowed the hypothesis;

(*) there exist $r > 0$ and K^* such that $|\varphi|_{[-r,0]} \le K^*|\varphi|_{\mathcal{B}}$

then there is a t_W such that $(t,x_t) \bar{\in} W$ for $t \in [t_W,\tau)$.

Proof. Suppose there is a ρ in $[\sigma,\tau)$ such that $(t,x_t) \in W$ for $\rho \le t < \tau$. Then, τ is finite, and $|\dot{x}(t)| \le B$ for $\rho \le t < \tau$, where $B = \sup|f(W)|$. Since W is a closed subset of Ω, it follows that (t,x_t) converges to a point (τ,ψ) in W as $t \to \tau-0$. Thus, the set $\{(t,x_t): t \in [\rho,\tau)\} \cup \{(\tau,\psi)\}$ is a compact subset of Ω, which contradicts Corollary 2.6.

Now, suppose that the second part is false for W. Then, there exists a sequence $\{t(k)\}$, $t(k) \to \tau-0$, such that $(t(k),x_{t(k)}) \in W$. We shall show that this fact implies that $\mathcal{Q} = \mathrm{cl}\{(t,x_t): t \in [\sigma,\tau)\}$ is a bounded set in Ω. At first, if $\mathcal{Q}$ is contained in Ω, which is always true when $\Omega = \mathbb{R} \times \mathcal{B}$, then it follows that $\mathcal{Q}$ is bounded. Indeed, if $\mathcal{Q}$ is unbounded, then there is a sequence $\{s(k)\}$ such that $s(k-1) < t(k) < s(k)$, $|x_t| < C$ for $t \in [t(k),s(k))$ and $|x_{s(k)}| = C$, where $C = \{K(0)H + \overline{\lim}_{\beta\to 0^+} M(\beta)\}C_1 + 1$ and $C_1 = \sup\{|\varphi|: (t,\varphi) \in W$ for some $t \in [\sigma,\tau)\}$. Set

$$\mathcal{Q}_0 = \mathrm{cl}\{(t,x_t): t \in J\},\ J = \bigcup_k [t(k),s(k)].$$

Since $\mathcal{Q}_0 \subset \mathcal{Q} \subset \Omega$ and $\mathcal{Q}_0$ is closed and bounded, $|f(t,x_t)| \le L$ for $t \in J$ and for some $L > 0$, and hence, from Axiom (A-iii) we have

$$\begin{aligned}|x_{s(k)}| &\le K(\beta(k))\{|x(t(k))| + L\beta(k)\} + M(\beta(k))|x_{t(k)}| \\ &\le \{K(\beta(k))H + M(\beta(k))\}C_1 + K(\beta(k))L\beta(k)\end{aligned}$$

for $\beta(k) = s(k) - t(k)$. From this inequality it follows that $|x_{s(k)}| < C$ for large k since $\beta(k) \to 0$ as $k \to \infty$. This is a contradiction. Therefore, now let $\mathcal{Q} \cap \partial\Omega \ne \phi$, and there exists a sequence $(\sigma(k),x_{\sigma(k)}) \to (\tau,\psi) \in \mathcal{Q} \cap \partial\Omega$. By hypothesis (*) $|x_{\sigma(k)} - \psi|_{[-r,0]} \to 0$ for an $r > 0$, $r < \tau - \sigma$. Thus, ψ is continuous on $[-r,0]$, and $x(t) = \psi(t-\tau)$ for $\tau-r \le t < \tau$. Hence, we have $\mathcal{Q} = \{(t,x_t) : t \in [\sigma,\tau)\} \cup \{(\tau,\psi)\}$, which means that $\mathcal{Q}$ is a compact set of $\mathbb{R} \times \mathcal{B}$. This implies that $W_0 = W \cap \mathcal{Q}$ is a compact subset of Ω. This is a contradiction since $(t(k),x_{t(k)}) \in W_0$ and $t(k) \to \tau-0$ as $k \to \infty$.

Consequently, (t,x_t) is in the closed bounded set $\mathcal{Q}$ of Ω for $t \in [\sigma,\tau)$, which contradicts the first part of the theorem.

The fundamental assumptions in Theorem 2.1 are that Axiom (A) and (A1) hold and that f and f^k are continuous functions on the open set Ω of $\mathbb{R}\times\mathcal{B}$. They may be natural ones to deal with the continuous dependence, as well as the existence and the uniqueness, etc., in the general situation, but they are not necessarily needed to treat the problem for equations which are given in the explicit form. In this direction, for the Volterra integrodifferential equation

$$\dot{x} = h(t,x) + \int_{-\infty}^{t} q(t,s,x(s))ds \qquad t \geq 0$$

Kaminogo [2] has given a partial answer to the problem, presented in Burton's book [2], on the continuous dependence of solutions on the initial condition

$$x(t) = \varphi(t) \qquad -\infty < t \leq 0.$$

He discussed the problem in the phase space C_g, which does not necessarily satisfy Axiom (A1), under very mild conditions. The main assumption is that the improper integral

$$\int_{-\infty}^{0} q(t,s,\varphi(s))ds = \lim_{R\to\infty} \int_{-R}^{0} q(t,s,\varphi(s))ds$$

exists and it is continuous in $t \in [0,T]$ for each fixed $\varphi \in C_g$. Under some additional assumptions it is proved that, if a bounded sequence $\{\varphi^k\}$ in C_g converges to φ in C_g a.e. on $(-\infty,0]$ and if $\{\varphi^k(0)\}$ converges to $\varphi(0)$, then $\{x(t,\varphi^k)\}$ converges to $x(t,\varphi)$ uniformly on $[0,T]$, where for each φ the unique existence of the solution $x(t,\varphi)$ of the above initial value problem is assumed. Furthermore, this result is extended to the case that the sequence of initial functions $\{\varphi^k\}$ is a general sequence of the space $C((-\infty,0],\mathbb{R}^n)$, and converges to a function φ compactly.

2.3. Kneser's property. In this section we shall give some results on Kneser's property for Equation (1.1), although we do not utilize them in the backward chapter of this book.

Let $\sigma < \tau$, and let $C([\sigma,\tau]),\mathbb{R}^n)$ be the Banach space of all continuous functions from the compact interval $[\sigma,\tau]$ into $\mathbb{R}^n$ with the usual supremum norm. For an $\mathbb{R}^n$-valued function u defined on $(-\infty,\tau]$, let $u|_{[\sigma,\tau]}$ be the restriction of u to the interval $[\sigma,\tau]$.

For $\varphi \in \mathscr{B}$, we denote by $S(\varphi)$ the set of all solutions of Equations (1.1) through (σ,φ). If $f : [\sigma,\tau] \times \mathscr{B} \to \mathbb{R}^n$ is bounded and continuous, then each element in $S(\varphi)$ is defined on the whole interval $[\sigma,\tau]$ by Theorem 2.7.

The following theorem is due to Kaminogo [1]. We omit the proof.

Theorem 3.1. If $f : [\sigma,\tau] \times \mathscr{B} \to \mathbb{E}$ is a bounded and continuous function, then for any $\varphi \in \mathscr{B}$, the sets

$$\{x|_{[\sigma,\tau]} : x \in S(\varphi)\},\ \{x(\tau) : x \in S(\varphi)\} \quad \text{and} \quad \{x_\tau : x \in S(\varphi)\}$$

are continuum (i.e., compact and connected) in $C([\sigma,\tau],\mathbb{R}^n)$, $\mathbb{R}^n$ an $\mathscr{B}$, respectively.

CHAPTER 3

STIELTJES INTEGRALS AND LINEAR OPERATORS ON $\mathscr{B}$

3.1. Riemann-Stieltjes integral and Darboux-Stieltjes integral.

Stieltjes integrals, relative to the Riesz representation theorem of linear operators, appear in the theory of linear functional differential equations. We deal with the integral of a function of the jump discontinuity. Such a function is not necessarily Riemann-Stieltjes integrable with respect to functions of bounded variation. To avoid this difficulty, the integral is extended to the Lebesgue-Stieltjes integral. For understanding this generalization, it is convenient to consider the Darboux-Stieltjes integral as an intermediate.

The theory of Stieltjes integrals may be well known. But we sometimes fall into confusion, which may be caused for the lack of the precise description of these three integrals. In this chapter, we first summarize the well known results about the Stieltjes integrals by rewriting the several theorems in the books of Hildebrandt [1] and Rudin [1]. Applying these results, we obtain a representation theorem of linear operators on the phase space $\mathscr{B}$.

Let $I = [a,b]$ be a finite interval, f and g be real or complex, bounded functions on I. A partition P of the interval I is a sequence such that

$$(1.1) \qquad P : a = t_0 < t_1 < \cdots < t_n = b.$$

We set

$$m(P) = \max\{t_i - t_{i-1} : i = 1,\cdots,n\}$$

$$I_i = [t_{i-1}, t_i], \quad \Delta_i g = g(t_i) - g(t_{i-1}) \qquad i = 1,\cdots,n,$$

and set

$$(1.2) \qquad S(f,g,P) = \sum_{i=1}^{n} f(s_i)\Delta_i g, \qquad s_i \in I_i.$$

If $\{S(f,g,P)\}$ converges to a constant, independent of the choice of $s_i \in I_i$ as $m(P) \to 0$, we say f is Riemann-Stieltjes (R-S) integrable with respect to g (or the form fdg is R-S integrable, in short), and write as

$$R\int_a^b fdg = \lim_{m(P)\to 0} S(f,g,P),$$

which is called the R-S integral of f w.r.t. g (or R-S integral of the form fdg).

Let $\mathcal{P} = \mathcal{P}(I)$ be the collection of all partitions of I. If P_1 and P_2 are in $\mathcal{P}$, and if every point of P_1 is a point of P_2, we say P_2 is a refinement of P_1, and write $P_1 < P_2$. The common refinement of P and Q in $\mathcal{P}$, denoted by $P \vee Q$, is a partition consisting of all points of P and Q; clearly, $P < P \vee Q$ and $Q < P \vee Q$. Hence, $\mathcal{P}$ becomes a direct set. The Darboux-Stieltjes integral of f w.r.t. g (D-S integral of fdg) is defined to be the limit of $\{S(f,g,P)\}$ along this direct set;

$$D\int_a^b fdg = \lim_{P\to\infty} S(f,g,P),$$

which means that, for every $\varepsilon > 0$, there exists a $Q \in \mathcal{P}$ such that, if $Q < P$, then

$$\left|D\int_a^b fdg - S(f,g,P)\right| < \varepsilon$$

for any choice of $s_i \in I_i$.

The definitions of two integrals imply the following.

<u>Theorem 1.1</u>. If fdg is R-S integrable, then fdg is D-S integrable, and

$$R\int_a^b fdg = D\int_a^b fdg.$$

<u>Example 1.2</u>. Set $I = [-1,1]$ and

$$g(t) = \begin{cases} 0 & t \le 0 \\ 1 & t > 0. \end{cases}$$

Then fdg is R-S integrable if and only if f is continuous at 0;

and if so, then

$$R\int_{-1}^{1} fdg = f(0).$$

On the other hand, fdg is D-S integrable if and only if f is right continuous at 0, and if so, then

$$D\int_{-1}^{1} fdg = f(0).$$

Theorem 1.3. If fdg is R-S integrable, then f and g have no common point of discontinuity. If fdg is D-S integrable, then f and g have no common point of discontinuity on the same side.

Proof. Suppose $a \le c < d \le b$, and $c \le s, s' \le d$. If $t_{j-1} = c$ and $t_j = d$ for the partition given by (1.1), and if we take s_i, $s_i' \in I_i$ as $s_i = s_i'$ if $i \neq j$ and $s_j = s$, $s_j' = s'$, then

$$(1.3) \qquad \left|\sum_{i=1}^{n} f(s_i)\Delta_i g - \sum_{i=1}^{n} f(s_i')\Delta_i g\right| = |f(s) - f(s')||g(d) - g(c)|.$$

Suppose fdg is D-S integrable. Then, for any $\varepsilon > 0$ there exists a $Q \in \mathcal{P}$ such that, if $P > Q$, the left hand side of (1.3) is less than ε. Suppose g is not continuous on the one side, say on the right side, of c; that is, $G := \limsup_{t \to c+0} |g(t) - g(c)| > 0$. Then, we can take a d such that $c < d < c+m(Q)$ and $|g(d) - g(c)| > G/2$. Taking a partition P consisting of c, d and the points of Q, we then know that $|f(s) - f(s')||g(d) - g(c)| < \varepsilon$ for $s, s' \in [c,d]$; hence

$$|f(s) - f(s')| \le 2\varepsilon/G \quad \text{for} \quad s, s' \in [c,d].$$

Therefore, f is continuous on the right side of c.

Suppose fdg is R-S integrable. Then, for any $\varepsilon > 0$ there exists a $\delta > 0$ such that, if $m(P) < \delta$, then the left hand side of (1.3) is less than ε. Since fdg is D-S integrable, it suffices to consider the case that g is discontinuous only on the one side of c. Suppose g is left continuous, but not right continuous at c. Then, for G defined above, we can take points c' and d such that $c' < c < d$, $d-c' < \delta$, $|g(d) - g(c')| \ge G/2$. Then, from the similar argument, we have

$$|f(s) - f(s')| \le 2\varepsilon/G \quad \text{for} \quad s, s' \in [c', d].$$

Therefore, f is continuous at c.

Theorem 1.4. fdg is R-S integrable if and only if fdg is D-S integrable and f and g have no common point of discontinuity.

Proof. The only if part follows from the previous theorem. We prove the if part.

Let $\varepsilon > 0$ be given. Since fdg is D-S integrable, there exists a partition

$$(1.4) \qquad Q : a = \tau_0 < \tau_1 < \cdots < \tau_q = b$$

such that, if $P, P' > Q$, then

$$(1.5) \qquad |S(f,g,P) - S(f,g,P')| < \varepsilon.$$

Taking a refinement if necessary, we can assume $q \ge 2$. Since f and g have no common point of discontinuity, there exists a $\delta_0 > 0$ such that

$$(1.6) \qquad |f(u_1) - f(\tau_j)||g(u_2) - g(u_3)| < \varepsilon/(q-1)$$

provided $|u_k - \tau_j| < \delta_0$ for $k = 1,2,3$ and for $j = 1,\cdots,q-1$. Set $\delta = \min\{\ell(Q), \delta_0\}$, where $\ell(Q) = \min\{\tau_k - \tau_{k-1} : k = 1,\cdots,n\}$. Suppose the partition (1.1) satisfies $m(P) < \delta$, and consider $S(f,g,P)$. Let $P' : a = t_0' < t_1' < \cdots < t_m' = b$ be a partition consisting of points of P and Q. Define $s_j' \in I_j' = [t_{j-1}', t_j']$ as follows. Let N be the set of index k such that τ_k is contained in an interior of $I_{i(k)}$ for some $i(k)$. Then, for $k \in N$, there exists a $j = j(k)$ such that $I_k = I_j' \cup I_{j+1}'$. Here t_{j-1}' and t_{j+1}' do not agree with any point of Q since $m(P) < \ell(Q)$. If $j = j(k)$ for some $k \in N$, we set

$$s_j' = \tau_k \quad \text{and} \quad s_{j+1}' = \tau_k.$$

Otherwise, I_j' agrees with some I_i: in this case, we set

$$s_j' = s_i.$$

Then, it follows that

$$S(f,g,P) - S(f,g,P') = \sum_{k\in N} [f(s_k) - f(\tau_k)][g(t_{i(k)}) - g(t_{i(k)-1})].$$

In view of (1.6), we have

$$|S(f,g,P) - S(f,g,P')| < \sum_{k\in N} \varepsilon/(q-1) \le \varepsilon.$$

On the other hand, it holds that

$$|S(f,g,P') - S(f,g,Q)| < \varepsilon$$

since $P', Q > Q$. Hence, we have $|S(f,g,P) - S(f,g,Q| < 2\varepsilon$ if $m(P) < \delta$, which implies

$$|S(f,g,P) - S(f,g,P')| < 4\varepsilon$$

if $m(P), m(P') < \delta$; that is, fdg is R-S integrable.

If parallel results hold for R-S integral and D-S integral, we omit the prefixes R-S and D-S, and write the integrals as

$$\int_a^b fdg.$$

Theorem 1.5. If fdg is integrable, then gdf is integrable and

$$\int_a^b gdf = f(b)g(b) - f(a)g(a) - \int_a^b fdg.$$

Proof. Let P be given by (1.1) and $s_i \in I_i$, $i = 1,\cdots,n$, $s_0 = a$, $s_{n+1} = b$. Then we can write

$$\sum_{i=1}^{n} g(s_i)[f(t_i) - f(t_{i-1})] = f(b)g(b) - f(a)g(a)$$

$$- \sum_{i=1}^{n} f(t_i)[g(s_{i+1}) - g(s_i)].$$

If we set $Q : a = s_0 < s_1 < \cdots < s_{n+1} = b$, then this becomes

$$S(g,f,P) = f(b)g(b) - f(a)g(a) - S(f,g,Q).$$

Since $m(Q) \to 0$ as $m(P) \to 0$, we have Theorem 1.5 for R-S integral. Since $f(t_i)\Big(g(s_{i+1}) - g(s_i)\Big) = f(t_i)\Big(g(s_{i+1}) - g(t_i)\Big) + f(t_i)\Big(g(t_i) - g(s_i)\Big)$, we can also write

$$S(g,f,P) = f(b)g(b) - f(a)g(a) - S(f,g,R),$$

where R is a partition consisting of all points t_i, s_j. Since $R > P$, we have Theorem 1.5 for D-S integral.

Example 1.6. Suppose that g is Riemann integrable on $[a,b]$. Then, for $a < c < b$, we have

$$\int_{c-1/n}^{c} n(t-c+1/n)dg(t) = g(c) - \int_{c-1/n}^{c} ng(t)dt.$$

Hence, if $g(c-0) = \lim_{t\to c-0} g(t)$ exists, then

$$\lim_{n\to\infty} \int_{c-1/n}^{c} n(t-c+1/n)dg(t) = g(c) - g(c-0).$$

Furthermore, if $g(c+0) = \lim_{t\to c+0} g(t)$ exists, we have the following formula:

$$\lim_{n\to\infty} \int_{c}^{c+1/n} [-n(t-c-1/n)]dg(t) = g(c+0) - g(c).$$

Theorem 1.7. The following hold:

(i) If $c^1, c^2 \in \mathbb{R}$ or $\mathbb{C}$ and if f^1dg and f^2dg are integrable, then $(c_1f^1 + c_2f^2)dg$ is integrable and

$$\int_a^b (c_1f^1 + c_2f^2)dg = c_1\int_a^b f^1dg + c_2\int_a^b f^2dg.$$

(ii) If fdg is integrable on $[a,b]$, then so is fdg on any $[c,d] \subset [a,b]$.

(iii) if fdg is D-S integrable on $[a,c]$ and on $[c,b]$, $a < c < b$, then so is fdg on $[a,b]$ and

$$D\int_a^b fdg = D\int_a^c fdg + D\int_c^b fdg.$$

This relation for R-S integral of fdg holds if and only if f or g is continuous at c.

Proof. The proof is easy.

A function g on I is said to be of bounded variation if the total variation of g defined by

$$\mathrm{Var}(g,[a,b]) = \sup\{\textstyle\sum_P |\Delta_i g| : P \in \mathcal{P}\}$$

is finite, where $\sum_P |\Delta_i g| = \sum_{i=1}^{n} |g(t_i) - g(t_{i-1})|$ for P given by (1.1). It is easy to see that

$$\mathrm{Var}(g,[a,b]) = \mathrm{Var}(g,[a,c]) + \mathrm{Var}(g,[c,b])$$

for $a < c < b$. Define the total variation function $T = T_g$ by

$$T(t) = \begin{cases} \mathrm{Var}(g,[a,t]) & a < t \le b \\ 0 & t = a. \end{cases}$$

Then $T(t)$ is nonnegative, nondecreasing and $T(t) - T(s) \ge |g(t) - g(s)|$ for $a \le s < t \le b$.

Suppose that g is a real function, and set

$$g^+(t) = \{T(t) + g(t) - g(a)\}/2$$

$$g^-(t) = \{T(t) - (g(t) - g(a))\}/2.$$

Then g^+ and g^- are nonnegative, nondecreasing functions, and we can write as follows:

$$g(t) - g(a) = g^+(t) - g^-(t) \tag{1.7}$$

$$T(t) = g^+(t) + g^-(t). \tag{1.8}$$

The formula (1.7) is called the Jordan decomposition of g.

Suppose that g is a complex function, and set

$$g_1(t) = \mathrm{Re}\, g(t), \qquad g_2(t) = \mathrm{Im}\, g(t).$$

Then g is of bounded variation if and only if g_1 and g_2 are of bounded variation.

Set

$$S^+(g) = \{t \in [a,b] : g \text{ is not right continuous at } t\}$$

$$S^-(g) = \{t \in [a,b] : g \text{ is not left continuous at } t\}$$

$$S(g) = S^+(g) \cup S^-(g).$$

Then the Jordan decomposition of g implies that these sets are at most countable sets.

Proposition 1.8. If g is of bounded variation, then

$$\mathrm{Var}(g,[a,b]) = |g(a+0) - g(a)| + \lim_{s\to a+0} \mathrm{Var}(g,[s,b])$$

$$\mathrm{Var}(g,[a,b]) = |g(b-0) - g(b)| + \lim_{t\to b-0} \mathrm{Var}(g,[a,t]).$$

Proof. We show the second relation; the first is also shown in the similar manner. Since $T(b) = T(t) + \mathrm{Var}(g,[t,b]) \ge T(t) + |g(t) - g(b)|$ for $a < t < b$, we have $T(b) \ge T(b-0) + |g(b-0) - g(b)|$. Let P be given by (1.1). If $t_{n-1} < t < t_n = b$, then

$$\sum_P |\Delta_i g| \le \sum_{i=1}^{n-1} |\Delta_i g| + |g(t) - g(t_{n-1})| + |g(b) - g(t)|$$

$$\le T(t) + |g(b) - g(t)|,$$

which implies $\sum_P |\Delta_i g| \le T(b-0) + |g(b) - g(b-0)|$. Since P is arbitrary, we have $T(b) \le T(b-0) + |g(b-0) - g(b)|$.

Corollary 1.9.

$$S^+(g) = S^+(T) = S^+(g^+) \cup S^+(g^-)$$

$$S^-(g) = S^-(T) = S^-(g^+) \cup S^-(g^-)$$

$$S(g) = S(T) = S(g^+) \cup S(g^-).$$

A complex function g of bounded variation on $I = [a,b]$ is said to be regular if $g(t)$ lies on the segment joining $g(t-0)$ and $g(t+0)$ for each t in (a,b); that is,

$$|g(t+0) - g(t-0)| = |g(t+0) - g(t)| + |g(t) - g(t-0)|$$

for each t in (a,b).

Theorem 1.10. If g is as in the above, then

$$\mathrm{Var}(g,[a,b]) = \lim_{m(P)\to 0} \sum_P |\Delta_i g|.$$

Proof. Since $\sum_Q |\Delta_j g| \le \sum_P |\Delta_i g|$ for $Q < P$, for a given $\varepsilon > 0$ there exists a Q, given by (1.4), such that $\sum_R |\Delta_k g| \ge \mathrm{Var}(g,[a,b]) - \varepsilon$ for $R > Q$. Since g is regular, there exists a $\delta_0 > 0$ such that, if $\tau_j - \delta_0 < s \le \tau_j \le t < \tau_j + \delta_0$, then

$$|g(t) - g(s)| \ge |g(t) - g(\tau_j)| + |g(\tau_j) - g(s)| - \varepsilon/(q-1)$$

for $j = 1,\cdots,q-1$.

Suppose that a partition P given by (1.1) satisfies $m(P) < \delta := \min\{\ell(Q),\delta_0\}$, where $\ell(Q)$ is the one considered in the proof of Theorem 1.4. Set $R = P \vee Q$. Then, for each $j = 1,\cdots,q-1$, there exists a unique $i(j)$ such that $t_{i(j)-1} \le \tau_j \le t_{i(j)}$; and, we have

$$\sum_P |\Delta_i g| - \sum_R |\Delta_k g|$$

$$= \sum_{j=1}^{q-1} \{|\Delta_{i(j)} g| - [|g(t_{i(j)}) - g(\tau_j)| + |g(\tau_j) - g(t_{i(j)-1})|]\}$$

$$\ge \sum_{j=1}^{q-1} (-\varepsilon)/(q-1) = -\varepsilon$$

since $m(P) < \delta_0$. Consequently, it follows that

$$\sum_P |\Delta_i g| \ge \sum_R |\Delta_k g| - \varepsilon \ge \mathrm{Var}(g,[a,b]) - 2\varepsilon$$

as long as $m(P) < \delta$.

Theorem 1.11. Suppose g is of bounded variation. If f is continuous, then fdg is R-S integrable. If fdg is D-S integrable, then

$$|D\int_a^b fdg| \le |f|\mathrm{Var}(g,[a,b]),$$

where $|f| = \sup\{|f(t)| : a \le t \le b\}$. Furthermore, if g is regular, then

$$(1.9) \qquad \mathrm{Var}(g,[a,b]) = \sup\{|\int_a^b f dg| : f \in C(I),\ |f| \le 1\}.$$

Proof. It is easy to show the first two results. Denote by L and M the left and right hand side of (1.9), respectively. It suffices to show that $L \le M$.

Let P be a partition given by (1.1). We show that there is a sequence $\{f^n\}$ in $C(I)$ such that $|f^n| \le 1$ and that

$$\sum_P |\Delta_i g| \le |\lim_{n\to\infty} \int_a^b f^n dg|.$$

Then it follows that $\sum_P |\Delta_i g| \le M$ for every P; hence $L \le M$.

For simplicity, we assume that P is given by three points $a < c < b$. Then, we have from the regularity of g that

$$\begin{aligned}
\sum_P |\Delta_i g| &= |g(c) - g(a)| + |g(b) - g(c)| \\
&\le |g(c-0) - g(a)| + |g(c) - g(c-0)| \\
&\qquad + |g(c+0) - g(c)| + |g(b) - g(c+0)| \\
&= |g(c-0) - g(a)| + |g(c+0) - g(c-0)| + |g(b) - g(c+0)| \\
&=: W_P
\end{aligned}$$

However, we can write

$$W_P = \sigma_1\{g(c-0) - g(a)\} + \sigma_2\{g(c+0) - g(c-0)\} + \sigma_3\{g(b) - g(c+0)\}$$

for some $\sigma_i \in \mathbb{C}$, $|\sigma_i| = 1$, $i = 1,2,3$.

Let u^n, v^n, w^n, $n = 1,2,\cdots$, be the continuous functions defined by

$$u^n(t) = \begin{cases} 1 & \text{on } (-\infty, c-1/n] \\ 0 & \text{on } [c, \infty) \\ \text{linear} & \text{on } (c-1/n, c], \end{cases}$$

$$v^n(t) = \begin{cases} 0 & \text{on } (-\infty, c] \\ 1 & \text{on } [c+1/n, \infty) \\ \text{linear} & \text{on } [c, c+1/n], \end{cases}$$

$$w^n(t) = 1 - u^n(t) - v^n(t).$$

Then, from Example 1.6 it follows that

$$\lim_{n\to\infty} \int_a^b u^n dg = g(c-0) - g(a)$$

$$\lim_{n\to\infty} \int_a^b v^n dg = g(b) - g(c+0)$$

$$\lim_{n\to\infty} \int_a^b w^n dg = g(c+0) - g(c-0).$$

Therefore, if we set $f^n = \sigma_1 u^n + \sigma_2 w^n + \sigma_3 v^n$, $n = 1,2,\cdots$, then $|f^n| \le 1$ and

$$W_P = \lim_{n\to\infty} \int_a^b f^n dg.$$

Since W_P is a positive number, it follows that

$$W_P = |\lim_{n\to\infty} \int_a^b f^n dg|,$$

which is the desired result.

Corollary 1.12. If g is a regular function of bounded variation, and if

$$\int_a^b fdg = 0 \quad \text{for all} \quad f \in C(I),$$

then $g(t) \equiv g(a)$ for all $t \in I$.

Example 1.13. Suppose $g(t) = 0$ for $t \ne 0$ and $g(0) = 1$. Then $\int_{-1}^1 fdg = 0$ for all $f \in C([-1,1])$, but $\mathrm{Var}(g,[-1,1]) = 2$.

A function g on $[a,b]$ is absolutely continuous if and only if it is differentiable almost everywhere, g' is Lebesgue integrable on $[a,b]$ and

$$g(t) = g(a) + \int_a^t g'(s)ds$$

for all $t \in [a,b]$. Hence, an absolutely continuous function is of bounded variation; in fact, we show in the next section that

$$\mathrm{Var}(g,[a,b]) = \int_a^b |g'(t)|dt.$$

Theorem 1.14. If f is continuous or of bounded variation, and if g is absolutely continuous, then fdg is R-S integrable, and

$$R\int_a^b fdg = \int_a^b f(t)g'(t)dt.$$

Proof. It follows from Theorems 1.5 and 1.11 that fdg is R-S integrable. For the partition (1.1), we can write

$$S(f,g,P) = \int_a^b f^P(t)g'(t)dt,$$

where $f^P(t) = f(s_i)$ for $t \in [t_{i-1},t_i)$, $i = 1,\cdots,n-1$, and $f^P(t) = f(s_n)$ for $t \in [t_{n-1},t_n]$. Obviously, $f^P(t) \to f(t)$ as $m(P) \to 0$ at $t \bar{\in} S(f)$ which is at most countable set. Then the conclusion follows from the dominated convergence theorem.

Suppose f is a real function, and α a nondecreasing function. For a $P \in \mathcal{P}$ in (1.1), set $M_i = \sup f(I_i)$, $m_i = \inf f(I_i)$, $i = 1,\cdots,n$, and define the upper sum and the lower sum as

$$(1.10)\qquad U(f,\alpha,P) = \sum_{i=1}^n M_i\Delta_i\alpha, \quad L(f,\alpha,P) = \sum_{i=1}^n m_i\Delta_i\alpha.$$

If $P < Q$, then $L(f,\alpha,P) \le L(f,\alpha,Q) \le U(f,\alpha,Q) \le U(f,\alpha,P)$; and for any P, R in $\mathcal{P}$ we have $L(f,\alpha,P) \le U(f,\alpha,R)$. This implies that the following relations hold:

$$\overline{\int_a^b} fd\alpha := \inf\{U(f,\alpha,P) : P \in \mathcal{P}\} = \lim_{P\to\infty} U(f,\alpha,P)$$

$$(1.11)\qquad \underline{\int_a^b} fd\alpha := \sup\{L(f,\alpha,P) : P \in \mathcal{P}\} = \lim_{P\to\infty} L(f,\alpha,P)$$

$$\underline{\int_a^b} fd\alpha \le \overline{\int_a^b} fd\alpha.$$

The first integral is called the upper integral of $fd\alpha$; the second integral the lower integral of $fd\alpha$. They agree with each other if and only if $fd\alpha$ is D-S integrable; and if so,

$$D\int_a^b fd\alpha = \overline{\int_a^b} fd\alpha = \underline{\int_a^b} fd\alpha.$$

This fact can be rewritten as follows.

Proposition 1.15. Let f and α be as in the above. Then $fd\alpha$ is D-S integrable if and only if, for every $\varepsilon > 0$, there exists a $P \in \mathcal{P}$ such that

$$\omega(f,\alpha,P) := U(f,\alpha,P) - L(f,\alpha,P) < \varepsilon.$$

Theorem 1.16. Suppose f and g are real functions, f is bounded, and g is of bounded variation. Then, fdg is integrable if and only if fdT_g is integrable.

Proof. Suppose fdg is D-S integrable. Let P be a partition given by (1.1), and u_i, u_i' are points in I_i, $i = 1,\cdots,n$. Then, we can write

$$\sum_{i=1}^n |f(u_i) - f(u_i')|\Delta_i T = \sum_{i=1}^n |f(u_i) - f(u_i')||\Delta_i g|$$

$$+ \sum_{i=1}^n |f(u_i) - f(u_i')|(\Delta_i T - |\Delta_i g|).$$

Set $s_i = u_i$ and $s_i' = u_i'$ if $(f(u_i) - f(u_i'))\Delta_i g \geq 0$; set $s_i = u_i'$ and $s_i' = u_i$ if $(f(u_i) - f(u_i'))\Delta_i g < 0$. Then, we have

$$\sum_{i=1}^n |f(u_i) - f(u_i')||\Delta_i g| = |\sum_{i=1}^n (f(s_i) - f(s_i'))\Delta_i g|.$$

On the other hand, we have

$$\sum_{i=1}^n |f(u_i) - f(u_i')|(\Delta_i T - |\Delta_i g|) \leq (M - m)(T(b) - \sum_{i=1}^n |\Delta_i g|),$$

where $M = \sup f(I)$ and $m = \inf f(I)$. Let $\varepsilon > 0$ be given. Since fdg is D-S integrable, there exists a $Q_1 \in \mathcal{P}$ such that $|\sum_{i=1}^n (f(s_i) - f(s_i'))\Delta_i g| < \varepsilon$ for $P > Q_1$. Since $T(b) =$

$\mathrm{Var}(g,[a,b])$, there exists a $Q_2 \in \mathcal{P}$ such that if $P > Q_2$, then $0 \le T(b) - \sum_{i=1}^{n} |\Delta_i g| < \varepsilon$. Therefore, if $P > Q_1 \vee Q_2$, we have

$$\sum_{i=1}^{n} |f(u_i) - f(u_i')| \Delta_i T < \varepsilon(1+M-m)$$

for $u_i, u_i' \in I_i$, $i = 1,\cdots,n$. This implies $\omega(f,T,P) < \varepsilon(1+M-m)$; hence, fdT is D-S integrable.

Suppose fdg is R-S integrable. Then, fdg is D-S integrable, and so is fdT. Since $S(f) \cap S(g) = \phi$ and $S(g) = S(T)$, it follows that $S(f) \cap S(T) = \phi$. Hence, fdT is R-S integrable by Theorem 1.3. We refer to such an argument as (*), which will be repeatedly used later on.

Suppose fdT is D-S integrable. Let P be the partition (1.1), and Q the one (1.4). Suppose P is a refinement of Q. Any subinterval $J_j = [\tau_{j-1},\tau_j]$, of Q, is a union of subintervals $I_k = [t_{k-1},t_k]$, $I_{k+1},\cdots, I_{k+m}$, of P. Hence we have

$$\left|\sum_{i=k}^{k+m} f(s_i)\Delta_i g - f(\tau_j)\Delta_j g\right| = \left|\sum_{i=k}^{k+m} (f(s_i) - f(\tau_j))\Delta_i g\right|$$

$$\le (M_j - m_j)\mathrm{Var}(g,[\tau_{j-1},\tau_j]) = (M_j - m_j)\Delta_j T,$$

which implies that

$$\left|S(f,g,P) - \sum_{j=1}^{q} f(\tau_j)\Delta_j g\right| \le \omega(f,T,Q).$$

Hence, if $P > Q$ and $P' > Q$, then

$$|S(f,g,P) - S(f,g,P')| \le 2\omega(f,T,Q).$$

Since fdT is D-S integrable, we have $\omega(f,T,Q) < \varepsilon$ for any $\varepsilon > 0$ by some $Q \in \mathcal{P}$: therefore, fdg is D-S integrable.

The similar result for R-S integral follows from the argument (*).

<u>Corollary 1.17</u>. Under the assumption of Theorem 1.16, fdg is integrable if and only if fdg^+ and fdg^- are integrable; and if so, then

$$\int_a^b fdg = \int_a^b fdg^+ - \int_a^b fdg^-.$$

Consider the case f and g are complex functions:

$$f = f_1 + \sqrt{-1}f_2, \qquad g = g_1 + \sqrt{-1}g_2,$$

where f_i, g_i are real functions. The function g is of bounded variation if and only if g_1 and g_2 are of bounded variation.

Theorem 1.18. Suppose g is a complex function of bounded variation. Then fdT is integrable if and only if $f_i dg_j$, i, j = 1, 2, are integrable; and if so, then

$$\int_a^b fdg = \int_a^b (fdg_1 + \sqrt{-1}fdg_2)$$

$$= \int_a^b \{f_1dg_1 - f_2dg_2 + \sqrt{-1}(f_1dg_2 + f_2dg_1)\},$$

and

$$\int_a^b f_i dg_j = \int_a^b f_i dg_j^{\,+} - \int_a^b f_i dg_j^{\,-}, \qquad i, j = 1, 2.$$

Proof. Since $S(f,T,P) = S(f_1,T,P) + \sqrt{-1}S(f_2,T,P)$, fdT is integrable if and only if f_1dT and f_2dT are integrable. Denote by T_1 and T_2 the total variation functions of g_1 and g_2, respectively. Set $|z| = |x| + |y|$ for $z = x + \sqrt{-1}y \in \mathbb{C}$. Then, it follows that $T = T_1 + T_2$, which implies

$$\omega(f_i,T,P) = \omega(f_i,T_1,P) + \omega(f_i,T_2,P).$$

Therefore, $f_i dT$ is D-S integrable if and only if $f_i dT_1$ and $f_i dT_2$ are D-S integrable. Then, from Theorem 1.16 and Corollary 1.17, we obtain the theorem for D-S integral. Using the argument (*) in the proof of Theorem 1.16, we have the result for R-S integral.

3.2. Lebesgue-Stieltjes integral. We summarize the theory of Lebesgue-Stieltjes integral by rewriting the Riesz representation theorem. We use the notations and the definitions of the book (Rudin [1]).

Suppose X is a locally compact, σ-compact, Hausdorff space.

A positive (real or complex, respectively) measure μ is a function, defined on a σ - algebra $\mathfrak{M}$ on X, whose range is in $[0,\infty]$ ($\mathbb{R}$ or $\mathbb{C}$, respectively), and which is countably additive. Define the space $L^p(\mu) = L^p(X,\mu)$ with the norm $|\cdot|_p$ for positive measure μ and for $1 \le p \le \infty$ as usual.

Let λ be a positive Borel measure on X. A set E is said to be λ-measurable if E is in $\mathfrak{M}$, the completion of S, the class of Borel sets, with respect to λ: that is, there are Borel sets A and B such that $A \subset E \subset B$ and $\lambda(B\setminus A) = 0$. A function f is said to be λ-measurable if it is measurable with respect to $\mathfrak{M}$. If $\lambda(X) < \infty$, we say λ is bounded; if $\lambda(K) < \infty$ for every compact set K, we say λ is locally finite; if

$$\lambda(E) = \inf\{\lambda(V): E \subset V,\ V \text{ open}\}$$

$$= \sup\{\lambda(K): K \subset E,\ K \text{ compact}\}$$

for every Borel set E, we say λ is regular. If X is a locally compact, σ-compact, Hausdorff space, any positive Borel measure on X is regular (Rudin [1, Theorem 2.18]).

The total variation $|\mu|$ of a complex measure μ is a function defined by

$$|\mu|(E) = \sup_i \Sigma|\mu(E_i)|,$$

where the supremum is taken over all partitions $\{E_i\}$ of E: E is the union of disjoint countable collection $\{E_i\}$ of $\mathfrak{M}$. It is a positive, bounded measure (Rudin [1, Theorem 6.4]). If μ is given by

$$\mu(E) = \int_E k d\lambda \qquad E \in \mathfrak{M}$$

for some positive measure λ on $\mathfrak{M}$ and for some λ-measurable function k, we write this equation as $d\mu = kd\lambda$. If $k \in L^1(\lambda)$, then $d|\mu| = |k|d\lambda$; see Rudin [1, Theorem 6.13].

From the Radon-Nikodim theorem, every complex measure μ on X is written as $d\mu = hd|\mu|$ for a measurable function h such that $|h(x)| = 1$ for all x in X (Rudin [1, Theorem 6.12]). We call the equation $d\mu = hd|\mu|$ the polar representation of μ.

A complex Borel measure μ is said to be regular if $|\mu|$ is regular. The Dirac measure δ_x, $x \in X$, is the measure defined by

$$\delta_x(E) = \begin{cases} 1 & \text{if } x \in E, \\ 0 & \text{if } x \bar{\in} E \end{cases}$$

It is a complex, regular Borel measure on X. The following claim will be used frequently.

Proposition 2.1. If λ is a positive, bounded, regular Borel measure on X, and if $k \in L^1(\lambda)$, then $d\mu = kd\lambda$ is a complex, regular Borel measure on X.

Proof. Since $d|\mu| = |k|d\lambda$ and $|k| \in L^1(\lambda)$, $|\mu|$ is absolutely continuous with respect to λ. Thus, for every $\varepsilon > 0$ there exists a $\delta > 0$ such that $\lambda(E) < \delta$ implies $|\mu|(E) < \varepsilon$. Since λ is bounded and regular, $|\mu|$ is regular.

Proposition 2.2. If μ_1 and μ_2 are complex, regular Borel measures on X, and if c_1, c_2 are complex numbers, then $\mu = c_1\mu_1 + c_2\mu_2$ is a complex, regular Borel measure on X.

Proof. It is easy to see that $\lambda := |\mu_1| + |\mu_2|$ is a positive bounded, regular Borel measure. Let k_i be the Radon-Nikodim derivative of μ_i with respect to λ, $i = 1,2$. Then, we have $d\mu = (c_1k_1 + c_2k_2)d\lambda$, and $c_1k_1 + c_2k_2$ is in $L^1(\lambda)$. Thus, Proposition 2.1 implies Proposition 2.2.

We define the integration with respect to a complex measure μ, with the polar representation $d\mu = hd|\mu|$, by the formula

$$\int f d\mu = \int fh d|\mu| \tag{2.1}$$

as long as fh is $|\mu|$-integrable. Set $f_1 = \mathrm{Re}\, f$, $f_2 = \mathrm{Im}\, f$, $h_1 = \mathrm{Re}\, h$, $h_2 = \mathrm{Im}\, h$, $\mu_1 = \mathrm{Re}\, \mu$ and $\mu_2 = \mathrm{Im}\, \mu$. Let c_1 and c_2 be the positive constants such that

$$c_1|z| \le |x| + |y| \le c_2|z| \tag{2.2}$$

for all $z = x + \sqrt{-1}y$ in $\mathbb{C}$, $x, y \in \mathbb{R}$.

Proposition 2.3. Let f, h, μ, μ_1, μ_2 be as cited above. Then

$$c_1|\mu|(E) \le |\mu_1|(E) + |\mu_2|(E) \le c_2|\mu|(E)$$

for every E in $\mathfrak{M}$; fh is $|\mu|$-integrable if and only if f_i is $|\mu_j|$-integrable for all $i,j = 1,2$; and if so,

$$\int f d\mu = \int [f_1 d\mu_1 - f_2 d\mu_2 + \sqrt{-1}(f_1 d\mu_2 + f_2 d\mu_1)].$$

Proof. Since $f = (fh)\bar{h}$ and $|h| = |\bar{h}| = 1$, fh is $|\mu|$-integrable if and only if f is $|\mu|$-integrable; that is, f_1 and f_2 are $|\mu|$-integrable.

Comparing the real and imaginary parts of the equation $d\mu = hd|\mu|$, we have $d\mu_i = h_i d|\mu|$, $i = 1,2$; hence $d|\mu_i| = |h_i| d|\mu|$, $i = 1,2$; consequently, $d(|\mu_1|+|\mu_2|) = (|h_1|+|h_2|)d|\mu|$. Since $|h| = 1$, we have the inequality in Proposition 2.3.

Therefore, a set in X is $|\mu|$-measurable if and only if it is $|\mu_1|$ and $|\mu_2|$-measurable; a function on X is $|\mu|$-integrable if and only if it is $|\mu_1|$ and $|\mu_2|$-integrable; and if so, then

$$\int f d\mu = \int fhd|\mu| = \int \{f_1 h_1 - f_2 h_2 + \sqrt{-1}(f_1 h_2 + f_2 h_1)\} d|\mu|$$

$$= \int \{f_1 d\mu_1 - f_2 d\mu_2 + \sqrt{-1}(f_1 d\mu_2 + f_2 d\mu_1)\}.$$

If μ is a real measure, then the measure $\mu^{\pm}$ defined by

$$\mu^+ = \frac{1}{2}(|\mu| + \mu), \qquad \mu^- = \frac{1}{2}(|\mu| - \mu)$$

are positive, bounded measures; and we have the equation

$$\mu = \mu^+ - \mu^-, \qquad |\mu| = \mu^+ + \mu^-.$$

The first equation is called the Jordan decomposition of μ. From the second equation, a function f is $|\mu|$-integrable if and only if it is μ^+ and μ^--integrable. Let $d\mu = hd|\mu|$ be the polar representation of μ. If we set

$$h^+ = \frac{1}{2}(|h| + h) = \frac{1}{2}(1 + h), \qquad h^- = \frac{1}{2}(|h| - h) = \frac{1}{2}(1 - h),$$

then $h = h^+ - h^-$, $d\mu^+ = h^+ d|\mu|$ and $d\mu^- = h^- d|\mu|$. Therefore, we have

$$\int f d\mu = \int fhd|\mu| = \int f d\mu^+ - \int f d\mu^-.$$

Consequently, the integration (2.1) is reduced to the integration with respect to positive bounded measures $\mu_i^{\pm}$, $i = 1,2$.

Let $C_c(X)$ be the space of complex, continuous functions on X with compact support, and $C_0(X)$ the completion of $C_c(X)$ with respect to the supremum norm: $|f| = \sup\{|f(x)| : x \in X\}$. A function f is in $C_0(X)$ if and only if f is continuous and it vanishes at infinity, cf. Rudin [1, Definition 3.16]. If X is compact, $C_c(X) = C_0(X) = C(X)$, the space of continuous functions on X.

Suppose that $\Phi : C_c(X) \to \mathbb{C}$ is a bounded linear operator in the sense that $\|\Phi\| := \sup\{|\Phi(f)| : f \in C_c(X), |f| \le 1\}$ is finite. It has a unique extension to a bounded linear operator on $C_0(X)$ with the same operator norm.

<u>Theorem 2.4 (Rudin [1, Theorems 2.14 and 2.16]).</u> To each bounded linear functional Φ on $C_0(X)$, there corresponds a unique, complex, regular Borel measure μ such that

$$\Phi(f) = \int_X f d\mu \qquad f \in C_0(X).$$

If Φ and μ are related in this manner, then

$$\|\Phi\| = |\mu|(X).$$

Moreover, μ is positive if Φ is positive in the sense that $\Phi(f) \ge 0$ for every $f \in C_0(X)$ with $f \ge 0$.

Suppose that g is a complex function on $I = [a,b]$, and that g is of bounded variation on I. The linear functional Φ defined by

$$\Phi(f) = \int_a^b f dg \qquad f \in C(I)$$

is a bounded linear functional on $C(I)$; hence, there exists a unique, complex, regular Borel measure μ on I which represents Φ in the sense of Theorem 2.4. We call μ the Lebesgue-Stieltjes (L-S) measure on I induced by g, and define the L-S integral of f with respect to g (or, L-S integral of fdg) by the formula

$$\text{(2.3)} \qquad L\int_a^b f dg = \int f d\mu .$$

This integral is defined for f in $L^1(|\mu|)$.

If $J = [c,d]$ is a subinterval of I, then the L-S measure μ_J on J induced by g is defined similarly from the linear functional

$$\Phi_J(f) = \int_c^d f dg \qquad f \in C(J),$$

and the L-S integral of fdg on J is defined as

$$L\int_c^d f dg = \int f d\mu_J.$$

The relationship of these measures and integrals will be given later.

If f is a bounded, Borel measurable function, the L-S integral of fdg is defined for every g of bounded variation. Notice that the following claim is obtained from the uniqueness of μ in Theorem 2.4.

Lemma 2.5. A given complex, Borel measure λ is the L-S measure on I induced by g if λ is regular, and

$$\int_a^b f dg = \int f d\lambda \qquad \text{for every } f \in C(I).$$

Example 2.6. For each $c \in (a,b)$, let $g_c: [a,b] \to \mathbb{R}$ be a function such that $g_c(b) - g_c(a) = 1$, $g_c(t) = g_c(a)$ for $a \leq t < c$, $g_c(t) = g_c(b)$ for $c < t \leq b$ and $g_c(a) \leq g_c(c) \leq g_c(b)$. Then δ_c is the L-S measure induced by g_c, and

$$L\int_a^b f dg_c = f(c)$$

for every function f on I such that $f(c)$ is finite.

Example 2.7. Suppose g is absolutely continuous on I. Since g' is integrable with respect to the Lebesgue measure, $d\mu = g'dt$ is the L-S measure induced by g due to Theorem 1.14, Proposition 2.1 and Lemma 2.5. Hence, the L-S integral of fdg is defined by

$$L\int_a^b f dg = \int f g' dt$$

for all functions f in $L^1(|\mu|)$, where $d|\mu| = |g'|dt$. Furthermore, it follows from Theorems 1.11 and 2.4 that

$$\mathrm{Var}(g,[a,b]) = \int_a^b |g'(t)| dt.$$

Theorem 2.8. If μ_1 and μ_2 are the L-S measures induced by g_1 and g_2, respectively, then $c_1\mu_1 + c_2\mu_2$, $c_1, c_2 \in \mathbb{C}$, is the L-S measure induced by $c_1g_1 + c_2g_2$.

Proof. Suppose that f is in $C(I)$. Then, we have

$$\int_a^b fd(c_1g_1 + c_2g_2) = c_1\int_a^b fdg_1 + c_2\int_a^b fdg_2$$

$$= c_1\int fd\mu_1 + c_2\int fd\mu_2 = \int fd(c_1\mu_1 + c_2\mu_2).$$

Since $c_1\mu_1 + c_2\mu_2$ is a complex, regular Borel measure, we have the conclusion from Lemma 2.5.

Corollary 2.9. Suppose $g = g_1 + \sqrt{-1}g_2$ is a complex function of bounded variation on I. Let $g_i = g_i^+ - g_i^-$ be the Jordan decomposition of g_i, and μ, μ_i, $\lambda_i^{\pm}$ the L-S measures induced by g, g_i, $g_i^{\pm}$. Then, $\mu = \mu_1 + \sqrt{-1}\mu_2 = (\lambda_1^+ - \lambda_1^-) + \sqrt{-1}(\lambda_2^+ - \lambda_2^-)$, and

$$L\int_a^b fdg = L\int_a^b (fdg_1 + \sqrt{-1}fdg_2)$$

$$= L\int_a^b [f(dg_1^+ - dg_1^-) + \sqrt{-1}f(dg_2^+ - dg_2^-)]$$

$$= L\int_a^b \{f_1(dg_1^+ - dg_1^-) - f_2(dg_2^+ - dg_2^-)$$

$$+ \sqrt{-1}[f_1(dg_2^+ - dg_2^-) + f_2(dg_1^+ - dg_1^-)]\}.$$

Proof. This is obtained from Proposition 2.3 and Theorem 2.8.

Corollary 2.10. Let g, g_i, $g_i^{\pm}$, $\lambda_i^{\pm}$ be as in Corollary 2.9, and let λ, λ_i be the L-S measures induced by T_g, T_{g_i}, respectively. Then, we have

$$\lambda_1 = \lambda_1^+ + \lambda_1^-, \qquad \lambda_2 = \lambda_2^+ + \lambda_2^-$$

and

$$c_1\lambda \le \lambda_1 + \lambda_2 \le c_2\lambda.$$

where c_1 and c_2 are the constants satisfying Inequality (2.2).

Proof. We only prove the third inequality. Assume $|z|$ satisfies (2.2), and set $|z|_\infty = |x| + |y|$. Denote by T^∞ the total variation function of g with respect to the norm $|\cdot|_\infty$. Then, Inequality (2.2) implies that $c_1 T \le T^\infty \le c_2 T$; or $T^\infty - c_1 T \ge 0$ and $c_2 T - T^\infty \ge 0$. Then, Theorem 2.4 implies that $\lambda^\infty - c_1\lambda \ge 0$ and $c_2\lambda - \lambda^\infty \ge 0$, where λ^∞ is the L-S measure induced by T^∞; hence, $c_1\lambda \le \lambda^\infty \le c_2\lambda$. On the other hand, we have $\lambda^\infty = \lambda_1 + \lambda_2$ since $T^\infty = T_{g_1} + T_{g_2}$.

From Definition (2.1), we have

$$|L\int_a^b f dg| \le \int |f| d|\mu|. \tag{2.4}$$

Of course, the dominated convergence theorem is valid: if $|f^n| \le \varphi$, if φdg, $f^n dg$, $n = 1, 2, \cdots$, are all L-S integrable, and if $\{f^n\}$ converges $|\mu|$-a.e., then

$$\lim_{n\to\infty} L\int_a^b f^n dg = L\int_a^b (\lim_{n\to\infty} f^n) dg.$$

Using these results, we prove an unsymmetric Fubini theorem. For a more general result, see Cameron & Martin [1].

Theorem 2.11. Let $[a,b]$ and $[c,d]$ be finite intervals in $\mathbb{R}$, and let $g : [a,b] \to \mathbb{C}$, $f : [c,d] \to \mathbb{C}$ and $h : [a,b] \times [c,d] \to \mathbb{C}$ be functions such that

(i) f is bounded, Borel measurable;

(ii) g is of bounded variation;

(iii) for each $s \in [c,d]$ the function $h(\cdot,s)$ is Borel measurable on $[a,b]$, and for each $t \in [a,b]$ the function $h(t,\cdot)$ is of bounded variation with

$$\mathrm{Var}(h(t,\cdot),[c,d]) \le v(t)$$

on $[a,b]$ for a Borel measurable function v.

If v and $h(\cdot,\tilde{s})$ for an $\tilde{s} \in [c,d]$ are L-S integrable with respect to g, then the following statements hold:

(iv) for each $s \in [c,d]$, the function $h(\cdot,s)$ is L-S integrable with respect to g over $[a,b]$, and the function

$L\int_a^b dg(t)h(t,s)$ of s is of bounded variation on $[c,d]$;

(v) the function $L\int_c^d [d_s h(t,s)]f(s)$ of t is Borel measurable on $[a,b]$ and L-S integrable with respect to g over $[a,b]$;

(vi) the following relation holds:

$$L\int_c^d d_s[L\int_a^b dg(t)h(t,s)]f(s) = L\int_a^b dg(t)\{L\int_c^d d_s h(t,s)f(s)\}.$$

<u>Proof</u>. The assertion (iv) immediately follows from Inequality (2.4) and the assumptions. In fact, we have

$$|L\int_a^b dg(t)h(t,s)| \le L\int_a^b (v(t) + |h(t,\tilde{s})|)d|\mu|(t)$$

and

$$\mathrm{Var}(L\int_a^b dg(t)h(t,\cdot),[c,d]) \le L\int_a^b v(t)d|\mu|(t).$$

Next, we shall prove the assertions (v) and (vi) in the case where $f = \chi_E$, the indicator of a Borel set E in $[c,d]$. Then, these assertions hold for any Borel simple function, and hold even for all functions f in (i) by the dominated convergence theorem since a function f is a limit of a sequence of Borel simple functions which do not exceed $|f|$ in the absolute value. Now, let $\mathfrak{B}$ denote the family of all Borel sets in $[c,d]$, and let $\mathfrak{U}$ denote the family of all elements F in $\mathfrak{B}$ such that the assertions (v) and (vi) hold as $f = \chi_F$. In what follows, we shall prove that $\mathfrak{U} = \mathfrak{B}$, which completes the proof of the theorem.

Let $F = [c_1,d_1] \subset [c,d]$. Then, for $f = \chi_F$, we have

$$L\int_c^d [d_s h(t,s)]f(s) = h(t,d_1+0) - h(t,c_1-0).$$

The function $h(t,d_1+0) - h(t,c_1-0)$ is Borel measurable as a limit function of a sequence of Borel measurable functions, and $|h(t,d_1+0) - h(t,c_1-0)| \le v(t)$. Hence, the assertion (v) holds for $f = \chi_F$ by (iii). Moreover, in this case, we have

$$L\int_a^b dg(t)\{L\int_c^d d_s h(t,s)f(s)\}$$

$$= L\int_a^b dg(t)h(t,d_1+0) - L\int_a^b dg(t)h(t,c_1-0)$$

$$= \lim_{s\to d_1+0} L\int_a^b dg(t)h(t,s) - \lim_{s\to c_1-0} L\int_a^b dg(t)h(t,s)$$

$$= L\int_c^d d_s\left(L\int_a^b dg(t)h(t,s)\right)\chi_{[c_1,d_1]}$$

$$= L\int_c^d d_s\left(L\int_a^b dg(t)h(t,s)\right)f(s)$$

by the dominated convergence theorem. Thus, $F = [c_1,d_1] \in \mathfrak{A}$. Hence, $\mathfrak{A}$ contains all of the closed intervals in $[c,d]$.

Moreover, the family $\mathfrak{A}$ has the following properties;

(vii) if $F \in \mathfrak{A}$, then $F^c \in \mathfrak{A}$;

(viii) if $F_1 \in \mathfrak{A}$, $F_2 \in \mathfrak{A}$ and $F_1 \cap F_2 = \phi$, then $F_1 \cup F_2 \in \mathfrak{A}$;

(ix) if $F_i \in \mathfrak{A}$, $G_i \in \mathfrak{A}$, $F_i \subset F_{i+1}$, $G_i \supset G_{i+1}$ for $i = 1,2,\cdots$, then $\bigcup_{i=1}^{\infty} F_i \in \mathfrak{A}$ and $\bigcap_{i=1}^{\infty} G_i \in \mathfrak{A}$.

For instance, we can easily verify Property (ix) by the dominated convergence theorem. Thus, $\mathfrak{A}$ is a monotone class and it contains all of the elementary sets in $[c,d]$; here a set $F \subset [c,d]$ is an elementary set if F can be represented as a finite union of disjoint intervals in $[c,d]$. Note that $\mathfrak{B}$ is the smallest monotone class which contains all elementary sets in $[c,d]$ (cf. Rudin [1, Theorem 7.3]). Therefore, $\mathfrak{B} \subset \mathfrak{A}$ and consequently $\mathfrak{B} = \mathfrak{A}$.

Notice that the L-S measure μ_J is not a restriction of μ on J, by definition. However, μ_J is induced from μ as follows. Put

$$j^+(t) = g(t+0) - g(t), \qquad j^-(t) = g(t) - g(t-0)$$

$$j(t) = g(t+0) - g(t-0).$$

<u>Lemma 2.12</u>. Let μ be the L-S measure on $[a,b]$ induced by g. Then

$$\mu(\{a\}) = j^+(a), \qquad \mu(\{b\}) = j^-(b)$$

$$\mu(\{c\}) = j(c) \quad \text{for } a < c < b.$$

<u>Proof.</u> Consider the case $a < c < b$. Let χ^n be the continuous function such that $\chi^n = 0$ on $(-\infty, c-1/n]$ and on

$[c+1/n,\infty)$, $\chi^n(c) = 1$, and χ^n is linear on $[c-1/n,n]$ and on $[c,c+1/n]$. Then, we have from Example 1.6 that

$$\mu(\{c\}) = \lim_{n\to\infty} \int \chi^n d\mu = \lim_{n\to\infty} \int_a^b \chi^n dg = j(c).$$

In the similar manner, we have that $\mu(\{a\}) = j^+(a)$, $\mu(\{b\}) = j^-(b)$.

<u>Theorem 2.13</u>. If E is a Borel set of $J = [c,d]$, then

$$\mu_J(E) = \mu(E \cap (c,d)) + j^+(c)\delta_c(E) + j^-(d)\delta_d(E)$$

$$|\mu_J|(E) = |\mu|(E \cap (c,d)) + |j^+(c)|\delta_c(E) + |j^-(d)|\delta_d(E).$$

<u>Proof</u>. Let f be a function in $C(J)$. Define a function f^n, $n = 1,2,\cdots$, in $C(I)$ as

$$f^n(t) = \begin{cases} 0 & \text{on } (-\infty,c-1/n],\ [d+1/n,\infty) \\ f(t) & \text{on } [c,d] \\ \text{linear} & \text{on } [c-1/n,c],\ [d,d+1/n]. \end{cases}$$

Consider the case $a < c < d < b$. Then, it follows that

$$\int f^n d\mu = \int_a^b f^n dg = \int_a^c f^n dg + \int_c^d f dg + \int_d^b f^n dg.$$

Taking the limits as $n \to \infty$, we obtain

$$\int_{[c,d]} f d\mu = f(c)j^-(c) + \int_c^d f dg + f(d)j^+(d).$$

If we denote by ν the restriction of μ on J, then we can write

$$\int_c^d f dg = \int f d(\nu - j^-(c)\delta_c - j^+(d)\delta_d).$$

In view of the definition of the total variation of a measure, we obtain that $|\nu|(E) = |\mu|(E)$ for every Borel set E of J; hence, ν is a complex, regular Borel measure on J. Therefore, $\mu_J = \nu - j^-(c)\delta_c - j^+(d)\delta_d$ by Proposition 2.2 and Lemma 2.5. Glancing at Lemma 2.12, we obtain the formula in the theorem.

From this formula, we again know that $|\mu_J|(E) = |\mu|(E)$ if E is a Borel set in (c,d). Since $|\delta_t|(E) = \delta_t(E)$ for every E, and since

$$|\mu_J|(E) = |\mu_J|(E \cap (c,d)) + |\mu_J|(E \cap \{c\}) + |\mu_J|(E \cap \{d\})$$

for Borel set E in J, we have the formula for $|\mu_J|$ as in the theorem.

Corollary 2.14. If $a \le c < d \le b$, then

$$L\int_c^d fdg = \int_{(c,d)} fd\mu + f(c)(g(c+0) - g(c)) + f(d)(g(d) - g(d-0))$$

$$= \int_{[c,d]} fd\mu - f(c)(g(c) - g(c-0)) - f(d)(g(d+0) - g(d)).$$

Theorem 2.15. Suppose $a < c < b$. Then, fdg is L-S integrable on $[a,b]$ if and only if fdg is L-S integrable on $[a,c]$ and $[c,b]$; and if so, then

$$L\int_a^b fdg = L\int_a^c fdg + L\int_c^b fdg.$$

Proof. Set $I_1 = [a,c]$, $I_2 = [c,b]$, and μ_i the L-S measure on I_i induced by g, $i = 1,2$. Since $|\mu_i|(A) = |\mu|(A)$ for every Borel set A of $I_i \setminus \{c\}$, $i = 1,2$, a subset B of $I_i \setminus \{c\}$ is $|\mu_i|$-measurable if and only if B is $|\mu|$-measurable. Let E be a subset of I, and set $E_1 = E \cap [a,c)$, $E_2 = E \cap (c,b]$, $E_3 = E \cap \{c\}$. Since $E = E_1 \cup E_2 \cup E_3$, and since E_3 is a Borel set of I, E is $|\mu|$-measurable if and only if E_i is $|\mu_i|$-measurable for $i = 1,2$. Hence, a function f on I is $|\mu|$-measurable if and only if $f|_{I_i}$ is $|\mu_i|$-measurable for $i = 1,2$; and if so, then

$$\int_I fd\mu = \int_{[a,c)} fd\mu + f(c)j(c) + \int_{(c,b]} fd\mu.$$

Since $j(c) = j^+(c) + j^-(c)$, this equation implies the formula in the theorem.

Applying Theorem 2.15 repeatedly, we obtain the following.

Corollary 2.16. Suppose that $a = t_0 < t_1 < \cdots < t_n = b$. Then, fdg is L-S integrable if and only if fdg is L-S integrable on $[t_{i-1}, t_i]$ for $i = 1,\cdots,n$; and if so, then

$$L\int_a^b f\,dg = \left(L\int_{t_0}^{t_1} + L\int_{t_1}^{t_2} + \cdots + L\int_{t_{n-1}}^{t_n}\right) f\,dg.$$

Theorem 2.17. Let μ and λ be the L-S measures induced by g and T_g, respectively. Then, we have $|\mu| \le \lambda$; hence, if fdT_g is L-S integrable, so is fdg, and

$$|L\int_a^b f\,dg| \le L\int_a^b |f|\,dT_g.$$

If g is regular as a function of bounded variation, then $|\mu| = \lambda$; hence, fdT_g is L-S integrable if and only if fdg is L-S integrable.

Proof. Let E be a Borel set of I. Since $|\mu|$ and λ are bounded, positive, regular Borel measures, for every $\varepsilon > 0$ there exist an open set V and a compact set K such that $K \subset E \subset V$, $|\mu|(V\setminus K) < \varepsilon$, $\lambda(V\setminus K) < \varepsilon$. Then, $|\mu(E)| \le |\mu(K)| + |\mu(E\setminus K)| \le |\mu(K)| + |\mu|(E\setminus K) \le |\mu(K)| + \varepsilon$. Take a function f in $C(I)$ such that $0 \le f \le 1$, $f = 1$ on K and supp f is in V. Let $d\mu = hd|\mu|$ be the polar representation. Then, we have

$$|\int_a^b f\,dg - \mu(K)| = |\int fhd|\mu| - \mu(K)| = |\int_{V\setminus K} fhd|\mu||$$

$$\le |\mu|(V\setminus K) < \varepsilon.$$

Similarly, we have $|\int_a^b f\,dT_g - \lambda(K)| < \varepsilon$. Hence, it follows that

$$|\mu(E)| \le |\mu(K)| + \varepsilon \le |\int_a^b f\,dg| + 2\varepsilon \le \int_a^b |f|\,dT_g + 2\varepsilon$$

$$\le \lambda(K) + 3\varepsilon \le \lambda(E) + 3\varepsilon,$$

or $|\mu(E)| \le \lambda(E)$. Since this holds for any Borel set E, we have $|\mu|(E) \le \lambda(E)$. The integral inequality follows from Inequality (2.4).

Suppose f is a function in $C(I)$. For a partition P given by (1.1), we consider the sum

$$S(f,T_g,P) = \sum_{i=1}^{n} f(t_i)\{T_g(t_i) - T_g(t_{i-1})\} = \sum_{i=1}^{n} f(t_i)\mathrm{Var}(g,I_i).$$

Set $\Psi_i = \Phi_{I_i}$. Suppose g is regular. Then, we have $\mathrm{Var}(g,I_i) =$

$\|\Psi_i\|$. Denote by μ_i the L-S measure on I_i induced by g. Then, $\|\Psi_i\| = |\mu_i|(I_i)$ by Theorem 2.4; $|\mu_i|(I_i) = |j^+(t_{i-1})| + |\mu|((t_{i-1},t_i)) + |j^-(t_i)|$ by Theorem 2.10. Since g is regular, $|\mu|(\{t_i\}) = |\mu(\{t_i\})| = |j(t_i)| = |j^+(t_i)| + |j^-(t_i)|$, or $|j^-(t_i)| = |\mu|(\{t_i\}) - |j^+(t_i)|$. Summarizing these results, we have

$$\mathrm{Var}(g,I_i) = |j^+(t_{i-1})| + |\mu|((t_{i-1},t_i]) - |j^+(t_i)|$$

for $i = 2,\cdots,n-1$, and

$$\mathrm{Var}(g,I_1) = |\mu|([a,t_1]) - |j^+(t_1)|$$

$$\mathrm{Var}(g,I_n) = |j^+(t_{n-1})| + |\mu|((t_{n-1},t_n]).$$

Therefore, we can write

$$S(f,T_g,P) = f(t_1)|\mu|([a,t_1]) + \sum_{i=2}^{n} f(t_i)\mu((t_{i-1},t_i])$$

$$+ \sum_{i=1}^{n-1} \{f(t_{i+1}) - f(t_i)\}|j^+(t_i)|$$

$$= \int f^P d|\mu| + \sum_{i=1}^{n-1} \Delta_i f|j^+(t_i)|,$$

where $\Delta_i f = f(t_{i+1}) - f(t_i)$ and f^P is defined by

$$f^P(t) = \begin{cases} f(t_1) & \text{on } [a,t_1] \\ f(t_i) & \text{on } (t_{i-1},t_i],\ i = 2,\cdots,n. \end{cases}$$

Set $s_i = t_i + \frac{1}{k}$, $i = 1,\cdots,n-1$, where $1/k < \min\{t_i-t_{i-1} : i = 1,\cdots,n\}$, and Q_k be the partition such that $a = t_0 < t_1 < s_1 < t_2 < s_2 < \cdots < t_{n-1} < s_{n-1} < t_n = b$. Then, we have

$$\sum_{i=1}^{n-1} |g(s_i) - g(t_i)| \le \sum_{Q_k} |\Delta_j g| \le \mathrm{Var}(g,[a,b]).$$

Taking the limit as $k \to \infty$, we obtain that

$$\sum_{i=1}^{n-1} |j^+(t_i)| \le \mathrm{Var}(g,[a,b]).$$

If we set $\omega(f,P) = \max\{|\Delta_i f| : i = 1,\cdots,n-1\}$, we therefore have that

$$|S(f,T_g,P) - \int f^P d|\mu|| \le \omega(f,P)\mathrm{Var}(g,[a,b]).$$

Since $f^P(t) \to f(t)$ and $\omega(f,P) \to 0$ as $m(P) \to 0$, it follows that

$$\int_a^b f dT_g = \int f d|\mu|.$$

Therefore, $|\mu|$ is the L-S measure induced by T_g.

Example 2.18. If g is the function in Example 1.13, then $\mu \equiv 0$ is the L-S measure induced by g, and $\lambda = 2\delta_0$ is the L-S measure induced by T_g.

Corollary 2.19. Let g, g_i, $g_i^{\pm}$, μ_i, $\lambda_i^{\pm}$ be as in Corollary 2.9, and let $\mu_i = \mu_i^+ - \mu_i^-$ be the Jordan decomposition of μ_i. Then, $\mu_i^{\pm} \le \lambda_i^{\pm}$; and $\mu_i^{\pm} = \lambda_i^{\pm}$ if g is regular as a function of bounded variation.

Proof. Let λ_i be the L-S measure induced by T_{g_i}. Then, $\lambda_i = \lambda_i^+ + \lambda_i^-$ from Corollary 2.10. On the other hand, $\mu_i = \lambda_i^+ - \lambda_i^-$ from Corollary 2.9. Since $\lambda_i \ge |\mu_i|$ by Theorem 2.17, we have

$$\mu_i^{\pm} = \frac{1}{2}\{|\mu_i| \pm \mu_i\} \le \frac{1}{2}\{\lambda_i \pm \mu_i\} = \lambda_i^{\pm}.$$

If g is regular, then g_1 and g_2 are regular, which implies $\lambda_i = |\mu_i|$. Therefore $\mu_i^{\pm} = \lambda_i^{\pm}$.

Example 2.20. Consider the function g in Example 2.18. Then, we have $\mu = |\mu| = 0$, hence $\mu^+ = \mu^- = 0$. On the other hand, $\lambda^+ = \lambda^- = \delta_0$.

Finally, we show that the L-S integral is an extension of D-S integral.

Theorem 2.21. Suppose α is a nondecreasing function on $I = [a,b]$. If $fd\alpha$ is D-S integrable, then it is L-S integrable, and

$$D\int_a^b fd\alpha = L\int_a^b fd\alpha.$$

Let λ be the L-S measure induced by α. Then, $fd\alpha$ is D-S integrable if and only if the following two conditions hold:

(i) f is right continuous at every point of $S^{+}(\alpha)$, and f is left continuous at every point of $S^{-}(\alpha)$.

(ii) f is continuous λ-a.e. in $I \setminus S(\alpha)$.

<u>Proof</u>. Since $D\int_a^b f d\alpha = D\int_a^b (f_1 d\alpha + \sqrt{-1} f_2 d\alpha)$, we can assume that f is a real function.

For a partition given by (1.1), define $U(f,\alpha,P)$ as in (1.11), and define a function $U = U(P)$ as

$$U(t) = \begin{cases} M_1 & \text{if } t_0 \le t < t_1, \\ M_i & \text{if } t_{i-1} < t < t_i,\ i = 2,\cdots,n-1, \\ M_n & \text{if } t_{n-1} < t \le t_n, \\ \max\{M_i, M_{i+1}\} & \text{if } t = t_i \text{ and } \lambda(\{t_i\}) = 0, \\ M_i\lambda^{-}(t_i) + M_{i+1}\lambda^{+}(t_i) & \text{if } t = t_i \text{ and } \lambda(\{t_i\}) > 0, \end{cases}$$

where $\lambda^{-}(\{t_i\}) = [\alpha(t_i) - \alpha(t_i-0)]/\lambda(\{t_i\})$ and $\lambda^{+}(\{t_i\}) = [\alpha(t_i+0) - \alpha(t_i)]/\lambda(\{t_i\})$. Then, we can write

$$U(f,\alpha,P) = \int U d\lambda.$$

Defining $L = L(P)$ with M_i replaced by m_i, we have

$$L(f,\alpha,P) = \int L d\lambda.$$

It is clear that $L(t) \le f(t) \le U(t)$ for all t in I.

Let $\{P_n\}$ be a sequence of partitions such that $P_1 < P_2 < \cdots$ and that

$$\underline{\int_a^b} f d\alpha = \lim_{n\to\infty} L(f,\alpha,P_n) \qquad \overline{\int_a^b} f d\alpha = \lim_{n\to\infty} U(f,\alpha,P_n).$$

Such a sequence $\{P_n\}$ can be chosen by Definition (1.11). Set $L_n = L(P_n)$, $U_n = U(P_n)$. If the refinement P_{n+1} of P_n is made by adding only one point to the points of P_n, it follows at once that

$$L_n(t) \le L_{n+1}(t) \le f(t) \le U_{n+1}(t) \le U_n(t) \quad \text{for } t \in I.$$

Since P_{n+1} is generally made by adding finite points successively to the points of P_n, these inequalities hold for $n = 1,2,\cdots$. Therefore, we have

$$L(t) := \lim_{n\to\infty} L_n(t) \le f(t) \le U(t) := \lim_{n\to\infty} U_n(t)$$

for all t in I, and

$$\bar{\int}_a^b f d\alpha = \lim_{n\to\infty} U(f,\alpha,P_n) = \lim_{n\to\infty} \int U_n d\lambda = \int U d\lambda$$

$$\underline{\int}_a^b f d\alpha = \lim_{n\to\infty} L(f,\alpha,P_n) = \lim_{n\to\infty} \int L_n d\lambda = \int L d\lambda.$$

Suppose $fd\alpha$ is D-S integrable. Then, the lower integral and the upper integral are the same, which implies that there exists a λ-null set N such that $L(t) = U(t)$ for t in $I \setminus N$. Since $L \le f \le U$, we have

$$L(t) = f(t) = U(t) \quad \text{for} \quad t \in I \setminus N, \tag{2.5}$$

and

$$L\int_a^b f d\alpha = \int f d\lambda = \int L d\lambda = \underline{\int}_a^b f d\alpha = D\int_a^b f d\alpha.$$

The condition (i) is mentioned in Theorem 1.3. Since every point in $S(\alpha)$ has a positive λ measure, the λ-null set N is contained in $K := I \setminus S(\alpha)$. Let t be in $K \setminus N$. Then, it follows from (2.5) that, for every $\varepsilon > 0$, there exists an $n(\varepsilon)$ such that

$$f(t) - \varepsilon < L_n(t) \le U_n(t) < f(t) + \varepsilon \tag{2.6}$$

for $n > n(\varepsilon)$. If t is not a point of P_n for any n, it is in the interior of some subinterval of P_n for any n. Inequality (2.6) then means that $|f(s) - f(t)| < \varepsilon$ for every s in this subinterval of P_n provided $n > n(\varepsilon)$; hence, f is continuous at t. Suppose the point t is a point of P_m for some m. Since $P_n > P_m$, it is then a point of P_n for $n \ge m$. Thus, for $n \ge m$, there exist adjacent intervals, say H_n^+ and H_n^-, of P_n, which have t as the common point. Notice that $\lambda(\{t\}) = 0$, since t is not in $S(\alpha)$. Glancing the definitions of U_n and L_n at such a point, we see that $|f(s) - f(t)| < \varepsilon$ for s in the union of H_n^+ and H_n^-. Therefore, f is continuous at t.

Conversely, suppose f satisfies Conditions (i) and (ii) in the theorem. Making a refinement, if necessary, we can assume that the sequence $\{P_n\}$ taken before satisfies that $m(P_n) \to 0$ as $n \to \infty$

and that $S(\alpha)$ is contained in the union of the sets of points of P_n. Let N be the λ-null set of K as before. Then, the first additional condition on $\{P_n\}$ assures that $U_n(t)$ and $L_n(t)$ converges to $f(t)$ as $n \to \infty$ for $t \in K \setminus N$. Let t be a point of $S(\alpha)$. From the second additional condition on $\{P_n\}$, there is an m such that t is a point of P_n for $n \geq m$. Let H_n^- and H_n^+ be the adjacent subintervals of P_n as before. We assume H_n^- is left to t and H_n^+ is right to t, and set

$$M_n^- = \sup f(H_n^-) \qquad M_n^+ = \sup f(H_n^+)$$

$$m_n^- = \inf f(H_n^-) \qquad m_n^+ = \sup f(H_n^+).$$

Then, we have

$$U_n(t) - L_n(t) = (M_n^- - m_n^-)\lambda^-(t) + (M_n^+ - m_n^+)\lambda^+(t).$$

If $\lambda^-(t) > 0$, then f is left continuous at t from Condition (i). This implies that $M_n^- - m_n^- \to 0$ as $n \to \infty$. Similarly, if $\lambda^+(t) > 0$, then $M_n^+ - m_n^+ \to 0$ as $n \to \infty$. Hence, we have $U_n(t) - L_n(t) \to 0$ as $n \to \infty$; that is, $L(t) = f(t) = U(t)$. Therefore, we obtain Relation (2.5); that is, $fd\alpha$ is D-S integrable.

<u>Corollary 2.22</u>. Suppose that α and λ be as in Theorem 2.20. Then, $fd\alpha$ is R-S integrable if and only if f is continuous λ-a.e. in $[a,b]$; and if so, then the both integrals are the same.

<u>Proof</u>. This follows from Theorems 1.4 and 2.20.

<u>Theorem 2.23</u>. Suppose that g is a complex function of bounded variation. If fdT_g is D-S integrable, then fdT_g and fdg are L-S integrable, and

$$D\int_a^b fdT_g = L\int_a^b fdT_g \qquad D\int_a^b fdg = L\int_a^b fdg.$$

Furthermore, fdT_g is D-S integrable if and only if f satisfies the following conditions:

(i) f is right continuous at every point of $S^+(g)$, and left continuous at every point of $S^-(g)$.

(ii) f is continuous λ-a.e. in $I \setminus S(g)$,

where λ is the L-S measure induced by T_g.

Proof. If fdT_g is D-S integrable, then $f_i dg_j^{\pm}$ are D-S integrable from Theorem 1.18. Therefore, it follows from Theorem 1.18, Corollary 2.9 and Theorem 2.21 that

$$D\int_a^b fdg = D\int_a^b f(dg_1^+ - dg_1^- + \sqrt{-1}(dg_2^+ - dg_2^-))$$

$$= L\int_a^b f(dg_1^+ - dg_1^- + \sqrt{-1}(dg_2^+ - dg_2^-))$$

$$= L\int_a^b fdg.$$

The other parts of the theorem follow from Theorem 2.21.

Corollary 2.24. If we replace the D-S integral by the R-S integral in Theorem 2.23, we can read Conditions (i) and (ii) as the condition

(iii) f is continuous λ-a.e. in I.

3.3. Radon Measures. Let X and $C_c(X)$ be as in Section 3.2, and $C_c^+(X)$ the class of nonnegative functions in $C_c(X)$, and $\Gamma(X)$ the linear space of bounded Borel measurable functions mapping X into $\mathbb{C}$ with compact support. Obviously, $C_c(X)$ is a linear subspace of $\Gamma(X)$. We consider extensions of linear operators on $C_c(X)$ which are not necessarily continuous with respect to the supremum norm.

A linear operator $\Phi : C_c(X) \to \mathbb{C}$ is called a Radon measure on X into $\mathbb{C}$ if, for each compact set K of X, there exists a constant c_K such that

$$|\Phi(f)| \le c_K \sup\{|f(x)| : x \in X\}$$

provided $\operatorname{supp} f :=$ closure of $\{x : f(x) = 0\}$ lies in K. A sequence $\{f^k\}$ in $\Gamma(X)$ is said to converge in Lebesgue to a function f in $\Gamma(X)$ if $\{f^k(x)\}$ are uniformly bounded, their supports are all contained in a compact set and $f^k(x) \to f(x)$ as $k \to \infty$ for each x in X. A linear operator $\nu : \Gamma(X) \to \mathbb{C}$ is said to be continuous in Lebesgue (in short, Lebesgue continuous) if the sequence $\{\nu(f^k)\}$ converges to $\nu(f)$ for any sequence $\{f^k\}$ in $\Gamma(X)$ which converges in Lebesgue to f.

If λ is a positive, locally finite, Borel measure on X and if $h \in L^{\infty}(\lambda)$, the integral

(3.1) $$\nu(f) = \int_X fh\,d\lambda$$

is well defined for $f \in \Gamma(X)$, and it is continuous in Lebesgue. We have the converse result.

<u>Theorem 3.1</u>. Suppose $\nu : \Gamma(X) \to \mathbb{C}$ is a linear operator which is continuous in Lebesgue. Then there exist a unique, positive, locally finite, Borel measure λ, and a unique function h in $L^\infty(\lambda)$ such that $|h(x)| = 1$ for all $x \in X$ and such that Relation (3.1) holds for every $f \in \Gamma(X)$.

<u>Proof</u>. We first show the uniqueness of λ and h. Suppose that

$$\int_X fh_1 d\lambda_1 = \int_X fh_2 d\lambda_2 \quad \text{for } f \in \Gamma(X),$$

and for positive, locally finite, Borel measures λ_1 and λ_2, and for some functions $h_1, h_2 \in L^\infty(\lambda)$ such that $|h_1(x)| = |h_2(x)| = 1$ for all $x \in X$. If E is a Borel set contained in a compact set, we then have

$$\lambda_1(E) = \int_X \chi_E \bar{h}_1 h_1 d\lambda_1 = \int_X \chi_E \bar{h}_1 h_2 d\lambda_2 \le \int_X \chi_E d\lambda_2 = \lambda_2(E),$$

and in the similar manner we have $\lambda_2(E) \le \lambda_1(E)$; that is, $\lambda_1(E) = \lambda_2(E)$. Since X is σ-compact, we can conclude that $\lambda_1 = \lambda_2$. Now, setting $\lambda = \lambda_1 = \lambda_2$, we have

$$\int_E h_1 d\lambda = \int_E h_2 d\lambda$$

for every E such that $\chi_E \in \Gamma(X)$. Since X is σ-compact, this equation shows that $h_1 = h_2$ a.e. $[\lambda]$.

To prove the existence, take a family of compact sets $X_1 \subset X_2 \subset X_3 \subset \cdots \subset X_n \subset \cdots$, whose union is X. For every Borel set E of X_n, define

$$\mu_n(E) = \nu(\chi_E).$$

Obviously, μ_n is a complex Borel measure on X_n, and $\mu_n(E) = \mu_m(E)$ if $E \subset X_n$ and $n \le m$. Set $\lambda_n = |\mu_n|$, and take the polar representation $d\mu_n = h_n d\lambda_n$. Then, it follows that, if $n \le m$, $\lambda_n(E) = \lambda_m(E)$ for $E \subset X_n$, and $h_n = h_m$ a.e. $[\lambda_n]$ in X_n. If we set Y_1

$= X_1$. $Y_n = X_n - X_{n-1}$ for $n = 2,3,\cdots$, then X is a union of disjoint countable Borel sets Y_n. Define

$$\lambda(E) = \sum_{n=1}^{\infty} \lambda_n(E \cap Y_n)$$

for any Borel set E of X, and define

$$h(x) = h_n(x) \quad \text{for} \quad x \in Y_n.$$

Then it follows immediately that λ is a positive Borel measure, h is a complex Borel measurable function, and that $|h(x)| = 1$ for $x \in X$.

Suppose E is a Borel set in X_n for some n. Then, we have

$$\nu(\chi_E) = \mu_n(E) = \int_E h_n d\lambda_n = \int_E hd\lambda = \int_X \chi_E hd\lambda.$$

Therefore, Relation (3.1) holds for every simple function f in $\Gamma(X)$, and so also for every nonnegative function in $\Gamma(X_n)$, for every real function in $\Gamma(X_n)$ and for every complex function in $\Gamma(X_n)$. Therefore, we have

$$\nu(\chi_E\varphi) = \int_X \chi_E \varphi hd\lambda$$

for any bounded Borel measurable function φ. In particular, setting $\varphi = \bar{h}$, we obtain

$$\nu(\chi_E\bar{h}) = \lambda(E)$$

if E is contained in X_n.

Next, let E be a Borel set contained in some compact set K. Then the sequence $\{\chi_{E\cap X_n}\}$ converges in Lebesgue to χ_E and hence the sequence $\{\chi_{E\cap X_n}\bar{h}\}$ also converges in Lebesgue to $\chi_E\bar{h}$. Thus,

$$\lambda(E) = \lim_{n\to\infty} \lambda(E \cap X_n) = \lim_{n\to\infty} \nu(\chi_{E\cap X_n}\bar{h}) = \nu(\chi_E\bar{h}).$$

Consequently, $\lambda(E) < \infty$, which shows that the measure λ is locally finite. Moreover, because $\lambda(E) < \infty$, the dominated convergence theorem shows that

$$\nu(\chi_E) = \lim_{n\to\infty} \nu(\chi_{E\cap X_n}) = \lim_{n\to\infty} \int_X (\chi_{E\cap X_n})hd\lambda = \int_X \chi_E hd\lambda.$$

Therefore, in the same manner as before, Relation (3.1) holds for every function in $\Gamma(X)$.

We denote by $|\nu|$ the Borel measure λ given in Theorem 3.1, and call it the total variation of ν. We also write Equation (3.1) as $d\nu = h d\lambda$. In case $\lambda = |\nu|$, we call it the polar representation of Lebesgue continuous operator $\nu : \Gamma(X) \to \mathbb{C}$.

<u>Theorem 3.2</u>. Suppose λ_1 and λ_2 are regular, positive, locally finite, Borel measures on X, and $h_i \in L^\infty(\lambda_i)$, $i = 1,2$. If Relation

$$(3.2) \qquad \int_X fh_1 d\lambda_1 = \int_X fh_2 d\lambda_2$$

holds for every f in $C_c(X)$, then it holds for every f in $\Gamma(X)$.

<u>Proof.</u> It suffices to show that Relation (3.2) holds for $f = \chi_E$ in $\Gamma(X)$. Since λ_1 and λ_2 are regular Borel measures, for every $\varepsilon > 0$, there exist a compact set K and an open set V such that $K \subset E \subset V$, and that $\lambda_i(V \setminus K) < \varepsilon$, $i = 1,2$. From Uryson's lemma, we can choose a function f in $C_c(X)$ such that $0 \le f \le 1$, $f = 1$ on K and $\operatorname{supp} f \subset V$. For this function, we have that, for $i = 1,2$,

$$\left|\int_X \chi_E h_i d\lambda_i - \int_X fh_i d\lambda_i\right| = \left|\int_{E\setminus K}(1-f)h_i d\lambda_i - \int_{V\setminus K} fh_i d\lambda_i\right|$$

$$\le |h_i|_\infty \lambda_i(E\setminus K) + |h_i|_\infty \lambda_i(V\setminus E)$$

$$= |h_i|_\infty \lambda_i(V\setminus K) \le |h_i|_\infty \varepsilon.$$

Since Relation (3.2) holds for this function f, we know that

$$\left|\int_X \chi_E h_1 d\lambda_1 - \int_X \chi_E h_2 d\lambda_2\right| \le (|h_1|_\infty + |h_2|_\infty)\varepsilon.$$

Since ε is an arbitrary positive number, we have the desired result

A Borel prolongation of a Radon measure Φ on X is a linear operator $\tilde{\Phi} : \Gamma(X) \to \mathbb{C}$, such that $\tilde{\Phi}$ is continuous in Lebesgue and such that

$$\tilde{\Phi}(f) = \Phi(f) \quad \text{for } f \in C_c(X).$$

If the total variation $|\tilde{\Phi}|$ is regular, we say that $\tilde{\Phi}$ is a regular Borel prolongation of Φ.

Theorem 3.3. Every Radon measure Φ on X have a unique, regular Borel prolongation $\tilde{\Phi}$; that is, there exist uniquely a Borel function h, $|h(x)| = 1$ for all $x \in X$, and a positive, regular, locally finite Borel measure λ such that

$$\Phi(f) = \int_X fh d\lambda \qquad f \in C_c(X).$$

We show the outline of the proof by following the argument given in the proof of Rudin [1, Theorem 6.19]. The uniqueness of such a prolongation is a direct consequence of Theorems 3.1 and 3.2.

To prove the existence, define

$$\Lambda(f) = \sup\{|\Phi(h)| : h \in C_c(X), |h| \le f\}$$

for f in $C_c^+(X)$. Then, Λ is extended on $C_c(X)$ as in Rudin [1, pp.141-142], so that Λ is a positive linear functional on $C_c(X)$ satisfying

$$|\Phi(f)| \le \Lambda(|f|) \qquad \text{for } f \in C_c(X).$$

Let λ be the corresponding positive, regular, Borel measure on X such that

$$\Lambda(f) = \int_X f d\lambda \qquad \text{for } f \text{ in } C_c(X),$$

cf. Rudin [1, Theorems 2.14 and 2.17]. Then, we have

$$|\Phi(f)| \le \Lambda(|f|) = \int_X |f| d\lambda = |f|_1 \qquad \text{for } f \text{ in } C_c(X).$$

Since $C_c(X)$ is dense in $L^1(\lambda)$, Φ has a unique extension $\tilde{\Phi}$ which is a linear functional on $L^1(\lambda)$ such that $|\tilde{\Phi}(f)| \le |f|_1$ for $f \in L^1(\lambda)$. Since λ is a σ-finite measure on the σ-compact space X, there is a unique function g in $L^\infty(\lambda)$ such that

$$\tilde{\Phi}(f) = \int_X fg d\lambda \qquad \text{for } f \text{ in } L^1(\lambda).$$

The dominated convergence theorem means that this extension $\tilde{\Phi}$ is continuous in Lebesgue. Furthermore, the above representation $\tilde{\Phi}$

$= g d\lambda$ is the polar representation of $\tilde{\Phi}$. Indeed, the inequality $|\tilde{\Phi}(f)| \le |f|_1$ for $f \in L^1(\lambda)$ implies $|g(x)| \le 1$ a.e. $[\lambda]$. Suppose V is an open set of X. Since

$$\lambda(V) = \sup\{\Lambda(f) : f \in C_c^+(X),\ 0 \le f \le 1,\ \operatorname{supp} f \subset V\},$$

cf. Rudin [1, p.43], it follows that

$$\lambda(V) = \sup\{|\Phi(h)| : h \in C_c(X, \mathbb{C}),\ |h| \le 1,\ \operatorname{supp} h \subset V\}.$$

This implies

$$\lambda(V) \le \int_V |g| d\lambda \le \int_V d\lambda = \lambda(V).$$

If $\lambda(V) < \infty$, these inequalities are compatible only if $|g| = 1$ a.e. $[\lambda]$ in V. Since λ is a regular, σ-finite measure, X is a union of countable open sets with finite λ measure. Therefore, we have $|g| = 1$ a.e. $[\lambda]$. Redefine g on $B = \{x : |g(x)| \ne 1\}$ so that $g(x) = 1$ on B. Since $\lambda(B) = 0$, the representation $\tilde{\Phi} = g d\lambda$ also holds for such a function g.

Now, we consider the Radon measure on $(-\infty, 0]$. If $g : (-\infty, 0] \to \mathbb{C}$ is of bounded variation over $[-r, 0]$ for every $r > 0$, we say g is locally of bounded variation on $(-\infty, 0]$. For such a g, we can define a linear operator Φ on $C_c((-\infty, 0], \mathbb{C})$ as

$$(3.3) \qquad \Phi(f) = R\int f dg \qquad f \in C_c((-\infty, 0], \mathbb{C}),$$

where the integral is over the interval $[-r, 0]$ containing $\operatorname{supp} f$. Clearly, Φ is a Radon measure on $C_c((-\infty, 0], \mathbb{C})$. The following theorem shows that the converse also holds.

<u>Theorem 3.4.</u> For every Radon measure Φ on $C_c((-\infty, 0], \mathbb{C})$, there corresponds a function $g : (-\infty, 0] \to \mathbb{C}$, locally of bounded variation, such that Relation (3.3) holds. Such a g is unique for Φ if g is normalized in the sense that $g(0) = 0$ and $g(\theta)$ is left continuous at $\theta < 0$; in fact, $g(\theta)$ is given by

$$(3.4) \qquad g(\theta) = \begin{cases} 0 & \text{if } \theta = 0, \\ -\tilde{\Phi}(\chi_{[\theta,0]}) = -\int \chi_{[\theta,0]} h d\lambda & \text{if } \theta < 0, \end{cases}$$

where $\tilde{\Phi}$ is the regular, Borel prolongation of Φ and $\tilde{\Phi} = hd\lambda$ is the polar representation of $\tilde{\Phi}$.

Proof. First of all, we will verify that the function g defined by (3.4) is locally of bounded variation and normalized. Clearly, g is left continuous at $\theta < 0$, because $\chi_{[s,0]}$ converges in Lebesgue to $\chi_{[\theta,0]}$ as $s \to \theta-0$. Let

$$(3.5) \qquad P : -r = t_0 < t_1 < \cdots < t_n = 0$$

be any partition of $[-r,0]$, and consider $V(g,P) := \sum_P |\Delta_i g|$. Taking complex numbers $\sigma(i)$ such that $|\sigma(i)| = 1$ and $|\Delta_i g| = \sigma(i)\Delta_i g$, we set

$$(3.6) \qquad f = -\sum_{i=1}^{n-1} \sigma(i)\{\chi_{[t_i,0]} - \chi_{[t_{i-1},0]}\} + \sigma(n)\chi_{[t_{n-1},0]}.$$

From (3.4), it follows that

$$V(g,P) = -\sum_{i=1}^{n-1} \sigma(i)\{\tilde{\Phi}(\chi_{[t_i,0]}) - \tilde{\Phi}(\chi_{[t_{i-1},0]})\} + \sigma(n)\tilde{\Phi}(\chi_{[t_{n-1},0]})$$

$$= \tilde{\Phi}(f).$$

Let f^m, $m = 1,2,\cdots$, be the function defined by (3.6) with $\chi_{[s,0]}$ replaced by $\chi^m{}_{[s,0]}$ which is given by

$$(3.7) \qquad \chi^m{}_{[s,0]}(t) = \begin{cases} 1 & \text{if } s \le t \le 0, \\ m(t - s + \frac{1}{m}) & \text{if } s - \frac{1}{m} \le t < s, \\ 0 & \text{if } t < s - \frac{1}{m} \end{cases}$$

for $s < 0$. Obviously, f^m is in $C_c((-\infty,0],\mathbb{C})$, $\sup_{t\le 0} |f^m(t)| \le 1$, and $f^m(\theta) \to f(\theta)$ as $m \to \infty$ for every $\theta \le 0$. Thus, we have $\Phi(f^m) = \tilde{\Phi}(f^m) \to \tilde{\Phi}(f) = V(g,P)$ as $m \to \infty$, because $\tilde{\Phi}$ is continuous in Lebesgue. Since $\sum_P |\Delta_i g| \ge 0$, this implies $V(g,P) = \lim_{m\to\infty} |\Phi(f^m)|$. Hence, $V(g,P) \le C_r$ for some constant C_r independent of P. This implies that $\mathrm{Var}(g,[-r,0]) \le C_r$.

Next, we establish Relation (3.3) for the above function g. For any partition P given by (3.5), let f^P be the function defined by (3.6) with $\sigma(i)$ replaced by $f(s_i)$, where $t_{i-1} \le s_i \le t_i$ for $i = 1,\cdots,n$. From the definition of g it follows that

$$\tilde{\Phi}(f^P) = \sum_{i=1}^{n} f(s_i)[g(t_i) - g(t_{i-1})] = S(f,g,P).$$

Since f^P converges in Lebesgue to f as $m(P) \to 0$, Relation (3.3) follows immediately from the above relation.

Finally, we prove the uniqueness of g which is normalized and satisfies Relation (3.3). Let $\tilde{g}$ be another function which possesses these properties. Then, we have

$$R\int \chi^m_{[s,0]} dg = \Phi(\chi^m_{[s,0]}) = R\int \chi^m_{[s,0]} d\tilde{g}.$$

Taking the limits as $m \to \infty$, we have, from Example 1.6, that

$$g(0) - g(s) + g(s) - g(s-0) = \tilde{g}(0) - \tilde{g}(s) + \tilde{g}(s) - \tilde{g}(s-0).$$

Since g and $\tilde{g}$ are normalized, it follows that $g(s) = \tilde{g}(s)$ for $s < 0$.

3.4. Representation of linear operators on the phase space $\mathscr{B}$. Due to the Riesz representation theorem (cf. Theorem 3.4), every continuous linear functional $L : C([-r,0],\mathbb{C}^n) \to \mathbb{C}^n$ is represented by a Stieltjes integral with respect to a matrix function of bounded variation in $[-r,0]$. In this section, we will try to make an analogous consideration for a linear continuous functional on the phase space $\mathscr{B}$.

Let $L : J \times \mathscr{B} \to \mathbb{C}^n$, where J is an interval, be a mapping of Caratheodory type, that is, $L(t,\varphi)$ is Borel measurable in $t \in J$ for each fixed $\varphi \in \mathscr{B}$ and is continuous in $\varphi \in \mathscr{B}$ for each fixed $t \in J$. Furthermore, let $L(t,\varphi)$ be linear in $\varphi \in \mathscr{B}$ for each $t \in J$. For each $t \in J$, set $|L(t)| = \sup\{|L(t,\varphi)| : |\varphi|_{\mathscr{B}} \le 1\}$. Clearly, $|L(t)| < \infty$ for $t \in J$. Let C_{00} and Γ be the product spaces of n-copies of $C_c((-\infty,0])$ and $\Gamma((-\infty,0])$, respectively. Clearly, C_{00} is a subspace of Γ. With these spaces C_{00} and Γ, we can define "Radon measures", "Borel prolongation" etc. as in Section 3.3. Note that $C_{00} \subset \mathscr{B}$ (cf. Section 1.2). For each fixed $t \in J$, we denote by $L_0(t,\cdot)$ the restriction of $L(t,\cdot)$ on C_{00}. Then, $L_0(t,\cdot)$ is a linear operator on C_{00} into $\mathbb{C}^n$, and

$$|L_0(t,\varphi)| \le |L(t)||\varphi|_{\mathscr{B}} \le |L(t)|K(r)|\varphi|_{[-r,0]},$$

where $|\varphi|_{[-r,0]} = \sup\{|\varphi(\theta)| : -r \le \theta \le 0\}$, by Axiom (A) provided supp φ lies in $[-r,0]$. Thus, $L_0(t,\cdot)$ is a "Radon" measure on

$(-\infty,0]$. Hence, by Theorem 3.3, $L_0(t,\cdot)$ has a unique, regular, Borel prolongation $\tilde{L}(t,\cdot) : \Gamma \to \mathbb{C}^n$.

Now, for any interval S in $(-\infty,0]$, let χ_S denote the indicator function of S, that is,

$$\chi_S(\theta) = \begin{cases} 1 & \text{if } \theta \in S, \\ 0 & \text{if } \theta \bar{\in} S. \end{cases}$$

For each $(t,\theta) \in J \times (-\infty,0]$, set

$$\eta(t,\theta) = \begin{cases} 0 & \text{for } \theta = 0 \\ -\tilde{L}(t,\chi_{[\theta,0]}I) & \text{for } \theta < 0, \end{cases} \tag{4.1}$$

where I is the $n \times n$ identity matrix. Since $\chi_{[\theta,0]} \in \Gamma((-\infty,0])$, the function $\eta(t,\theta)$ is well defined and $\eta : J \times (-\infty,0] \to \mathcal{M}$, where $\mathcal{M}$ is the space of all complex $n \times n$ matrices with operator norm induced by the usual vector norm on $\mathbb{C}^n$. As shown in Theorem 3.4, the above function $\eta(t,\theta)$ is locally of bounded variation and normalized in θ for each fixed $t \in J$.

<u>Lemma 4.1.</u> The function $\eta(t,\theta)$ defined by (4.1) is Borel measurable for (t,θ) in $J \times (-\infty,0]$.

<u>Proof</u>. Let $\chi^m{}_{[s,0]}$, $m = 1,2,\cdots$, be the function defined by (3.7). As seen in the proof of Theorem 3.4, we have

$$\eta(t,\theta) = \lim_{m\to\infty} L(t,-\chi^m{}_{[s,0]}I), \qquad s < 0.$$

Since $L(t,\varphi)$ is a mapping of Caratheodory type and $\chi^m{}_{[s,0]}I$ is continuous in $s < 0$ as a function with values in $\mathcal{B}$, the function $L(t,-\chi^m{}_{[s,0]}I)$ is Borel measurable in $(t,s) \in J \times (-\infty,0)$. Hence the function η is Borel measurable on $J \times (-\infty,0)$ as a limit function of the sequence of Borel measurable functions. Consequently, η is also Borel measurable on $J \times (-\infty,0]$, since $\eta(t,0) \equiv 0$ for t in J.

The following theorem is due to Naito [6], which is a generalization of Riesz representation theorem for linear operators on the space $\mathcal{B}$.

<u>Theorem 4.2</u>. Suppose $L : J \times \mathcal{B} \to \mathbb{C}^n$, where J is an interval,

is a mapping of Caratheodory type such that $L(t,\varphi)$ is linear in $\varphi \in \mathcal{Z}$ for each t in J. Then there exists a Borel measurable function $\eta : J \times (-\infty,0] \to \mathcal{M}$ such that for each $t \in J$ the function η is locally of bounded variation for θ in $(-\infty,0]$ and that

$$(4.2) \qquad L(t,\varphi) = \int [d_\theta \eta(t,\theta)]\varphi(\theta) \qquad \text{for } \varphi \in C_{00}$$

$$(4.3) \qquad \operatorname{Var}(\eta(t,\cdot),[-r,-s]) \le c|L(t)|K(r-s)M(s) \quad \text{for } r > s \ge 0,$$

where the integral is over the interval containing $\operatorname{supp}\varphi$, and c is a constant dependent only on the norm of $\mathbb{C}^n$. If $\eta(t,0)$ in Relation (4.2) is normalized in θ, then it is determined uniquely by L.

Proof. Except the estimate (4.3), we have already seen that the results in Theorem 4.2 are true by Theorem 3.4 and Lemma 4.1. We shall verify the estimate (4.3). To do so, it suffices to show that the similar estimate is valid for each component of η. Thus, without restricting the generality, we can assume $n = 1$.

Let P be any partition of an interval $[-r,-s]$ such that $-r = \theta_0 < \theta_1 < \cdots < \theta_n = -s$, and set

$$V^P(t) = \sum_{i=1}^{d} |\eta(t,\theta_i) - \eta(t,\theta_{i-1})|.$$

For each $i = 1,\cdots,d$, take a complex number $\sigma(i)$ such that $|\sigma(i)| = 1$ and that $|\eta(t,\theta_i) - \eta(t,\theta_{i-1})| = \sigma(i)\Big(\eta(t,\theta_i) - \eta(t,\theta_{i-1})\Big)$. In case $\theta_d = -s < 0$, we set

$$(4.4) \qquad \varphi = -\sum_{i=1}^{d} [\chi_{[\theta_i,0]} - \chi_{[\theta_{i-1},0]}]\sigma(i),$$

and in case $\theta_d = -s = 0$, we set

$$(4.5) \qquad \varphi = -\sum_{i=1}^{d} [\chi_{[\theta_i,0]} - \chi_{[\theta_{i-1},0]}]\sigma(i) + \chi_{[\theta_{d-1},0]}\sigma(d).$$

Then, from the definition of η it follows that $V^P(t) = \tilde{L}(t,\varphi)$. Let φ^m, $m = 1,2,\cdots$, be the function defined by Relation (4.4) or (4.5) with $\chi_{[s,0]}$ replaced by $\chi^m_{[s,0]}$ which was defined by Relation (3.4). By the same reason as in the proof of Theorem 3.4, we have $V^P(t) = \lim_{m\to\infty} |L(t,\varphi^m)|$. From Axiom (A) it follows that

$$|\varphi^m|_{\mathscr{B}} \le M(s)|(\varphi^m)_{-s}|_{\mathscr{B}} \le K(r-s+\frac{1}{m})M(s)|\varphi^m|_{[-r,0]} \le K(r-s+\frac{1}{m})M(s).$$

Thus,

$$|L(t,\varphi^m)| \le |L(t)||\varphi^m|_{\mathscr{B}} \le |L(t)|K(r-s+\frac{1}{m})M(s)$$

for $m = 1,2,\cdots$. This relation yields that

$$V^P(t) \le |L(t)|K(r-s)M(s).$$

Since the partition P is arbitrary, this proves the desired estimate.

In the above theorem, we have proved Relation (4.2) for each $\varphi \in C_{00}$. Does the relation hold for all elements in $\mathscr{B}$? Unfortunately, this is not always the case. The following theorem shows this fact (cf. Hagemann & Naito [1] and Naito [6]).

Theorem 4.3. Let $\mathscr{B} = C_\gamma$, which is the space introduced in Section 1.3, and let L be the one in Theorem 4.2. Then there are functions $\Lambda : J \to \mathscr{M}$ and $\eta : J \times (-\infty,0] \to \mathscr{M}$ such that

(i) Λ and η are Borel measurable on J and $J \times (-\infty,0]$, respectively, and for each $t \in J$ the function $\eta(t,\cdot) : (-\infty,0] \to \mathscr{M}$ is normalized and locally of bounded variation, and

$$|\Lambda(t)| \le c|L(t)|$$

$$\mathrm{Var}(\eta(t,\cdot),[-r,-s]) \le c \max\{e^{-\gamma r}, e^{-\gamma s}\}|L(t)| \quad \text{for} \quad r > s \ge 0,$$

where c is a constant dependent only on the norm of $\mathbb{C}^n$;

(ii) $$L(t,\varphi) = \Lambda(t)\tilde{\varphi}(-\infty) + \lim_{r\to\infty}\int_{-r}^{0}[d_\theta \eta(t,\theta)]\varphi(\theta)$$

for all $(t,\varphi) \in J \times C_\gamma$, where $\tilde{\varphi}(-\infty) = \lim_{\theta\to-\infty} e^{\gamma\theta}\varphi(\theta)$. If $\eta(t,\theta)$ in (ii) is normalized in $\theta \in (-\infty,0]$, then Λ and η are uniquely determined by L.

Proof. First, we observe that C_γ and the space $C([-1,0])$ of $\mathbb{C}^n$-valued continuous functions on $[-1,0]$ with the supremum norm are isometric; in fact, the mapping $\iota : C_\gamma \to C([-1,0])$ defined by

$$i(\varphi)(s) = \begin{cases} e^{\gamma s/(1+s)}\varphi(s/(1+s)), & s \in (-1,0] \\ \tilde{\varphi}(-\infty), & s = -1, \end{cases}$$

for $\varphi \in C_\gamma$ is linear, bijective and norm-preserving: $|\varphi| = |i(\varphi)|$. The function $N : J \times C([-1,0]) \to \mathbb{C}^n$ defined by $N(t,\psi) = L(t,i^{-1}(\psi))$ for $(t,\psi) \in J \times C([-1,0])$ is a mapping of Caratheodory type and $N(t,\cdot) : C([-1,0]) \to \mathbb{C}^n$ is linear for each $t \in J$. By the Riesz representation theorem (cf. Theorems 1.11 and 3.4) there is a function $\mu : J \times [-1,0] \to \mathcal{M}$ such that $\mu(t,\cdot)$ is normalized and of bounded variation on $[-1,0]$ with

$$N(t,\psi) = \int_{-1}^{0} [d_s\mu(t,s)]\psi(s) \quad \text{for all} \quad (t,\psi) \in J \times C([-1,0])$$

and $\mathrm{Var}(\mu(t,\cdot),[-1,0]) = |N(t)| = |L(t)|$ in the case $n = 1$; in the general case note that all norms in $\mathbb{C}^n$ are equivalent, so that there is a constant c dependent only on the norm of $\mathbb{C}^n$ such that

$$\mathrm{Var}(\mu(t,\cdot),[-1,0]) \le c|L(t)|.$$

Set

$$\Lambda(t) = \mu(t,-1+0) - \mu(t,-1), \quad t \in \mathbb{R}. \tag{4.6}$$

Then

$$\begin{aligned} |\Lambda(t)| &- |\mu(t,-1+0) - \mu(t,-1)| \\ &\le \mathrm{Var}(\mu(t,\cdot),[-1,0]) \\ &\le c|L(t)|. \end{aligned}$$

Next, set $\zeta(t,\theta) = \mu(t,\theta/(1-\theta))$ for $(t,\theta) \in J \times (-\infty,0]$ and

$$\eta(t,\theta) = \begin{cases} \int_0^\theta e^{\gamma\alpha} d_\alpha \zeta(t,\alpha) & \text{if } \theta < 0, \\ 0 & \text{if } \theta = 0 \end{cases} \tag{4.7}$$

for $t \in J$. Since $\mu(t,\cdot)$, $t \in J$, is normalized and of bounded variation on $[-1,0]$, $\zeta(t,\cdot)$ is clearly normalized and of bounded variation on $(-\infty,0]$; hence $\eta(t,\cdot)$ is normalized and locally of bounded variation on $(-\infty,0]$. Moreover, we have

$$\begin{aligned}\mathrm{Var}(\zeta(t,\cdot),[-r,-s]) &\le \mathrm{Var}(\mu(t,\cdot),[-r/(1+r),-s/(1+s)])\\ &\le \mathrm{Var}(\mu(t,\cdot),[-1,0])\\ &\le c|L(t)|.\end{aligned}$$

Thus,

$$\begin{aligned}&\mathrm{Var}(\eta(t,\cdot),[-r,-s])\\ &\le \int_{-r}^{-s} e^{\gamma\alpha}|d_\alpha\zeta(t,\alpha)|\\ &\le \max\{e^{-\gamma r}, e^{-\gamma s}\}\mathrm{Var}(\zeta(t,\cdot),[-r,-s])\\ &\le c\max\{e^{-\gamma r}, e^{-\gamma s}\}|L(t)|,\end{aligned}$$

which proves (i) except the assertion on the Borel measurability of Λ and η.

Next, we prove (ii). For any $(t,\varphi) \in J \times C_\gamma$, one has

$$\begin{aligned}L(t,\varphi) &= N(t,\iota(\varphi))\\ &= \int_{-1}^{0}[d_s\mu(t,s)]\iota(\varphi)(s)\\ &= \Lambda(t)\iota(\varphi)(-1) + \lim_{\ell\to1-0}\int_{-\ell}^{0}[d_s\mu(t,s)]\iota(\varphi)(s)\\ &= \Lambda(t)\tilde{\varphi}(-\infty) + \lim_{\ell\to1-0}\int_{-\ell}^{0}[d_s\mu(t,s)]e^{\gamma s/(1+s)}\varphi(s/(1+s))\\ &= \Lambda(t)\tilde{\varphi}(-\infty) + \lim_{r\to\infty}\int_{-r}^{0}[d_\theta\zeta(t,\theta)]e^{\gamma\theta}\varphi(\theta)\\ &= \Lambda(t)\tilde{\varphi}(-\infty) + \lim_{r\to\infty}\int_{-r}^{0}[d_\theta\eta(t,\theta)]\varphi(\theta),\end{aligned}$$

which proves (ii).

The remainder of the assertions follows from Theorem 4.2. In fact, let $\Lambda(t)$ and $\eta(t,\theta)$ be any functions satisfying Relation (ii) such that $\eta(t,\theta)$ is normalized in $\theta \in (-\infty,0]$. Since

$$L(t,\varphi) = \int [d_\theta \eta(t,\theta)]\varphi(\theta) \qquad \text{for} \quad \varphi \in C_{00}$$

by (ii), Theorem 4.2 says that $\eta(t,\theta)$ is uniquely determined by L and it is Borel measurable in $(t,\theta) \in J \times (-\infty,0]$. Set $(e^{-\gamma\cdot})(\theta) = e^{-\gamma\theta}$, $\theta \le 0$. Clearly, $e^{-\gamma\cdot}a \in C_\gamma$ for all $a \in \mathbb{C}^n$. So, one can compute $\Lambda(t)a$ for every a in $\mathbb{C}^n$; consequently

$$\Lambda(t) = L(t,e^{-\gamma\cdot}I) - \lim_{r\to\infty} \int_{-r}^{0} [d_\theta \eta(t,\theta)]e^{-\gamma\theta}$$

for t in $\mathbb{R}$. This also shows that $\Lambda(t)$ is unique for L and it is Borel measurable in $t \in J$. At the same time, this fact yields that the functions Λ and η in Relations (4.6) and (4.7) are Borel measurable.

CHAPTER 4

GENERAL LINEAR SYSTEMS

4.1. Fundamental matrix and variation of constants formula.

We consider systems of linear functional differential equations with infinite delay

$$(1.1)\qquad \dot{x}(t) = L(t,x_t)$$

$$(1.2)\qquad \dot{x}(t) = L(t,x_t) + h(t),$$

where $L : \mathbb{R} \times \mathscr{B} \to \mathbb{C}^n$ is a continuous mapping such that $L(t,\varphi)$ is linear in $\varphi \in \mathscr{B}$ for each t in $\mathbb{R}$ and $h : \mathbb{R} \to \mathbb{C}^n$ is locally integrable. Since the operator norm $|L(t)| = \sup\{|L(t,\varphi)| : |\varphi|_{\mathscr{B}} \le 1\}$ is a lower semi-continuous function for t in $\mathbb{R}$, it is Borel measurable. If $\mathscr{B}$ is complete, the Banach Steinhauss theorem implies that $|L(t)|$ is locally bounded for t in $\mathbb{R}$. Even if $\mathscr{B}$ is not necessarily complete, we can obtain the following result.

<u>Lemma 1.1 (Sawano [1])</u>. There exists a continuous function $n(t)$ such that

$$|L(t)| \le n(t) \quad \text{on } \mathbb{R}.$$

In particular, $|L(t)|$ is locally bounded for t in $\mathbb{R}$.

<u>Proof</u>. Let J be a compact interval in $\mathbb{R}$. We show that $|L(t,\varphi)| \le \ell|\varphi|_{\mathscr{B}}$ on $J \times \mathscr{B}$ for some constant $\ell > 0$. Otherwise there would exist a sequence $\{(t_k,\varphi^k)\}$ such that $t_k \in J$, $|\varphi_k|_{\mathscr{B}} = 1$ and $|L(t_k,\varphi^k)| \ge k$. Set $\psi^k = k^{-1/2}\varphi^k$. Since $|\psi^k|_{\mathscr{B}} = k^{-1/2} \to 0$ as $k \to \infty$, the set $S = \{\psi^k : k = 1,2,\cdots\} \cup \{0\}$ is compact. Therefore $L(t,\varphi)$ should be bounded on $J \times S$, but

$$|L(t_k,\psi^k)| = k^{-1/2}|L(t_k,\varphi^k)| \ge k^{1/2},$$

which implies $|L(t_k,\psi^k)| \to \infty$ as $k \to \infty$, a contradiction. Then it is not difficult to construct a desirable $n(t)$.

Theorem 2.1.1 yields that for any (σ,φ) in $\mathbb{R} \times \mathcal{B}$, there exists a (noncontinuable) function x with the property that $x_\sigma = \varphi$, x is continuous on $[\sigma,\delta)$ and

$$x(t) = \varphi(0) + \int_\sigma^t L(s,x_s)ds + \int_\sigma^t h(s)ds$$

for $t \in [\sigma,\delta)$. Since x_t is a continuous function for t in $[\sigma,\delta)$ by Axiom (A_1), from the assumptions on L and h it follows that $x(t)$ is locally absolutely continuous on $[\sigma,\delta)$ and satisfies Equation (1.2) for a.e. t in $[\sigma,\delta)$. Henceforth, we call $x(t)$ a (locally absolutely continuous) solution of Equation (1.2) on $[\sigma,\delta)$ and denote it by $x(t,\sigma,\varphi,h)$.

Theorem 1.2. For any $(\sigma,\varphi) \in \mathbb{R} \times \mathcal{B}$, there exists a unique solution of Equation (1.2) on $[\sigma,\infty)$ such that $x_\sigma = \varphi$.

Proof. Let $[\sigma,\delta)$, $\sigma < \delta$, be the existence interval of the solution $x(t) := x(t,\sigma,\varphi,h)$. Since the right hand side of Equation (1.2) satisfies the local Lipschitz condition by Lemma 1.1, the uniqueness of such a solution follows from Theorem 2.1.2. We must prove the global existence i.e., $\delta = \infty$. Suppose $\delta < \infty$. From Axiom (A) and Lemma 1.1, it follows that

$$|x_t|_\mathcal{B} \le A(t-\sigma,\sigma,h,\varphi) + K(t-\sigma)\int_\sigma^t n(s)|x_s|_\mathcal{B}ds$$

for $t \in [\sigma,\delta)$, where

$$A(u,\sigma,h,\varphi) = [HK(u) + M(u)]|\varphi|_\mathcal{B} + K(u)\int_0^u |h(\sigma+\tau)|d\tau.$$

Hence, by Gronwall's inequality we have

$$(1.3) \qquad |x_t|_\mathcal{B} \le A(t-\sigma,\sigma,h,\varphi)$$
$$+ K(t-\sigma)\int_\sigma^t n(s)A(s-\sigma,\sigma,h,\varphi)\exp[\int_s^t n(u)K(u-\sigma)du]ds$$

for $t \in [\sigma,\delta)$. The right hand side in the above inequality is locally bounded for $t \in [\sigma,\infty)$. Hence, we obtain $\sup_{\sigma\le t<\delta} |x_t|_\mathcal{B} =: N < \infty$. Thus $\lim_{t\to\delta^-} x_t$ exists, since x is uniformly continuous on $[\sigma,\delta)$ by the inequality

$$|x(t) - x(t')| \le N\int_t^{t'} n(s)ds + \int_t^{t'} |h(s)|\, ds, \qquad \sigma \le t < t' < \delta.$$

Therefore, applying Theorem 2.1.1, one sees that the solution $x(t)$ must be extended beyond δ. This is a contradiction.

Let $\eta(t,\theta)$ be the normalized function defined by Relation (4.1) in Section 3.4. By Theorem 3.4.2. and Lemma 1.1, it is locally bounded for (t,θ) in $\mathbb{R} \times (-\infty,0]$; in fact

$$(1.4) \qquad |\eta(t,\theta)| \le c|L(t)|K(-\theta) \quad \text{for} \quad (t,\theta) \in \mathbb{R} \times (-\infty,0].$$

For the following discussion, we set $\eta(t,\theta) = 0$ for $\theta > 0$.

To introduce the fundamental matrix of Equation (1.1), we consider the equation

$$(1.5) \qquad \begin{aligned} \dot{x}(t) &= \int_{\sigma-t}^{0} d_\theta \eta(t,\theta)x_t(\theta) + g(t), \qquad \sigma \le t, \\ x(\sigma) &= a, \end{aligned}$$

where $g : [\sigma,\infty) \to \mathbb{C}^n$ is locally integrable.

<u>Theorem 1.3 (Naito [6])</u>. Under the above assumptions for L, η and g, Equation (1.5) is reduced to Equation (1.2) with initial condition $x_\sigma = 0$. Thus for any a in $\mathbb{C}^n$ there exists uniquely a locally absolutely continuous function $x(t)$ for $t \ge \sigma$ such that $x(\sigma) = a$ and the first relation of (1.5) holds a.e. in $[\sigma,\infty)$.

<u>Proof</u>. Suppose $x(t)$ is a solution having the above properties. If we set $y(t) = 0$ for $t < \sigma$ and $y(t) = x(t) - a$ for $t \ge \sigma$, then $y(t)$ satisfies

$$\dot{y}(t) = \int_{\sigma-t}^{0} d_\theta \eta(t,\theta)y_t(\theta) - \eta(t,\sigma-t)a + g(t)$$

a.e. in $t \ge \sigma$. Since y_t lies in C_{00} with $\operatorname{supp} y_t \subset [\sigma-t,0]$ for $t \ge \sigma$, this relation is reduced to the equation $\dot{y}(t) = L(t,y_t) - \eta(t,\sigma-t)a + g(t)$ a.e. in $t \ge \sigma$. Since $\eta(t,\theta)$ is Borel measurable and locally bounded for (t,θ) in $\mathbb{R}^2$, the function $\eta(t,\sigma-t)$, as a function of t, is also Borel measurable and locally bounded on $\mathbb{R}$. Hence, by Theorem 1.2 the equation for y has a unique solution such that $y_\sigma = 0$; this implies that $x(t) = y(t) + a$ for $t \ge \sigma$ is a unique solution of Equation (1.5) having the desired properties.

For σ in $\mathbb{R}$, let $X(t,\sigma)$ be the matrix solution of the equation

$$\frac{\partial X}{\partial t}(t,\sigma) = \int_{\sigma-t}^{0} d_\theta \eta(t,\theta)X(t+\theta,\sigma) \quad \text{a.e. in} \quad t \geq \sigma$$

$$X(\sigma,\sigma) = I \quad \text{and} \quad X(t,\sigma) = 0 \quad \text{for} \quad t < \sigma.$$

We call $X(t,\sigma)$ the fundamental matrix of Equation (1.1). If $g(t) \equiv 0$ in Equation (1.5), it follows from the uniqueness property that the solution of Equation (1.5) is given by

$$x(t) = X(t,\sigma)a \quad \text{for} \quad t \geq \sigma.$$

Next, let $(\mathbb{C}^n)^*$ be the space of n-dimensional row-vectors and consider the equation

$$(1.6) \qquad y(s) + \int_{s}^{t} y(\alpha)\eta(\alpha,s-\alpha)d\alpha = b(s), \qquad s \leq t,$$

which is referred to as the adjoint equation of Equation (1.1). Here $y(s)$ is in $(\mathbb{C}^n)^*$ and $b : (-\infty,t] \to (\mathbb{C}^n)^*$ is locally of bounded variation.

<u>Theorem 1.4.</u> Given t in $\mathbb{R}$ and $b : (-\infty,t] \to (\mathbb{C}^n)^*$ locally of bounded variation, Equation (1.6) has a unique solution $y(s)$ for s in $(-\infty,t]$ which is locally of bounded variation. The total variation of y satisfies

$$(1.7) \qquad \mathrm{Var}(y,[s,t]) \leq \mathrm{Var}(b,[s,t]) + b^*(s)\{\exp(\int_s^t c|L(\alpha)|K^*(\alpha-s)d\alpha) - 1\},$$

where $b^*(s) = \sup\{|b(u)| : s \leq u \leq t\}$ and $K^*(r) = \sup\{K(u) : 0 \leq u \leq r\}$. In particular, if b is normalized, then so is y.

<u>Proof.</u> Suppose $y(s)$ is Borel measurable and locally bounded for s in $(-\infty,t]$ and designate by $(\Omega y)(s)$ the integral of Equation (1.6). Since $\eta(\alpha,s-\alpha) = 0$ for $\alpha \leq s$, one has

$$(\Omega y)(s) = \int_\sigma^t y(\alpha)\eta(\alpha, s-\alpha)d\alpha \quad \text{for } \sigma \le s \le t,$$

and $\mathrm{Var}(\eta^\alpha, [\sigma, t]) = \mathrm{Var}(\eta^\alpha, [\sigma, \alpha]) \le c|L(\alpha)|K(\alpha-\sigma)$ for $\sigma \le \alpha \le t$ by Theorem 3.4.2, where $\eta^\alpha(s) = \eta(\alpha, s-\alpha)$. This leads to

$$(1.8) \qquad \mathrm{Var}(\Omega y, [\sigma, t]) \le \int_\sigma^t |y(\alpha)|c|L(\alpha)|K(\alpha-\sigma)d\alpha,$$

which implies Ωy is locally of bounded variation on $(-\infty, t]$. Such a function is also Borel measurable and locally bounded on $(-\infty, t]$.

From this remark, one can define successive approximations $y^m(s)$ for $m = 0, 1, 2, \cdots$ as $y^0(s) = b(s)$ and $y^m(s) = b(s) - (\Omega y^{m-1})(s)$ for $s \le t$. Then, from Inequality (1.4), one has successively

$$|y^1(s) - y^0(s)| \le \int_s^t c|b(\alpha)||L(\alpha)|K(\alpha-s)d\alpha$$

$$\le cb^*(s)\int_s^t |L(\alpha)|K^*(\alpha-s)d\alpha$$

and

$$|y^2(s) - y^1(s)| \le \int_s^t c|L(\alpha)|K(\alpha-s)\{\int_\alpha^t cb^*(u)|L(u)|K^*(u-\alpha)du\}d\alpha$$

$$\le b^*(s)\int_s^t c|L(\alpha)|K^*(\alpha-s)\{\int_\alpha^t c|L(u)|K^*(u-s)du\}d\alpha$$

$$\le \frac{b^*(s)}{2}\{\int_s^t c|L(\alpha)|K^*(\alpha-s)d\alpha\}^2$$

for $s \le t$, since $0 \le K(u) \le K^*(u) \le K^*(r)$ for $0 \le u \le r$ and $0 \le |b(\alpha)| \le b^*(\alpha) \le b^*(s)$ for $s \le \alpha \le t$. Repeating this procedure, we easily obtain (by induction on m)

$$(1.9) \qquad |y^m(s) - y^{m-1}(s)| \le \frac{b^*(s)}{m!}\{\int_s^t c|L(\alpha)|K^*(\alpha-s)d\alpha\}^m, \qquad s \le t,$$

for $m = 1, 2, \cdots$. Therefore $y^m(s)$ converges to a function $y(s)$ compactly on $(-\infty, t]$, and

$$(1.10) \qquad |y(s)| \le b^*(s)\exp\{\int_s^t c|L(\alpha)|K^*(\alpha-s)d\alpha\}, \qquad s \le t.$$

This implies that $y(s) = \lim_{m\to\infty}[b(s) - (\Omega y^{m-1})(s)] = b(s) - (\Omega y)(s)$; that is, $y(s)$ is the solution of Equation (1.6). Since $y^m(s)$ is Borel measurable for $s \le t$, $y(s)$ is also Borel measurable for $s \le t$. Moreover, (1.7) follows from (1.8) and (1.10).

Suppose $z(s)$ is a solution of Equation (1.6) with $b = 0$, and set $A_\sigma = \sup\{|z(s)| : \sigma \le s \le t\}$. Then, following the argument similar to the proof of (1.9), one can show that, for $\sigma \le s \le t$ and $m = 1,2,\cdots$, $|z(s)|$ is not greater than the right hand side of Inequality (1.9) with $b^*(s)$ replaced by A_σ. Therefore, $z(s) = 0$; in other words, the solution of Equation (1.6) is unique for the function b.

Finally, consider the case b is normalized. Obviously, $y(t) = 0$. Moreover, by (1.4), applying the bounded convergence theorem, one sees that $(\Omega y)(s)$ is continuous to the left for $s < t$, since $\eta(t,s)$ is normalized and y is locally bounded. Therefore, $y(s)$ is also continuous to the left for $s < t$ because of the normality for b.

Let $Y(s,t)$ be the matrix solution of the system

$$(1.11) \qquad Y(s,t) + \int_s^t Y(\alpha,t)\eta(\alpha,s-\alpha)d\alpha = I \quad \text{for} \quad s \le t.$$

If $b(t) \equiv b$, a constant vector, in Equation (1.6), it follows from the uniqueness property that the solution y is given by

$$y(s) = Y(s,t)b \quad \text{for} \quad s \le t.$$

We call $Y(s,t)$ the fundamental matrix of Equation (1.6). By Theorem 1.4, $Y(s,t)$ is locally of bounded variation in $s \in (-\infty,t]$.

Next, we shall give an intimate relation between $X(t,\sigma)$ and $Y(\sigma,t)$.

<u>Theorem 1.5.</u> If $x(t)$ and $y(t)$ are solutions of Equations (1.5) and (1.6), respectively, then

$$y(t)x(t) - y(\sigma)x(\sigma) = \int_\sigma^t db(s)x(s) + \int_\sigma^t y(s)g(s)ds \quad \text{for} \quad \sigma < t.$$

<u>Proof.</u> Suppose $x(t)$ is the solution of Equation (1.5). By Theorem 3.1.2, one has

$$(1.12)\qquad \int_\sigma^t [d_\alpha y(\alpha)]x(\alpha) + \int_\sigma^t y(\alpha)d_\alpha x(\alpha) = y(t)x(t) - y(\sigma)x(\sigma).$$

By Theorem 3.1.14, the second term on the left hand side becomes

$$\int_\sigma^t y(\alpha)x'(\alpha)d\alpha = \int_\sigma^t y(\alpha)\{\int_{\sigma-\alpha}^0 d_\theta\eta(\alpha,\theta)x_\alpha(\theta)\}d\alpha + \int_\sigma^t y(\alpha)g(\alpha)d\alpha$$

$$= \int_\sigma^t y(\alpha)\{\int_\sigma^\alpha d_s\eta(\alpha,s-\alpha)x(s)\}d\alpha + \int_\sigma^t y(\alpha)g(\alpha)d\alpha$$

$$= \int_\sigma^t y(\alpha)\{\int_\sigma^t d_s\eta(\alpha,s-\alpha)x(s)\}d\alpha + \int_\sigma^t y(\alpha)g(\alpha)d\alpha,$$

since $\eta(\alpha,s-\alpha) = 0$ for $s \geq \alpha$. Furthermore, applying Theorem 3.2.11 to the first integral in the last equation, one obtains

$$\int_\sigma^t y(\alpha)d_\alpha x(\alpha) = \int_\sigma^t d_s[\int_\sigma^t y(\alpha)\eta(\alpha,s-\alpha)d\alpha]x(s) + \int_\sigma^t y(\alpha)g(\alpha)d\alpha$$

$$= \int_\sigma^t d_s[\int_s^t y(\alpha)\eta(\alpha,s-\alpha)d\alpha]x(s) + \int_\sigma^t y(\alpha)g(\alpha)d\alpha.$$

Since y satisfies Equation (1.6), Relation (1.12) yields the desired relation.

The following corollaries follow immediately from Theorem 1.5.

<u>Corollary 1.6.</u> If $g(t) \equiv 0$ and $b(t) \equiv$ constant, then the conclusion of Theorem 1.5 becomes

$$y(t)x(t) = y(\sigma)x(\sigma) \quad \text{for} \quad \sigma < t.$$

In particular,

$$(1.13)\qquad X(t,\sigma) = Y(\sigma,t) \quad \text{for} \quad \sigma \leq t.$$

<u>Corollary 1.7.</u> The solution $x(t)$ of Equation (1.5) and the solution $y(s)$ of Equation (1.6) are given by

$$(1.14)\qquad x(t) = X(t,\sigma)x(\sigma) + \int_\sigma^t X(t,s)g(s)ds, \qquad \sigma \leq t$$

$$(1.15)\qquad y(s) = y(t)X(t,s) - \int_s^t db(u)X(u,s), \qquad s < t.$$

Now, we can easily establish the following relation:

$$(1.16)\qquad x(t,\sigma,\varphi,h) = x(t,\sigma,\varphi,0) + x(t,\sigma,0,h) \quad \text{for} \quad t \geq \sigma,$$

which is called the superposition principle. By using this formula we can derive the variation of constants formula as follows.

Theorem 1.8 (Naito [6]). Suppose $L : \mathbb{R} \times \mathscr{B} \to \mathbb{C}^n$ is continuous, $L(t,\varphi)$ is linear for φ in $\mathscr{B}$ and $h : [\sigma,\infty) \to \mathbb{C}^n$ is locally integrable. Then for every φ in $\mathscr{B}$ the solution $x(t,\sigma,\varphi,h)$ of Equation (1.2) is given by

$$(1.17)\qquad x(t,\sigma,\varphi,h) = \varphi(0) + \int_\sigma^t X(t,s)L(s,S(s-\sigma)\varphi)ds + \int_\sigma^t X(t,s)h(s)ds \quad \text{for} \quad t \geq \sigma,$$

where $S(t)$ is the operator defined in Section 1.2.

Proof. Since $x(t,\sigma,0,h)$ is a solution of Equation (1.5) with $a = 0$ and $g = h$, it is equal to the third term on the right hand side of Relation (1.17). Now, consider the function $z(t) = x(t,\sigma,\varphi,0) - y(t)$, where $y(t) = \varphi(t-\sigma)$ for $t \leq \sigma$ and $y(t) = \varphi(0)$ for $t > \sigma$. Then $z(t)$ satisfies $\dot{z}(t) = L(t,z_t) + L(t,y_t)$ a.e. in $t \geq \sigma$, and $z_\sigma = 0$. Since $y_t = S(t-\sigma)\varphi$ for $t \geq \sigma$, from Formula (1.14) it follows that $z(t)$ is equal to the second term on the right hand side of Relation (1.17). Therefore, Relation (1.17) holds, since $x(t,\sigma,\varphi,h) = \varphi(0) + z(t) + x(t,\sigma,0,h)$ for $t \geq \sigma$ by (1.16).

Corollary 1.9. For any $\varphi \in C_{00}$, the solution $x(t,\sigma,\varphi,h)$ of Equation (1.2) is given by

$$(1.18)\qquad x(t,\sigma,\varphi,h) = Y(\sigma,t)\varphi(0) + \int_{-r}^0 d_\xi\Big[\int_\sigma^t Y(u,t)\eta(u,\sigma+\xi-u)du\Big]\varphi(\xi) + \int_\sigma^t Y(u,t)h(u)du \quad \text{for} \quad t \geq \sigma,$$

where $\operatorname{supp}\varphi \subset [-r,0]$.

Proof. Let $\varphi \in C_{00}$, and $\operatorname{supp}\varphi \subset [-r,0]$. Note that $S(t)\varphi \in C_{00}$ for $t \geq 0$. From Formula (4.2) in Section 3.4 it follows

$$L(u,S(u-\sigma)\varphi) = \int_{-r+\sigma-u}^{0} d_\theta\eta(u,\theta)[S(u-\sigma)\varphi](\theta)$$

$$= \int_{\sigma-u}^{0} d_\theta\eta(u,\theta)\varphi(0) + \int_{-r+\sigma-u}^{\sigma-u} d_\theta\eta(u,\theta)\varphi(u-\sigma+\theta)$$

$$= -\eta(u,\sigma-u)\varphi(0) + \int_{-r}^{0} d_\xi\eta(u,\sigma+\xi-u)\varphi(\xi)$$

for $u \geq \sigma$. Hence, in virtue of Theorem 3.2.11, we have

$$\int_\sigma^t X(t,u)L(u,S(u-\sigma)\varphi)du$$

$$= \int_\sigma^t Y(u,t)[-\eta(u,\sigma-u)\varphi(0) + \int_{-r}^{0} d_\xi\eta(u,\sigma+\xi-u)\varphi(\xi)]du$$

$$= [Y(\sigma,t) - I]\varphi(0) + \int_{-r}^{0} d_\xi[\int_\sigma^t Y(u,t)\eta(u,\sigma+\xi-u)du]\varphi(\xi)$$

by (1.11) and (1.13). Substituting to (1.17), one obtains Formula (1.18).

When $\mathscr{B} = C_\gamma$, in Theorem 3.4.3 we established the representation theorem on L which is valid for all elements in C_γ. Combining it with Formula (1.17), we can obtain a result similar to Formula (1.18), which is also valid for all elements in C_γ.

Theorem 1.10 (Hagemann & Naito [1]). Let $\mathscr{B} = C_\gamma$. Then, for any $\varphi \in C_\gamma$ the solution $x(t,\sigma,\varphi,h)$ of Equation (1.2) is represented as

$$(1.19) \qquad x(t,\sigma,\varphi,h) = X(t,\sigma)\varphi(0) + [\int_\sigma^t X(t,s)\Lambda(s)e^{-\gamma(s-\sigma)}ds]\tilde{\varphi}(-\infty)$$

$$+ \lim_{r\to\infty}\int_{-r}^{0} d_\theta[\int_\sigma^t X(t,s)\eta(s,\sigma+\theta-s)ds]\varphi(\theta)$$

$$+ \int_\sigma^t X(t,s)h(s)ds \qquad \text{for} \quad t \geq \sigma.$$

<u>Proof.</u> It suffices to show that Relation (1.17) can be rewritten as Relation (1.19). Since $[S(t)\varphi]^{\sim}(-\infty) = e^{-\gamma t}\tilde{\varphi}(-\infty)$, by Theorem 3.4.3 the second term of the right hand side of (1.17) becomes

$$(1.20) \qquad [\int_{\sigma}^{t} X(t,s)\Lambda(s)e^{-\gamma(s-\sigma)}ds]\tilde{\varphi}(-\infty)$$

$$+ \int_{\sigma}^{t} X(t,s)\{\lim_{r\to\infty}\int_{-r}^{0}[d_{\theta}\eta(s,\theta)][S(s-\sigma)\varphi](\theta)\}ds.$$

The first term of (1.20) appears as the second term of the right hand side of Relation (1.19).

Observe the second term of (1.20). Suppose $r > s-\sigma$ and divide the integration interval of the integral in the braces as $[-r,0] = [-r,\sigma-s] \cup [\sigma-s,0]$. The integral on the second subinterval becomes $[\eta(s,0) - \eta(s,\sigma-s)]\varphi(0) = -\eta(s,\sigma-s)\varphi(0)$ since $[S(s-\sigma)\varphi](\theta) = \varphi(0)$ for $s-\sigma+\theta \geq 0$. This result leads to

$$(1.21) \qquad \int_{\sigma}^{t} X(t,s)\{\int_{\sigma-s}^{0}[d_{\theta}\eta(s,\theta)][S(s-\sigma)\varphi](\theta)\}ds = [X(t,\sigma) - I]\varphi(0)$$

by (1.11) and (1.13). Since the limit operation in the braces of Relation (1.20) acts on the integral on $[-r,\sigma-s]$, the sum of $\varphi(0)$ and the left hand side of (1.21) yield the term $X(t,\sigma)\varphi(0)$ in Relation (1.19).

Since $[S(s-\sigma)\varphi](\theta) = \varphi(s-\sigma+\theta)$ for $\theta \leq \sigma-s$, the above limit operation is rewritten as

$$(1.22) \qquad \int_{\sigma}^{t} X(t,s)\{\lim_{r\to\infty}\int_{-r}^{\sigma-s}[d_{\theta}\eta(s,\theta)][S(s-\sigma)\varphi](\theta)\}ds$$

$$= \int_{\sigma}^{t} X(t,s)\{\lim_{r\to\infty}\int_{s-\sigma-r}^{0}[d_{\xi}\eta(s,\sigma+\xi-s)]\varphi(\xi)\}ds$$

$$= \int_{\sigma}^{t} X(t,s)\{\lim_{r\to\infty}\int_{-r}^{0}[d_{\xi}\eta(s,\sigma+\xi-s)]\varphi(\xi)\}ds.$$

The integral $I_r(s)$ in the last braces is the limit of the sequence of Riemann sums

$$A^{p}(s) = \sum_{j=1}^{N}[\eta(s,\sigma+\xi_j-s) - \eta(s,\sigma+\xi_{j-1}-s)]\varphi(\tau_j),$$

as $m(p) \to 0$, where $P : -r = \xi_0 < \xi_1 < \cdots < \xi_N = 0$, $\xi_{j-1} \leq \tau_j \leq \xi_j$, $j = 1,\cdots,N$, and $m(P) = \max\{|\xi_j - \xi_{j-1}| : j = 1,\cdots,N\}$. Since

$\eta(t,\theta)$ is a Borel measurable function in (t,θ), the function $\eta(s,\sigma+\xi-s)$ is a Borel measurable in $s \in [\sigma,t]$ for each $\xi \in [-r,0]$. Thus the integral $I_r(s)$ is a Borel measurable function in $s \in [\sigma,t]$ as a limit of the sequence of the Borel measurable functions $\Lambda^p(s)$. Furthermore, Relation (4.8) in Section 3.4 implies that

$$|I_r(s)| \le \int_{-r}^{0} |d_\xi \zeta(s,\sigma+\xi-s)| e^{\gamma(\sigma+\xi-s)} |\varphi(\xi)|$$

$$\le c e^{\gamma(\sigma-s)} |\varphi|_{C_\gamma} |L(s)|$$

for $\sigma \le s \le t$, $r > 0$. From the dominated convergence theorem and Theorem 3.2.11 it follows that

$$(1.23) \qquad \int_\sigma^t X(t,s)\{\lim_{r\to\infty} \int_{-r}^{0} [d_\xi \eta(s,\sigma+\xi-s)]\varphi(\xi)\}ds$$

$$= \lim_{r\to\infty} \int_\sigma^t X(t,s)\{\int_{-r}^{0} [d_\xi \eta(s,\sigma+\xi-s)]\varphi(\xi)\}ds$$

$$= \lim_{r\to\infty} \int_{-r}^{0} d_\xi [\int_\sigma^t X(t,s)\eta(s,\sigma+\xi-s)ds]\varphi(\xi).$$

Summarizing (1.22) and (1.23), and replacing ξ by θ, we obtain the third term in the right hand side of Relation (1.19).

4.2. Formulation of the variation of constants formula in the second dual space.

In the previous section, we have established some representation theorems of solutions of Equation (1.2) in $\mathbb{C}^n$. In this section, we treat Equation (1.2) again and will try to obtain a representation of solutions of the equation in the phase space $\mathcal{B}$.

Now, let $x(t,\sigma,\varphi,h)$ be the solution of Equation (1.2) such that $x_\sigma = \varphi$, where $(\sigma,\varphi) \in \mathbb{R} \times \mathcal{B}$ and $h : \mathbb{R} \to \mathbb{C}^n$ is locally integrable. As was shown in Theorem 1.2, this solution is uniquely determined and is defined for all $t \ge \sigma$. Define an operator $T(t,\sigma) : \mathcal{B} \to \mathcal{B}$, $t \ge \sigma$, by

$$T(t,\sigma)\varphi = x_t(\sigma,\varphi,0), \qquad \varphi \in \mathcal{B},$$

which is usually referred to as the solution operator of Equation (1.1). Furthermore, for $t \ge \sigma$ we define an operator $K(t,\sigma)$: $L^1([\sigma,t],\mathbb{C}^n) \to \mathcal{B}$ by

$$K(t,\sigma)h = x_t(\sigma,0,h), \qquad h \in L^1([\sigma,t],\mathbb{C}^n).$$

From Relation (1.16), it follows that

$$x_t(\sigma,\varphi,h) = T(t,\sigma)\varphi + K(t,\sigma)h$$

for $t \geq \sigma$, $\varphi \in \mathscr{B}$ and $h \in L^1([\sigma,t],\mathbb{C}^n)$. Moreover, by Formulas (1.13) and (1.17), the operator $K(t,\sigma)$ can be written more explicitly:

$$[K(t,\sigma)h](\theta) = \begin{cases} \int_\sigma^{t+\theta} Y(s,t+\theta)h(s)ds, & \theta \in [\sigma-t,0] \\ 0, & \theta \in (-\infty,\sigma-t) \end{cases}$$

for $h \in L^1([\sigma,t],\mathbb{C}^n)$.

Theorem 2.1. For $t \geq \sigma$, the operators $T(t,\sigma) : \mathscr{B} \to \mathscr{B}$ and $K(t,\sigma) : L^1([\sigma,t],\mathbb{C}^n) \to \mathscr{B}$ are bounded linear operators. Furthermore, the following estimates holds:

$$|T(t,\sigma)| \leq M(t-\sigma) + K(t-\sigma)\Big(H + \int_0^{t-\sigma} n(\sigma+\tau)(HK(\tau) + M(\tau))\exp[\int_{\sigma+\tau}^t n(u)K(u-\sigma)du]d\tau\Big)$$

and

$$|K(t,\sigma)| \leq K(t-\sigma)\exp[\int_\sigma^t n(u)K(u-\sigma)du],$$

where $n(t)$ is the function arising in Lemma 1.1.

Proof. The linearity of the operators is obvious. Furthermore, the estimates on $|T(t,\sigma)|$ and $|K(t,\sigma)|$ immediately follow from the inequality (1.3) in the proof of Theorem 1.2.

Let $\mathscr{B}^*$ be the space of bounded linear functions from $\mathscr{B}$ into $\mathbb{C}$. Since $\mathscr{B}$ is a semi-normed space, $\mathscr{B}^*$ is a Banach space with the usual operator norm (see, e.g., Edwards [1, p.87]). $\mathscr{B}^*$ is called the dual space of $\mathscr{B}$. Denote by BV the set of all functions mapping $(-\infty,0]$ into $(\mathbb{C}^n)^*$ which are locally of bounded variation and are normalized. Then, for any $\psi \in \mathscr{B}^*$ there corresponds a unique $\widetilde{\psi} \in BV$ such that

$$(2.2)\qquad \langle\psi,\varphi\rangle = \int[d\widetilde{\psi}(\theta)]\varphi(\theta), \qquad \varphi \in C_{00}$$

and

$$(2.3)\qquad Var(\widetilde{\psi},[-r,0]) \le cK(r)|\psi|, \qquad r > 0,$$

by Theorem 3.4.2, where $\langle x^*,x\rangle = x^*(x)$ for $x \in \mathscr{B}$, $x^* \in \mathscr{B}^*$.

The adjoints of operators $T(t,\sigma)$ and $K(t,\sigma)$ are defined by

$$T^*(t,\sigma) : \mathscr{B}^* \to \mathscr{B}^*, \qquad K^*(t,\sigma) : \mathscr{B}^* \to L^\infty([\sigma,t],(\mathbb{C}^n)^*)$$

$$\langle T^*(t,\sigma)\psi,\varphi\rangle = \langle\psi,T(t,\sigma)\varphi\rangle$$

$$\int_\sigma^t[K^*(t,\sigma)\psi](s)h(s)ds = \langle\psi,K(t,\sigma)h\rangle$$

where $\varphi \in \mathscr{B}$, $\psi \in \mathscr{B}^*$, $h \in L^1([\sigma,t],\mathbb{C}^n)$ and $t \ge \sigma$. For a $\psi \in \mathscr{B}^*$ and a bounded linear operator $T : \mathscr{B} \to \mathscr{B}$, $[T^*\psi]^\sim \in BV$ is uniquely determined by the values $\langle T^*\psi,\cdot\rangle$ on C_{00}. For example, $[S^*(t)\psi]^\sim = \widetilde{S}(-t)\widetilde{\psi}$, where $\widetilde{S}(\beta) : BV \to BV$ is given by

$$[\widetilde{S}(\beta)\chi](\theta) = \begin{cases} 0 & \text{if } \theta = 0, \\ \chi(\beta+\theta) & \text{if } \theta < 0 \end{cases}$$

for $\beta \le 0$ and $\chi \in BV$.

Consider the adjoint equation

$$(2.5)\qquad y(s) + \int_s^t y(\alpha)\eta(\alpha,s-\alpha)d\alpha = \widetilde{\psi}(s-t), \qquad s \le t,$$

for each $(y,\widetilde{\psi}) \in \mathbb{R} \times BV$. By Corollaries 1.6 and 1.7, the unique solution $y(\cdot,t,\widetilde{\psi})$ of Equation (2.5) is given by

$$(2.6)\qquad y(s,t,\widetilde{\psi}) = \begin{cases} 0 & s = t, \\ -\int_{s-t}^0 [d\widetilde{\psi}(\theta)]Y(s,t+\theta) & s < t, \end{cases}$$

and moreover $y(\cdot,t,\widetilde{\psi})$ is a normalized function by Theorem 1.4.

Rewrite Equation (2.5) as

$$y(t+\theta) + \int_{\theta}^{0} y(t+\beta)\eta(t+\beta,\theta-\beta)d\beta = \tilde{\psi}(\theta), \qquad \theta \le 0.$$

Hence, if we use a linear operator $\Omega(t)$ on BV defined by

$$(2.7) \qquad [\Omega(t)\chi](\theta) = \int_{\theta}^{0} \chi(\beta)\eta(t+\beta,\theta-\beta)d\beta, \qquad \theta \le 0,\ \chi \in BV,$$

for each $t \in \mathbb{R}$, then Equation (2.5) is reduced to the equation $[I + \Omega(t)]y_t = \tilde{\psi}$, where $y_t(\theta) = y(t+\theta)$ for $\theta \le 0$ and I is the identity operator on BV. Thus we obtain

$$(2.8) \qquad \begin{aligned} &(I + \Omega(t))y_t(\cdot,t,\tilde{\psi}) = \tilde{\psi} \\ &\text{or} \\ &y_t(\cdot,t,\tilde{\psi}) = (I + \Omega(t))^{-1}\tilde{\psi}. \end{aligned}$$

<u>Theorem 2.2.</u> For any $\psi \in \mathcal{B}^*$ and $t \ge \sigma$, we have

$$[T^*(t,\sigma)\psi]^{\sim} = (I + \Omega(\sigma))\tilde{S}(\sigma-t)(I + \Omega(t))^{-1}\tilde{\psi}.$$

Moreover,

$$[T^*(t,\sigma)\psi]^{\sim}(0^-) = y(\sigma,t,\tilde{\psi}) \quad \text{for} \quad t > \sigma.$$

<u>Proof.</u> Let $\varphi \in C_{00}$ and $\psi \in \mathcal{B}^*$ be given, and let $x_t = T(t,\sigma)\varphi$ for $t \ge \sigma$. Let $\operatorname{supp} \varphi \subset [-r,0]$. Since $x_t \in C_{00}$, from Corollary 1.9 and (2.2) it follows that

$$\begin{aligned} \langle T^*(t,\sigma)\psi,\varphi\rangle &= \langle\psi,x_t\rangle \\ &= \int_{-r-t}^{0} d\tilde{\psi}(\theta)x(t+\theta) \\ &= \int_{-r-t}^{\sigma-t} d\tilde{\psi}(\theta)\varphi(t+\theta-\sigma) + \int_{\sigma-t}^{0} [d\tilde{\psi}(\theta)Y(\sigma,t+\theta)]\varphi(0) \\ &\quad + \int_{\sigma-t}^{0} d\tilde{\psi}(\theta)[\int_{-r}^{0} d_{\xi}\{\int_{\sigma}^{t} Y(\alpha,t+\theta)\eta(\alpha,\sigma+\xi-\alpha)d\alpha\}\varphi(\xi)]. \end{aligned}$$

Apply Theorem 3.2.11 to the last integral. Then the last integral is equal to

$$\int_{-r}^{0} d_{\xi}\{\int_{\sigma-t}^{0} d\tilde{\psi}(\theta)\int_{\sigma}^{t} Y(\alpha,t+\theta)\eta(\alpha,\sigma+\xi-\alpha)d\alpha\}\varphi(\xi).$$

Moreover, applying Fubini's theorem to the integral in the braces, the above integral becomes

$$\int_{-r}^{0} d_\xi \{\int_\sigma^t [\int_{\sigma-t}^{0} d\tilde{\psi}(\theta)Y(\alpha,t+\theta)]\eta(\alpha,\sigma+\xi-\alpha)d\alpha\}\varphi(\xi).$$

Since $Y(\alpha,t+\theta) = 0$ for $\theta < \alpha-t$ and $Y(\alpha,t+\theta) = I$ for $\theta = \alpha-t$, by Corollary 3.2.14, Theorem 3.2.15 and (2.6) we have

$$\int_{\sigma-t}^{0} d\tilde{\psi}(\theta)Y(\alpha,t+\theta) = \int_{\alpha-t}^{0} d\tilde{\psi}(\theta)Y(\alpha,t+\theta) + \int_{\sigma-t}^{\alpha-t} d\tilde{\psi}(\theta)Y(\alpha,t+\theta)$$

$$= \int_{\alpha-t}^{0} d\tilde{\psi}(\theta)Y(\alpha,t+\theta) + \int_{[\sigma-t,\alpha-t]} Y(\alpha,t+\theta)d\mu$$

$$- [\tilde{\psi}(\alpha-t+0) - \tilde{\psi}(\alpha-t)]Y(\alpha,\alpha)$$

$$- [\tilde{\psi}(\sigma-t) - \tilde{\psi}(\sigma-t-0)]Y(\alpha,\sigma)$$

$$= \int_{\alpha-t}^{0} d\tilde{\psi}(\theta)Y(\alpha,t+\theta)$$

$$+ \int_{\{\alpha-t\}} Y(\alpha,t+\theta)d\mu - [\tilde{\psi}(\alpha-t+0) - \tilde{\psi}(\alpha-t)]$$

$$= \int_{\alpha-t}^{0} d\tilde{\psi}(\theta)Y(\alpha,t+\theta)$$

$$+ [\tilde{\psi}(\alpha-t+0) - \tilde{\psi}(\alpha-t-0)]Y(\alpha,\alpha)$$

$$- [\tilde{\psi}(\alpha-t+0) - \tilde{\psi}(\alpha-t)]$$

$$= \int_{\alpha-t}^{0} d\tilde{\psi}(\theta)Y(\alpha,t+\theta)$$

$$= -y(\alpha,t,\tilde{\psi})$$

for $\alpha \in (\sigma,t)$. Consequently,

$$\langle T^*(t,\sigma)\psi,\varphi\rangle = \int_{-r}^{0} d_\xi[\tilde{\psi}(\xi+\sigma-t) - \int_\sigma^t y(\alpha,t,\tilde{\psi})\eta(\alpha,\sigma+\xi-\alpha)d\alpha]\varphi(\xi)$$

$$- y(\sigma,t,\tilde{\psi})\varphi(0)$$

for all $t \geq \sigma$. Set

$$\chi(\xi) = \tilde{\psi}(\xi+\sigma-t) - \int_\sigma^t y(\alpha,t,\tilde{\psi})\eta(\alpha,\sigma+\xi-\alpha)d\alpha$$

for $\xi \le 0$. Then χ is locally of bounded variation on $(-\infty,0]$ with $\chi(0) = y(\sigma,t,\tilde{\psi})$, and moreover it is left continuous on $(-\infty,0]$. In particular, if $t > \sigma$, χ is continuous even at $\xi = 0$. Hence the function ν defined by

$$\nu(\xi) = \begin{cases} \chi(\xi) & \text{if } \xi < 0, \\ 0 & \text{if } \xi = 0, \end{cases}$$

is in BV, and moreover we have

$$\int_{-r}^0 d\nu(\xi)\varphi(\xi) = \int_{-r}^0 d\chi(\xi)\varphi(\xi) + \int_{-r}^0 d[\nu-\chi](\xi)\varphi(\xi)$$

$$= \langle T^*(t,\sigma)\psi,\varphi\rangle + y(\sigma,t,\tilde{\psi})\varphi(0) - \chi(0)\varphi(0)$$

$$= \langle T^*(t,\sigma)\psi,\varphi\rangle .$$

Since $[T^*(t,\sigma)\psi]^\sim$ is determined uniquely by the values $\langle T^*(t,\sigma)\psi,\cdot\rangle$ on C_{00}, we have $[T^*(t,\sigma)\psi]^\sim = \nu$, and consequently $[T^*(t,\sigma)\psi]^\sim(\xi) = \chi(\xi)$ for all $\xi < 0$. In particular, if $t > \sigma$, then

$$[T^*(t,\sigma,\psi)]^\sim(0^-) = \chi(0) = y(\sigma,t,\tilde{\psi}).$$

On the other hand, from (2.4) - (2.8) it follows

$$\chi(\xi) = y(\sigma+\xi,t,\tilde{\psi}) + \int_{\sigma+\xi}^t y(\alpha,t,\tilde{\psi})\eta(\alpha,\sigma+\xi-\alpha)d\alpha$$

$$- \int_\sigma^t y(\alpha,t,\tilde{\psi})\eta(\alpha,\sigma+\xi-\alpha)d\alpha$$

$$= y(\sigma+\xi,t,\tilde{\psi}) + \int_\xi^0 y(\sigma+\beta,t,\tilde{\psi})\eta(\sigma+\beta,\xi-\beta)d\beta$$

$$= [(I + \Omega(t))^{-1}\tilde{\psi}](\sigma+\xi-t)$$

$$+ \int_\xi^0 [(I + \Omega(t))^{-1}\tilde{\psi}](\sigma+\beta-t)\eta(\sigma+\beta,\xi-\beta)d\beta$$

$$= [(I + \Omega(\sigma))\tilde{S}(\sigma-t)(I + \Omega(t))^{-1}\tilde{\psi}](\xi)$$

for all $\xi \le 0$ and $t \ge \sigma$. Thus $[T^*(t,\sigma)\psi]^\sim = (I + \Omega(\sigma))\tilde{S}(\sigma-t)(I + \Omega(t))^{-1}\tilde{\psi}$ for $t \ge \sigma$.

<u>Theorem 2.3.</u> For any $\psi \in \mathscr{B}^*$ we have

$$[K^*(t,\sigma)\psi](\alpha) = -[T^*(t,\alpha)\psi]^\sim(0^-)$$

for a.e. α in $[\sigma,t]$.

<u>Proof.</u> Let $\psi \in \mathscr{B}^*$ and $h \in L^1([\sigma,t],\mathbb{C}^n)$ be given. From (2.1) it follows that $K(t,\sigma)h \in C_{00}$. Then,

$$\int_\sigma^t [K^*(t,\sigma)\psi](\alpha)h(\alpha)d\alpha = \langle\psi, K(t,\sigma)h\rangle = \int d\tilde{\psi}(\theta)[K(t,\sigma)h](\theta)$$

$$= \int_{\sigma-t}^0 d\tilde{\psi}(\theta)\{\int_\sigma^{t+\theta} Y(\alpha,t+\theta)h(\alpha)d\alpha\}$$

$$= \int_{\sigma-t}^0 d\tilde{\psi}(\theta)\{\int_\sigma^t Y(\alpha,t+\theta)h(\alpha)d\alpha\}$$

$$= \int_\sigma^t \{\int_{\sigma-t}^0 d\tilde{\psi}(\theta)Y(\alpha,t+\theta)\}h(\alpha)d\alpha$$

$$= \int_\sigma^t \{\int_{\alpha-t}^0 d\tilde{\psi}(\theta)Y(\alpha,t+\theta)\}h(\alpha)d\alpha$$

$$= -\int_\sigma^t y(\alpha,t,\tilde{\psi})h(\alpha)d\alpha$$

$$= -\int_\sigma^t [T^*(t,\alpha)\psi]^\sim(0^-)h(\alpha)d\alpha$$

by (2.6) and Theorem 2.2. Since $h \in L^1([\sigma,t],\mathbb{C}^n)$ is arbitrarily given, we obtain the desired relation.

Next, we shall formulate the variation-of-constants formula for $x_t(\sigma,\varphi,h)$, the t-section of the solution of Equation (2.2). To do this, at first, we introduce some notations and definitions.

Let X be a complex linear space with a semi-norm $|\cdot|_X$, and consider the quotient space $\hat{X} := X/|\cdot|_X$. For any $x \in X$ we denote by $\hat{x}$ the equivalence class of x. $\hat{X}$ is a normed space with the norm $|\cdot|_{\hat{X}}$ naturally induced by $|\cdot|_X$. Let X^* and $\hat{X}^*$ be the dual spaces of X and $\hat{X}$, respectively. X^* and $\hat{X}^*$ are Banach spaces.

In addition, X^* is isometrically isomorphic to $\hat{X}^*$ by the canonical mapping $x^* \in X^* \to \hat{x}^* \in \hat{X}^*$, where $\hat{x}^* \in \hat{X}^*$ is defined by $\langle \hat{x}^*, \hat{x} \rangle = \langle x^*, x \rangle$, $x \in \hat{x}$. In what follows, we identify $\hat{X}^*$ with $\hat{X}$ since there would no arise confusion.

Let $T : X \to Y$ be a bounded linear operator, where Y is a semi-normed linear space. For each $\hat{x} \in \hat{X}$, set $\hat{T}\hat{x} = (Tx)^\wedge$, $x \in \hat{X}$. Then $\hat{T}$ defines a bounded linear operator from $\hat{X}$ into $\hat{Y}$. We call $\hat{T}$ the operator induced by T. Clearly, the adjoint operator $\hat{T}^*$ of $\hat{T}$ can be identified with the adjoint operator T^* of T by the canonical identifications between X^*, $\hat{X}^*$ and Y^*, $\hat{Y}^*$, respectively.

Let X be a semi-normed space and let f be a function defined on an interval J in $\mathbb{R}$ with values in X^*. The function f is said to be measurable on J in the weak-star topology of X^* (w^*-measurable on J, for short), if the function $\langle f(t), x \rangle$ of t is measurable on J for each $x \in X$. Furthermore, the function f is said to be integrable on J in the weak-star topology of X^* (w^*-integrable on J, for short), if it is w^*-measurable on J, and if the function $\langle f(t), x \rangle$ of t is Lebesgue integrable on J for each $x \in X$. Afterwards, we shall employ the following lemmas as $X = \mathscr{B}^*$. So, in Lemma 2.4 throughout Lemma 2.6, we assume that X is complete.

<u>Lemma 2.4.</u> If a function $f : J \to X^*$ is w^*-integrable on J, then there exists a unique element $x^* \in X^*$ such that

$$\langle x^*, x \rangle = \int_J \langle f(t), x \rangle dt$$

for all $x \in X$.

<u>Proof.</u> Employing the same argument as in the proof of Hille & Phillips [1, Theorem 3.7.1.], we can prove this lemma. Indeed, since the function f is w^*-integrable on J, we can define a linear operator $F : X \to L^1(J, \mathbb{C})$ by

$$[F(x)](t) = \langle f(t), x \rangle, \qquad t \in J,\ x \in X.$$

Let $x_n \to x$ in X and $F(x_n) \to g$ in $L^1(J, \mathbb{C})$ as $n \to \infty$. Then $g(t) = [F(x)](t)$ for a.e. t in J, because of $[F(x_n)](t) \to [F(x)](t)$ as $n \to \infty$ for each $t \in J$. This shows that F is a closed linear operator. Since F is defined on the entire of X, F is bounded by the closed graph theorem. Thus

$$\left|\int_J \langle f(t),x\rangle dt\right| \le |F(x)|_{L^1(J,\mathbb{C})} \le |F||x|_X,$$

and consequently the linear functional $x \in X \to \int_J \langle f(t),x\rangle dt$ is bounded. This is the desired result.

Henceforth, we denote by $\int_{*J} f(t)dt$ the element $x^* \in X^*$ assured in Lemma 2.4, and hence the following relation holds:

$$(2.9) \qquad \langle \int_{*J} f(t)dt, x\rangle = \int_J \langle f(t),x\rangle dt, \qquad x \in X.$$

Lemma 2.5. If a function $f : J \to X^*$ is w^*-integrable and T is bounded linear operator on X, then $T^*f : J \to X^*$ is w^*-integrable and

$$(2.10) \qquad T^*\int_{*J} f(t)dt = \int_{*J} T^*f(t)dt.$$

Proof. The w^*-integrability of T^*f follows from the w^*-integrability of f and the relation $\langle T^*f(t),x\rangle = \langle f(t),Tx\rangle$, $x \in X$. Moreover,

$$\begin{aligned}\langle T^*\int_{*J} f(t)dt, x\rangle &= \langle \int_{*J} f(t)dt, Tx\rangle \\ &= \int_J \langle f(t),Tx\rangle dt \\ &= \int_J \langle T^*f(t),x\rangle dt \\ &= \langle \int_{*J} T^*f(t)dt, x\rangle, \qquad x \in X,\end{aligned}$$

by (2.9), from which Relation (2.10) follows.

Lemma 2.6. If $f : J \to X^*$ is Bochner integrable on J, then it is w^*-integrable on J and the Bochner integral $\int_J f(t)dt$ is equal to $\int_{*J} f(t)dt$.

Proof. If f is Bochner integrable on J, then there exists a sequence of countably-valued Bochner integrable functions $\{f_n\}$ such that $f_n(t) \to f(t)$ as $n \to \infty$ a.e. in $t \in J$ and

$\lim_{n\to\infty} \int_J |f_n(t)-f(t)|_{X^*} dt = 0$. For each $x \in X$, $\{<f_n(t),x>\}$ is a Cauchy sequence in $L^1(J,\mathbb{C})$ and $<f_n(t),x> \to <f(t),x>$ as $n \to \infty$ a.e. in $t \in J$. Hence $<f(t),x>$ is in $L^1(J,\mathbb{C})$, which shows that f is w^*-integrable on J. Moreover, for each $x \in X$,

$$
\begin{aligned}
<\int_J f(t)dt,x> &= <\lim_{n\to\infty} \int_J f_n(t)dt,x> \\
&= \lim_{n\to\infty} \int_J <f_n(t),x>dt \\
&= \int_J <f(t),x>dt \\
&= <\int_{*J} f(t)dt,x>
\end{aligned}
$$

by (2.9). This yields $\int_J f(t)dt = \int_{*J} f(t)dt.$

Henceforth, we identify a normed linear space $\hat{X}$ with the subspace of the second dual space $X^{**} := (X^*)^*$ in the usual manner. Furthermore, we denote by $<x^*,x^{**}>$ the duality pairing between $x^* \in X^*$ and $x^{**} \in X^{**}$.

Now, consider the space $\hat{\mathscr{B}}$, which we shall refer to as the quotient phase space. $\hat{\mathscr{B}}$ is a normed linear space, and the Banach spaces $\mathscr{B}^*$ and $\hat{\mathscr{B}}^*$ can be identified by the canonical mapping. Note that the value $\varphi(0)$ is independent of the particular choice of φ in $\hat{\varphi} \in \hat{\mathscr{B}}$ by Axiom (A-ii). Sometimes, we denote by $\hat{\varphi}(0)$ the value $\varphi(0)$, $\varphi \in \hat{\varphi}$. Next, for each $i = 1,2,\cdots,n$, we consider a function $\gamma_i : \mathscr{B}^* \to \mathbb{C}$ defined by

$$\gamma_i(\psi) = -\tilde{\psi}_i(0^-)$$

for $\psi \in \mathscr{B}^*$, where $\tilde{\psi}_i$ denotes the i-th component of $\tilde{\psi} \in BV$ uniquely determined by $\psi \in \mathscr{B}^*$. Then $\gamma_i \in \mathscr{B}^{**}$ for $i = 1,2,\cdots,n$, because

$$
\begin{aligned}
|\tilde{\psi}_i(0^-)| &\le \mathrm{Var}(\tilde{\psi}(\cdot),[-1,0]) \\
&\le cK(1)|\psi|_{\mathscr{B}^*}
\end{aligned}
$$

by Estimate (2.3). Moreover, we consider a functional $\Gamma : \mathscr{B}^* \to (\mathbb{C}^n)^*$ defined by

$$\langle\psi,\Gamma\rangle = (\langle\psi,\gamma_1\rangle, \cdots, \langle\psi,\gamma_n\rangle) = -\tilde{\psi}(0^-)$$

for $\psi \in \mathscr{B}^*$, and for any operator T on $\mathscr{B}^{**}$ we set

$$T\Gamma = (T\gamma_1, \cdots, T\gamma_n).$$

Let $\hat{T}(t,\sigma)$ be the operator induced by $T(t,\sigma)$. The following theorem gives a formulation of the variation-of-constants formula in the second dual space $\mathscr{B}^{**}$.

Theorem 2.7 (Murakami [4]). Let $x(t) = x(t,\sigma,\varphi,h)$ be the solution of Equation (1.2). Then $T^{**}(t,\cdot)\Gamma h(\cdot) : [\sigma,t] \to \mathscr{B}^{**}$ is w^*-integrable and

$$\hat{x}_t = \hat{T}(t,\sigma)\hat{x}_\sigma + {}_*\!\int_\sigma^t T^{**}(t,s)\Gamma h(s)ds, \quad t \geq \sigma, \tag{2.11}$$

where $T^{**}(t,s) = (T^*(t,s))^*$.

Proof. From Theorem 2.3 it follows that

$$\langle\psi,T^{**}(t,s)\Gamma h(s)\rangle = \langle T^*(t,s)\psi,\Gamma\rangle h(s) = -[T^*(t,s)\psi]^{\sim}(0^-)h(s)$$

$$= [K^*(t,\sigma)\psi](s)h(s) \quad \text{a.e. } s \in [\sigma,t]$$

for each $\psi \in \mathscr{B}^*$. Hence, $T^{**}(t,s)\Gamma h(s)$ is w^*-integrable for s on $[\sigma,t]$. Furthermore, we have ${}_*\!\int_\sigma^t T^{**}(t,s)\Gamma h(s)ds = \hat{K}(t,\sigma)h$, since

$$\langle\psi,{}_*\!\int_\sigma^t T^{**}(t,s)\Gamma h(s)ds\rangle = \int_\sigma^t \langle\psi,T^{**}(t,s)\Gamma h(s)\rangle ds$$

$$= \int_\sigma^t [K^*(t,\sigma)\psi](s)h(s)ds$$

$$= \langle\psi,K(t,\sigma)h\rangle$$

$$= \langle\psi,\hat{K}(t,\sigma)h\rangle$$

for all $\psi \in \mathscr{B}^*$ by (2.9). On the one hand, $x_t = T(t,\sigma)x_\sigma + K(t,\sigma)h$ and hence $\hat{x}_t = \hat{T}(t,\sigma)\hat{x}_\sigma + \hat{K}(t,\sigma)h$. Therefore, Relation (2.11) holds.

In the next chapter, we will treat the case where $\hat{\mathscr{B}}$ is decomposed as the direct sum of closed subspaces $\mathscr{B}_1(s)$ and $\mathscr{B}_2(s)$, $s \in \mathbb{R}$, satisfying

$$\hat{\mathscr{B}} = \mathscr{B}_1(s) \oplus \mathscr{B}_2(s)$$

(2.12)

$$\hat{T}(t,s)\mathscr{B}_i(s) \subset \mathscr{B}_i(t), \quad s \le t, \quad i = 1, 2.$$

Let $\pi_i(s)$ be the projection on $\mathscr{B}_i(s)$. Then the following relations hold:

$$\pi_i(t)\hat{T}(t,s) = \hat{T}(t,s)\pi_i(s),$$

(2.13)

$$\pi_i^{**}(t)T^{**}(t,s) = T^{**}(t,s)\pi_i^{**}(s), \quad i = 1, 2.$$

<u>Theorem 2.8 (Murakami [4])</u>. Suppose $\hat{\mathscr{B}}$ is decomposed as in (2.12). Then $x_t = x_t(\sigma,\varphi,h)$ satisfies

$$(2.14) \qquad \pi_i(t)\hat{x}_t = \hat{T}(t,\sigma)\pi_i(\sigma)\hat{x}_\sigma + {}_*\!\int_\sigma^t T^{**}(t,s)\pi_i^{**}(s)\Gamma h(s)ds$$

for $\sigma \le t$ and $i = 1,2$.

<u>Proof.</u> Observe that $\pi_i(t) = \pi_i^{**}(t)$ on $\hat{\mathscr{B}}$, since $\langle\psi,\pi_i(t)\hat{\varphi}\rangle = \langle\pi_i^*(t)\psi,\hat{\varphi}\rangle = \langle\psi,\pi_i^{**}(t)\hat{\varphi}\rangle$ for $\hat{\varphi} \in \hat{\mathscr{B}}$ and $\psi \in \mathscr{B}^*$. Then Relation (2.14) follows from Relations (2.11), (2.12) and Lemma 2.5. Indeed,

$$\pi_i(t)\hat{x}_t = \pi_i(t)[\hat{T}(t,\sigma)\hat{x}_\sigma + {}_*\!\int_\sigma^t T^{**}(t,s)\Gamma h(s)ds]$$

$$= \hat{T}(t,\sigma)\pi_i(\sigma)\hat{x}_\sigma + {}_*\!\int_\sigma^t \pi_i^{**}(t)T^{**}(t,s)\Gamma h(s)ds$$

$$= \hat{T}(t,\sigma)\pi_i(\sigma)\hat{x}_\sigma + {}_*\!\int_\sigma^t T^{**}(t,s)\pi_i^{**}(s)\Gamma h(s)ds.$$

Let h be a (fixed) locally integrable function, and consider the operator $\hat{\Pi}(t,\sigma) : \hat{\mathscr{B}} \to \hat{\mathscr{B}}$, $t \ge \sigma$, defined by

$$(2.15) \qquad \hat{\Pi}(t,\sigma)\hat{\varphi} = \hat{T}(t,\sigma)\hat{\varphi} + {}_*\!\int_\sigma^t T^{**}(t,s)\Gamma h(s)ds$$

for $\hat{\varphi} \in \hat{\mathscr{B}}$. If $x(t)$ satisfies Equation (1.2), the function $\hat{\xi} : \mathbb{R} \to$

$\hat{\mathscr{B}}$ defined by $\hat{\xi}(t) = \hat{x}_t$ for $t \in \mathbb{R}$ satisfies

$$(2.16) \qquad \hat{\Pi}(t,\sigma)\hat{\xi}(\sigma) = \hat{\xi}(t) \quad \text{for all} \quad t \geq \sigma.$$

Under the additional assumption on $\mathscr{B}$, the converse holds:

<u>Theorem 2.9.</u> Suppose $\mathscr{B}$ satisfies Axiom (C1). If $\hat{\xi} : \mathbb{R} \to \hat{\mathscr{B}}$ is a function which satisfies Relation (2.16), then the function $x(t)$ defined by $x(t) = [\hat{\xi}(t)](0)$ for $t \in \mathbb{R}$ is a solution of Equation (1.2) on $\mathbb{R}$ and satisfies $x_t \in \hat{\xi}(t)$ for all $t \in \mathbb{R}$.

<u>Proof.</u> For each $\sigma \in \mathbb{R}$, choose an element $\varphi(\sigma)$ in $\hat{\xi}(\sigma)$. From (2.11), (2.15) and (2.16) it follows that $\hat{x}_t(\sigma,\varphi(\sigma),h) = \hat{\Pi}(t,\sigma)\hat{\varphi}(\sigma) = \hat{\Pi}(t,\sigma)\hat{\xi}(\sigma) = \hat{\xi}(t)$ for all $t \geq \sigma$. In particular,

$$(2.17) \qquad x(t,\sigma,\varphi(\sigma),h) = x(t) \quad \text{for all} \quad t \geq \sigma.$$

Let a $t \in \mathbb{R}$ be fixed, and set $\varphi^n = x_t(t-n,\varphi(t-n),h)$ for $n = 1,2,\cdots$. Since $x_t(t-n,\varphi(t-n),h) \in \hat{\xi}(t)$, we have $\varphi^n \in \hat{\xi}(t)$ for $n = 1,2,\cdots$, and hence $|\varphi^n - \varphi^m|_{\mathscr{B}} = 0$ for all $n,m = 1,2,\cdots$. On the other hand, $\varphi^n(\theta) \to x(t+\theta)$ as $n \to \infty$ compactly on $(-\infty,0]$, because $\varphi^n(\theta) = x(t+\theta,t-n,\varphi(t-n),h) = x(t+\theta)$, $-n \leq 0 \leq 0$, by (2.17). Axiom (C1) implies that $x_t \in \mathscr{B}$ and $\varphi^n \to x_t$ as $n \to \infty$. Hence, $x_t \in \hat{\xi}(t)$ because of $\varphi^n \in \hat{\xi}(t)$. Let $t \geq \sigma$. Then $\hat{x}_t(\sigma,\varphi(\sigma),h) = \hat{\xi}(t) = \hat{x}_t$. From this fact and (2.17) it follows that

$$\begin{aligned}(d/dt)x(t) &= (d/dt)x(t,\sigma,\varphi(\sigma),h) \\ &= L(t,x_t(\sigma,\varphi(\sigma),h)) + h(t) \\ &= L(t,x_t) + h(t)\end{aligned}$$

for a.e. t in $[\sigma,\infty)$. Therefore $x(t)$ is a solution of Equation (1.2) defined on $\mathbb{R}$, since σ is arbitrarily given so that $t \geq \sigma$.

Moreover, the similar transform is valid for a general equation

$$(2.18) \qquad \dot{y}(t) = f(t,y_t),$$

where f is continuous on an open set of $\mathbb{R} \times \mathscr{B}$. Denote by $y(\sigma,\varphi,f)$ a solution of this equation such that $y_\sigma = \varphi$. We regard

Equation (2.18) as a perturbed equation of the trivial equation
$\dot{y}(t) = 0$.

Theorem 2.10 (Naito [7]). The t-segment $\eta(t) = y_t(\sigma,\varphi,f)$
satisfies an integral equation in $\hat{\mathscr{B}}$ such that

$$(2.19) \qquad \hat{\eta}(t) = \hat{S}(t)\hat{\varphi} + \int_\sigma^t S^{**}(t-s)\Gamma f(s,\eta(s))ds \qquad t \geq \sigma.$$

Conversely, if $\hat{\eta}(t)$ is a solution of Equation (2.19), then $\hat{\eta}(t)$
$= \hat{y}_t(\sigma,\varphi,f)$ for some solution $y(\sigma,\varphi,f)$.

Proof. It suffices to show the second part. If $\hat{\eta}(t)$
satisfies Equation (2.19), then $f(t,\eta(t))$ is continuous for $t \geq \sigma$
as long as $\hat{\eta}(t)$ is defined. The weak*-integral in (2.19)
represents the t-segment of the solution of

$$\dot{z}(t) = 0 + f(t,\eta(t))$$

with the initial condition $z_\sigma = 0 \in \mathscr{B}$. If we set $y(t) = \varphi(0) +$
$z(t)$ for $t \geq \sigma$ and $y(t) = \varphi(t-\sigma)$ for $t \leq \sigma$, it is easy to
see that $\hat{y}_t = \hat{\eta}(t)$ for $t \geq \sigma$. Hence, $\dot{y}(t) = \dot{z}(t) = f(t,y_t)$.

4.3. The essential spectral radius of solution operators of linear
equations. In this section, we suppose $\mathscr{B}$ is a complex linear space
satisfying Axioms (A), (A$_1$) and (B). Hence, the quotient phase space
$\hat{\mathscr{B}} = \mathscr{B}/|\cdot|_{\mathscr{B}}$ is a Banach space.

We consider a linear functional differential equation

$$(3.1) \qquad\qquad \dot{x}(t) = L(t,x_t),$$

where $L : \mathbb{R} \times \mathscr{B} \to \mathbb{C}^n$ is a continuous mapping such that $L(t,\varphi)$ is
linear in $\varphi \in \mathscr{B}$. As was stated in the preceding section, the
solution operator $T(t,\sigma)$, $t \geq \sigma$, of Equation (3.1) is a bounded
linear operator on $\mathscr{B}$ defined by

$$T(t,\sigma)\varphi = x_t(\sigma,\varphi),$$

where $x(\cdot,\sigma,\varphi)$ denotes the solution of Equation (3.1) such that x_σ
$= \varphi$, which exists uniquely on $[\sigma,\infty)$. The following properties for
$T(t,\sigma)$ immediately follows from the definition of $T(t,\sigma)$ and the
uniqueness of solutions of Equation (3.1) for the initial value

problem:

(a) $T(\sigma,\sigma) = I$; the identity operator on $\mathscr{B}$;

(b) $T(t,\tau)T(\tau,\sigma) = T(t,\sigma)$ for $t \geq \tau \geq \sigma$.

Moreover, from Axiom (A1), it follows

(c) $T(t,\sigma)\varphi$ is continuous for t on $[\sigma,\infty)$ for each fixed φ in $\mathscr{B}$.

Let us consider the case where $L(t+\omega,\varphi) = L(t,\varphi)$ for all $t \in \mathbb{R}$, $\varphi \in \mathscr{B}$ and some constant $\omega > 0$. In this case, Equation (3.1) is called a periodic equation with period ω (an ω-periodic equation, for short). If Equation (3.1) is ω-periodic, from the uniqueness of solutions it follows that $x_{t+\omega}(\sigma+\omega,\varphi) = x_t(\sigma,\varphi)$ for $t \geq \sigma$, because $x(\cdot+\omega,\sigma+\omega,\varphi) = y(\cdot)$ is also a solution of Equation (3.1) by the fact that

$$(d/dt)y(t) = (d/dt)x(t+\omega,\sigma+\omega,\varphi) = L(t+\omega,x_{t+\omega}(\sigma+\omega,\varphi)) = L(t,y_t)$$

and $y_\sigma = \varphi$. From this fact it follows that

(d) $T(t+\omega,\sigma+\omega) = T(t,\sigma)$ for $t \geq \sigma$ if Equation (3.1) is ω-periodic.

In particular, if Equation (3.1) is autonomous, that is, the value $L(t,\varphi)$ does not depend on t, then Property (d) implies that $T(t+\omega,\sigma+\omega) = T(t,\sigma)$ for all $t \geq \sigma$ and all ω in $\mathbb{R}$. Thus,

(e) if Equation (3.1) is autonomous, then $T(t,\sigma) = T(t-\sigma,0)$ for $t \geq \sigma$.

In the autonomous case, we henceforth write $T(t,0)$ as $T(t)$. By Properties (a)-(c), $\{T(t)\}_{t\geq 0}$ is a strongly continuous semigroup on $\mathscr{B}$. We call $\{T(t)\}_{t\geq 0}$ the solution semigroup corresponding to the autonomous linear equation (3.1). For more details of the solution semigroup, we shall study in the next chapter.

Our aim in this section is to investigate the essential spectral radius of the operator $\hat{T}(t,\sigma)$ induced by the solution operator $T(t,\sigma)$ of Equation (3.1). To do this, we first introduce some notations and definitions.

Let X be a Banach space and let $T : X \to X$ be a closed linear operator with dense domain $\mathscr{D}(T)$ in X. We denote by $\rho(T)$, $\sigma(T)$ and $P_\sigma(T)$ the resolvent set, spectrum and point spectrum of T, respectively. Furthermore, we denote by ess(T) the set of all λ in $\sigma(T)$, for which at least one of the following holds:

(i) $\mathcal{R}(T-\lambda I) =: \{(T-\lambda I)\varphi : \varphi \in X\}$ is not closed;

(ii) the point λ is a limit point of $\sigma(T)$;

(iii) the generalized eigenspace $\bigcup_{k\geq 1} \mathcal{N}((T-\lambda I)^k)$ for λ is infinite dimensional;

where I is the identity operator on X and $\mathcal{N}((T-\lambda I)^k)$ denotes the null set of the operator $(T-\lambda I)^k$. The set ess(T) is called the essential spectrum of T (cf. Browder[1]). We define the essential spectral radius of T by

$$r_e(T) = \sup\{|\lambda| : \lambda \in \mathrm{ess}(T)\}.$$

A complex number $\lambda \in \sigma(T)\setminus \mathrm{ess}(T)$ is called a normal eigenvalue of T. A point λ is called a normal point of T if it is either a normal eigenvalue of T or in $\rho(T)$. If λ is a normal eigenvalue, then it is in $P_\sigma(T)$ with finite dimensional generalized eigenspace $\mathcal{N}((T-\lambda I)^k)$ and X can be represented as the direct sum of $\mathcal{N}((T-\lambda I)^k)$ and $\mathcal{R}((T-\lambda I)^k)$ (see Appendix, Corollary 2.2).

Next, let Ω be a bounded set in a Banach space X. Kuratowski's measure of noncompactness $\alpha(\Omega)$ of Ω is defined as

$$\alpha(\Omega) = \inf\{d > 0 : \Omega \text{ has a finite cover of diameter} < d\}$$

(cf. Darbo [1], Kuratowski [1] and Sadovskii [1]). From the definition of $\alpha(\Omega)$, it easily follows that, for bounded sets Ω, Ω_1 and Ω_2 in X,

(i) $\alpha(\Omega) \geq 0$ and $\alpha(\bar{\Omega}) = \alpha(\Omega)$, where $\bar{\Omega}$ is the closure of Ω;

(ii) $\alpha(\Omega_1) \leq \alpha(\Omega_2)$ if $\Omega_1 \subset \Omega_2$;

(iii) $\alpha(\Omega_1 \cup \Omega_2) = \max\{\alpha(\Omega_1), \alpha(\Omega_2)\}$;

(iv) $\alpha(\Omega_1 + \Omega_2) \leq \alpha(\Omega_1) + \alpha(\Omega_2)$, where $\Omega_1 + \Omega_2 = \{x_1 + x_2 : x_i \in \Omega_i,\ i = 1,2\}$;

(v) $\alpha(\Omega) = 0$ if and only if Ω is relatively compact in X.

For instance, Property (v) follows from the fact that a complete metric space is relatively compact if and only if it is totally bounded.

Let $T : X \to X$ be a bounded linear operator. We set

$$\alpha(T) = \inf\{k \geq 0 : \alpha(T\Omega) \leq k\alpha(\Omega) \text{ for all bounded sets } \Omega \text{ in } X\}.$$

$\alpha(T)$ is called Kuratowski's measure of T or α-measure of T. We can easily check the following properties:

(vi) $0 \le \alpha(T) \le |T| := \sup\{|Tx| : |x| \le 1\}$;

(vii) $\alpha(T) = 0$ if and only if T is completely continuous;

(viii) $\alpha(T_1 + T_2) \le \alpha(T_1) + \alpha(T_2)$;

(ix) $\alpha(T_1T_2) \le \alpha(T_1)\alpha(T_2)$;

for any bounded linear operators T, T_1, T_2 on X.

By Nussbaum's theorem (for the proof, see Edmunds & Evans [1, pp. 39-45]), the essential spectral radius of a bounded linear operator T is computed by the formula

$$r_e(T) = \lim_{n\to\infty}[\alpha(T^n)]^{1/n}, \tag{3.2}$$

which plays an important role in the following.

Now, we shall estimate the essential spectral radius of the solution operator $T(t,\sigma)$ of Equation (3.1). First of all, we begin with the easiest case. Let $\{S(t)\}_{t\ge0}$ be the solution semigroup corresponding to the equation

$$\dot{x}(t) = 0,$$

and let $S_0(t)$ be the restriction of $S(t)$ on the closed subspace

$$\mathscr{B}_0 = \{\varphi \in \mathscr{B} : \varphi(0) = 0\}.$$

Clearly, $S_0(t)$ is a bounded linear operator on $\mathscr{B}_0$, and it satisfies

$$|S_0(t)| \le M(t), \qquad t \ge 0,$$

where $M(t)$ is the function in Axiom (A-iii). Let $\hat{S}(t)$ and $\hat{S}_0(t)$ be the operators induced by $S(t)$ and $S_0(t)$, respectively.

The essential idea in the proof of the following result is seen in Hale [3] and Shin [3].

<u>Lemma 3.1.</u> $\alpha(\hat{S}_0(t)) \le \alpha(\hat{S}(t)) \le |S_0(t)|$ for $t \ge 0$.

<u>Proof.</u> The first part of the inequality is clear. We shall verify the second part. Let $\hat{\Omega}$ be any bounded set of $\hat{\mathscr{B}}$ such that $\alpha(\hat{\Omega}) = \rho$. Then, for any $\varepsilon > 0$ there exists a finite cover $\{\hat{\Omega}_i\}_{i=1}^n$ of $\hat{\Omega}$ satisfying $d(\hat{\Omega}_i) < \rho+\varepsilon$, $i = 1,\cdots,n$, where $d(\hat{\Omega}_i)$ denotes the diameter of $\hat{\Omega}_i$. Now, for any $a \in \mathbb{E}$, we set

$$\hat{U}_a = \{\varphi+\psi : \varphi, \psi \in \mathscr{B} \text{ with } \varphi(0) = a \text{ and } |\psi|_{\mathscr{B}} < \varepsilon\}.$$

Since the set $\{\varphi(0) : \hat{\varphi} \in \hat{\Omega}\}$ is bounded in $\mathbb{E}$ by Axiom (A-ii), there exists a finite sequence $\{a_j\}_{j=1}^m$ in $\mathbb{E}$ such that $\hat{\Omega} \subset \bigcup_{j=1}^m \hat{U}_{a_j}$. Set

$$\hat{\mathscr{Q}}_{ij} = \hat{\Omega}_i \cap \hat{U}_{a_j}, \quad i = 1,\cdots,n; \quad j = 1,\cdots,m.$$

Then $\{\hat{\mathscr{Q}}_{ij}\}$ is a finite cover of $\hat{\Omega}$ satisfying $d(\hat{\mathscr{Q}}_{ij}) < \rho+\varepsilon$. Let $\hat{x}^k = \hat{\varphi}^k + \hat{\psi}^k \in \hat{\mathscr{Q}}_{ij}$ with $\varphi^k(0) = a_j$ and $|\psi^k|_{\mathscr{B}} < \varepsilon$ for $k = 1, 2$. Then

$$\begin{aligned}|\hat{S}(t)\hat{x}^1 - \hat{S}(t)\hat{x}^2| &\le |\hat{S}_0(t)(\hat{\varphi}^1-\hat{\varphi}^2)| + |\hat{S}(t)||\hat{\psi}^1-\hat{\psi}^2| \\ &\le |\hat{S}_0(t)|(|\hat{x}^1-\hat{x}^2| + |\hat{\psi}^1-\hat{\psi}^2|) + 2\varepsilon|S(t)| \\ &\le |S_0(t)|d(\hat{\mathscr{Q}}_{ij}) + 4\varepsilon|S(t)| \\ &\le |S_0(t)|\rho + 5\varepsilon|S(t)|;\end{aligned}$$

consequently,

$$d(\hat{S}(t)\hat{\mathscr{Q}}_{ij}) \le |S_0(t)|\alpha(\hat{\Omega}) + 5\varepsilon|S(t)|$$

for all $i = 1,\cdots,n$ and $j = 1,\cdots,m$. Since $\{\hat{S}(t)\hat{\mathscr{Q}}_{ij}\}$ is a finite cover of $\hat{S}(t)\hat{\Omega}$, we have $\alpha(\hat{S}(t)\hat{\Omega}) \le |S_0(t)|\alpha(\Omega) + 5\varepsilon|S(t)|$. Thus, $\alpha(\hat{S}(t)\hat{\Omega}) \le |S_0(t)|\alpha(\hat{\Omega})$, because $\varepsilon > 0$ is arbitrary. Hence, $\alpha(\hat{S}(t)) \le |S_0(t)|$, which is the desired one.

<u>Example 3.2.</u> <u>(i) The case $\mathscr{B} = C([-r,0]) \times L^p(g)$.</u> By the same argument as in the proof of Theorem 1.4.11, we see that

$$\alpha(S(t)) \le |S_0(t)| = \left(\operatorname*{ess\,sup}_{\theta\le -r} \frac{g(\theta-t)}{g(\theta)}\right)^{1/p}$$

for $t > r$.

<u>(ii) The case $\mathscr{B} = C_g$.</u> By Theorem 1.4.3, we have

$$\alpha(S(t)) \le |S_0(t)| = \sup\{\frac{g(\theta+t)}{g(\theta)} : -\infty < \theta \le -t\}.$$

In particular, if $\mathscr{B} = C_\gamma$, we have

$$\alpha(S(t)) = e^{-\gamma t},$$

cf. Shin [6]. Indeed, to certify the above relation, it suffices to verify by Lemma 3.1 that $e^{-\gamma t} \le \alpha(S_0(t))$. The inequality can be proved in the following way. Take a family Ω in $C((-\infty,0],E)$ with the properties that

(a) each element in Ω has the support contained in $[-2,-1]$;

(b) the family Ω is uniformly bounded, but not equicontinuous on $[-2,-1]$.

Clearly, Ω is a bounded set in C_γ, but it is not a relatively compact set in C_γ; hence $\alpha(\Omega) > 0$. Let $\{\mathcal{Q}_i\}$ be any finite cover $\{\mathcal{Q}_i\}$ of $S_0(t)\Omega$. Notice that any function in $S_0(t)\Omega$ is identically zero on the interval $[-t,0]$. Therefore, if we set $\Omega_i = \{x_{-t} : x \in \mathcal{Q}_i\}$, then $\Omega_i \subset (C_\gamma)_0$, and the family $\{\Omega_i\}$ is a finite cover of the set Ω. Moreover,

$$\begin{aligned} d(\Omega_i) &= \sup\{|x_{-t} - y_{-t}|_{C_\gamma} : x,\ y \in \mathcal{Q}_i\} \\ &= \sup\{\sup_{s\le -t} |x(s) - y(s)|e^{\gamma(s+t)} : x,\ y \in \mathcal{Q}_i\} \\ &= e^{\gamma t}\sup\{\sup_{s\le 0}|x(s) - y(s)|e^{\gamma s} : x,\ y \in \mathcal{Q}_i\} \\ &= e^{\gamma t} d(\mathcal{Q}_i). \end{aligned}$$

Consequently, we have $\alpha(\Omega) \le e^{\gamma t}\alpha(S_0(t)\Omega)$. Thus,

$$e^{-\gamma t} \le \alpha(S_0(t)\Omega)/\alpha(\Omega) \le \alpha(S_0(t)),$$

which is the desired one.

Property (ix) of α-measure implies that $\alpha(\hat{S}(t+s)) = \alpha(\hat{S}(t)\hat{S}(s)) \le \alpha(\hat{S}(t))\alpha(\hat{S}(s))$ for $t,\ s \ge 0$. Thus $\alpha(\hat{S}(t))$ is a locally bounded and submultiplicative function of t in $[0,\infty)$. So, we can define a constant β by the relation

$$\beta = \lim_{t\to\infty} \frac{\log \alpha(\hat{S}(t))}{t} = \inf_{t>0} \frac{\log \alpha(\hat{S}(t))}{t} \tag{3.3}$$

where $-\infty \le \beta < \infty$, cf. Hille & Phillips [1].

Lemma 3.3. $r_e(\hat{S}(t)) = \exp(t\beta) \leq |S_0(t)|$ for $t > 0$.

Proof. By Formula (3.2), we obtain

$$r_e(\hat{S}(t)) = \lim_{n\to\infty}[\alpha(\hat{S}(t)^n)]^{1/n} = \lim_{n\to\infty}[\alpha(\hat{S}(nt))]^{1/n}$$

$$= \lim_{n\to\infty} \exp[(1/n)\log \alpha(\hat{S}(nt))] = \exp(t\beta)$$

by (3.3). On the other hand, (3.3) and Lemma 3.1 imply that

$$\exp(t\beta) \leq \alpha(\hat{S}(t)) \leq |S_0(t)| \quad \text{for} \quad t > 0.$$

This completes the proof.

Define operators $U(t,\sigma)$, $t \geq \sigma$, by

$$T(t,\sigma)\varphi = U(t,\sigma)\varphi + S(t-\sigma)\varphi \quad \text{for} \quad \varphi \in \mathscr{B}.$$

Lemma 3.4 (Hale & Kato [1]). For any $t \geq \sigma$, the operator $U(t,\sigma)$ on $\mathscr{B}$ is completely continuous.

Proof. Notice that $U(t,\sigma)\varphi$ can be represented as

$$[U(t,\sigma)\varphi](\theta) = \begin{cases} 0 & \text{if} \quad t+\theta < \sigma, \\ \int_\sigma^{t+\theta} L(s,T(s,\sigma)\varphi)ds & \text{if} \quad t+\theta \geq \sigma. \end{cases}$$

Also, Lemma 1.1 and Theorem 2.1 imply that the functions $|L(s)|$ and $|T(s,\sigma)|$ of s on $[\sigma,\infty)$ are locally bounded. Hence, by Axiom (A-iii) and Ascoli-Arzéla's theorem one can easily obtain the desired conclusion.

Now, we shall give some informations on the magnitude of α-measure and the essential spectral radius of solution operator $T(t,\sigma)$.

Theorem 3.5. Let $T(t,\sigma)$, $t \geq \sigma$, be the solution operator of Equation (3.1). Then

$$\alpha(\hat{T}(t,\sigma)) \leq |S_0(t-\sigma)|$$

and

$$r_e(\hat{T}(t,\sigma)) = e^{\beta(t-\sigma)} \le |S_0(t-\sigma)|$$

for $t > \sigma$, where β is the number defined by Relation (3.3). In particular, if $|S_0(t_1)| < 1$ for some $t_1 > 0$, then $\beta < 0$.

Proof. Since $\hat{U}(t,\sigma)$ is completely continuous on $\hat{\mathcal{B}}$ by Lemma 3.4, from Properties (vii) and (viii) it follows that $\alpha(\hat{T}(t,\sigma)) = \alpha(\hat{S}(t-\sigma))$. Thus, by Formula (3.2) we obtain $r_e(\hat{T}(t,\sigma)) = r_e(\hat{S}(t-\sigma))$. Hence, the conclusion of the theorem immediately follows from Lemmas 3.1 and 3.3.

CHAPTER 5

LINEAR AUTONOMOUS SYSTEMS

5.1. The perturbation of solution semigroups. In the following, we assume that the space $\mathscr{B}$ satisfies Axioms (B) and (C1) with $\mathbb{E} = \mathbb{C}^n$.

Consider a linear autonomous functional differential equation

$$\dot{x}(t) = L(x_t), \tag{1.1}$$

where L is in $\mathscr{L}(\mathscr{B}, \mathbb{C}^n)$, the space of all bounded linear operators from $\mathscr{B}$ into $\mathbb{C}^n$. Denote by $x(\varphi)$ or $x(\varphi, L)$ the solution of Equation (1.1) such that $x_0 = \varphi$ in $\mathscr{B}$. Let $T(t)$, $t \geq 0$, be the solution semigroup of Equation (1.1); that is, $T(t)\varphi = x_t(\varphi)$ for $t \geq 0$ and $\varphi \in \mathscr{B}$. From Axiom (A), $T(t)$ is a strongly continuous semigroup of bounded linear operators on $\mathscr{B}$. The infinitesimal generator A of $T(t)$ is defined as

$$A\varphi = \lim_{t \to 0^+} (1/t)\{T(t)\varphi - \varphi\}$$

whenever this limit exists in $\mathscr{B}$. Let $\hat{T}(t)$ and $\hat{A}$ be the operators on $\hat{\mathscr{B}}$ induced by $T(t)$ and A, respectively;

$$\hat{T}(t)\hat{\varphi} = (T(t)\varphi)\hat{}, \qquad \varphi \in \mathscr{B}$$

$$\hat{A}\hat{\varphi} = (A\varphi)\hat{}, \qquad \varphi \in \mathscr{D}(A)$$

where $\hat{\varphi}$ is the equivalence class of $\varphi \in \mathscr{B}$ and $\mathscr{D}(A)$ denote the domain of A.

Obviously, $\hat{T}(t)$ is a strongly continuous semigroup of bounded linear operators on the Banach space $\hat{\mathscr{B}}$, and $\hat{A}$ is the infinitesimal generator of $\hat{T}(t)$. For such a semigroup, it is well known that the relation

$$\alpha = \alpha_L := \lim_{t \to \infty} \frac{1}{t} \log|\hat{T}(t)| = \inf_{t>0} \frac{1}{t} \log|\hat{T}(t)| \tag{1.2}$$

holds, $-\infty \leq \alpha_L < \infty$, and that the spectral radius $r_\sigma(\hat{T}(t))$ is given by

$$(1.3)\qquad r_\sigma(\hat{T}(t)) = e^{t\alpha}.$$

The constant α is called the type number of $T(t)$. Of course, if α is given by (1.2), then for every $\varepsilon > 0$ there exists a $c(\varepsilon)$ such that

$$(1.4)\qquad |\hat{T}(t)| \le c(\varepsilon)e^{(\alpha+\varepsilon)t} \qquad \text{for } t \ge 0.$$

Furthermore, the resolvent $R(\lambda;\hat{A}) = (\lambda I-\hat{A})^{-1}$ exists for $\mathrm{Re}\,\lambda > \alpha$, and it is given by the Laplace transform;

$$R(\lambda;\hat{A})\hat{\varphi} = \mathcal{L}(\hat{T}(t)\hat{\varphi})(\lambda) = \int_0^\infty e^{-\lambda t}\hat{T}(t)\hat{\varphi}dt$$

for φ in $\mathcal{B}$ and for $\mathrm{Re}\,\lambda > \alpha$. These results are found in Butzer & Berens [1], Hille & Phillips [1], Yosida [1], etc..

Now, we consider the perturbation of the solution semigroup. Let L_1 and L_2 be in $\mathcal{L}(\mathcal{B},\mathbb{C}^n)$, $T_i(t)$, $i = 1,2$, the solution semigroup of the equation

$$\dot{x}(t) = L_i(x_t),$$

$i = 1,2$. Recall the variation-of-constants formula in Theorem 4.2.7.

<u>Theorem 1.1</u>. Let $T_1(t)$ and $T_2(t)$ be as above. Then, we have

$$(1.5)\qquad \hat{T}_1(t)\hat{\varphi} = \hat{T}_2(t)\hat{\varphi} + \int_{*0}^t T_2^{**}(t-s)\Gamma\{(L_1-L_2)(T_1(s)\varphi)\}ds$$

for $\varphi \in \mathcal{B}$ and for $t \ge 0$.

<u>Proof</u>. This result follows directly from Theorem 4.2.7, since the solution $x(t) = x(t,0,\varphi,L_1)$ satisfies the equation

$$\dot{x}(t) = L_2(x_t) + L_1(x_t) - L_2(x_t) = L_2(x_t) + (L_1-L_2)(T_1(t)\varphi)$$

for $t \ge 0$ and for φ in $\mathcal{B}$.

Clément et al. [1] obtained the similar result by solving directly the equation of the variation-of-constants formula. But we need not to do so due to Theorem 4.2.7.

Before taking the Laplace transform of the both sides in (1.5), we observe the Laplace transform of $T_2^{**}(t)\Gamma$. Generally speaking, for the semigroup $T(t)$ of Equation (1.1), we set

$$\Psi(t,L,\lambda) = \int_0^t{}_{*} e^{-\lambda s}T^{**}(s)\Gamma ds.$$

Substituting the integral variable as $s = t - u$, the j-th component Ψ_j of Ψ is given by

$$\Psi_j(t) := \Psi_j(t,L,\lambda) = e^{-\lambda t}\int_0^t{}_{*} T^{**}(t-u)\gamma_j e^{\lambda u}du.$$

Hence, $e^{\lambda t}\Psi_j$ is an element in $\hat{\mathscr{B}}$, which is the equivalence class of the t-segment of the solution of the equation

$$\dot{x}(t) = L(x_t) + e^{\lambda t}e_j \tag{1.6}$$

with the initial condition $x_0 = 0$, where

$$e_j = \mathrm{col}(\delta_{1j},\cdots,\delta_{nj})$$

and δ_{ij} is Kronecker's δ.

<u>Proposition 1.2</u>. Suppose that α is the constant given by (1.2). Then, each component $\Psi_j(t,L,\lambda)$ converges in the norm of $\hat{\mathscr{B}}$ as $t \to \infty$ for $\mathrm{Re}\,\lambda > \alpha$.

<u>Proof.</u> From the definition of the weak* integral, it follows that, for any φ^* in $\mathscr{B}^*$,

$$\langle\varphi^*,\Psi_j(t)\rangle = \int_0^t e^{-\lambda s}\langle T^*(s)\varphi^*,\gamma_j\rangle ds.$$

Hence, we have that, if $0 \le u \le t$,

$$|\langle\varphi^*,\Psi_j(t)-\Psi_j(u)\rangle| \le \int_u^t e^{-(\mathrm{Re}\,\lambda)s}|T^*(s)||\varphi^*||\gamma_j|ds.$$

Since φ^* is an arbitrary element of $\mathscr{B}^*$, the Hahn-Banach theorem implies that

$$|\Psi_j(t)-\Psi_j(u)| \le |\gamma_j| |\int_u^t e^{-(\mathrm{Re}\,\lambda)s} |T^*(s)| ds.$$

Thus, if $\mathrm{Re}\,\lambda > \alpha$, then $|\Psi_j(t)-\Psi_j(u)| \to 0$ as $t, u \to \infty$, which implies the conclusion of the theorem.

From this proposition, we can define the weak* Laplace transform of $T^{**}(t)\Gamma$ as

$$\mathcal{L}_*(T^{**}(t)\Gamma)(\lambda) = \lim_{t\to\infty} \int_0^t e^{-\lambda s} T^{**}(s)\Gamma ds,$$

which converges in the norm of $\hat{\mathcal{B}}$ for $\mathrm{Re}\,\lambda > \alpha$. The j-th component, $\mathcal{L}_*(T^{**}(t)\gamma_j)(\lambda)$, is the limit of $e^{-\lambda t}\hat{x}_t$ as $t \to \infty$ in the norm of $\hat{\mathcal{B}}$, where x_t is the t-segment of the solution x of Equation (1.6) with $x_0 = 0$.

<u>Theorem 1.3.</u> Let $T_1(t)$ and $T_2(t)$ be as in Theorem 1.1. Then, $R(\lambda;\hat{A}_1)$ and $R(\lambda;\hat{A}_2)$ exist for $\mathrm{Re}\,\lambda > \alpha := \max\{\alpha_{L_1}, \alpha_{L_2}\}$ and

$$R(\lambda;\hat{A}_1)\varphi = R(\lambda;\hat{A}_2)\varphi + \mathcal{L}_*(T_2{}^{**}(t)\Gamma)(\lambda)[(L_1-L_2)R(\lambda;\hat{A}_1)\varphi)]$$

for φ in $\mathcal{B}$.

<u>Proof</u>. Denote by $V(t)\varphi$ the integral term in the right hand side of Equation (1.5), and assume that $\mathrm{Re}\,\lambda > \alpha$. Since the Laplace transforms $\mathcal{L}(\hat{T}_i(t)\varphi)$, $i = 1,2$, are convergent, it follows from Equation (1.5) that $\mathcal{L}(V(t)\varphi)(\lambda)$ is also convergent, and

$$\mathcal{L}(V(t)\varphi)(\lambda) = R(\lambda;\hat{A}_1)\varphi - R(\lambda;\hat{A}_2)\varphi.$$

Let φ^* be an element in $\mathcal{B}^*$. Since $\mathcal{L}(V(t)\varphi)(\lambda)$ is convergent in $\hat{\mathcal{B}}$, we have

$$<\varphi^*, \int_0^\infty e^{-\lambda t} V(t)\varphi dt> = \int_0^\infty e^{-\lambda t} <\varphi^*, V(t)\varphi> dt.$$

The definition of the weak* integral means that

$$<\varphi^*, V(t)\varphi> = \int_0^t <\varphi^*, T_2{}^{**}(t-s)\Gamma[(\Delta L)(T_1(s)\varphi)]> ds$$

$$= \int_0^t <\varphi^*, T_2^{**}(t-s)\Gamma>\cdot \Delta L(T_1(s)\varphi)ds$$

$$= \int_0^t f(t-s)g(s)ds,$$

where $f(t) = <\varphi^*, T_2^{**}(t)\Gamma>$, $g(t) = \Delta L(T_1(t)\varphi)$ and $\Delta L = L_1 - L_2$. Since the j-th component of f satisfies $|f_j(t)| = |<\varphi^*, T_2^{**}(t)\gamma_j>| \le |\varphi^*||\gamma_j||T_2(t)|$, and since $|g(t)| \le |\Delta L||T_1(t)||\varphi|$, the Laplace transform of the convolution of f and g is convergent for $\operatorname{Re}\lambda > \alpha$; hence, we have

$$\int_0^\infty e^{-\lambda t}<\varphi^*, V(t)\varphi>dt = \int_0^\infty e^{-\lambda t}f(t)dt \int_0^\infty e^{-\lambda t}g(t)dt$$

for $\operatorname{Re}\lambda > \alpha$. On the other hand, we can write

$$\int_0^\infty e^{-\lambda t}f(t)dt = <\varphi^*, \mathcal{L}_*(T_2^{**}(t)\Gamma)(\lambda)>$$

by the definition of the weak* Laplace transform. Since ΔL is in $\mathcal{L}(\mathcal{B}, \mathbb{C}^n)$, we can also write

$$\int_0^\infty e^{-\lambda t}g(t)dt = \Delta L(\int_0^\infty e^{-\lambda t}T_1(t)\varphi dt).$$

These results are summarized as

$$<\varphi^*, \mathcal{L}(V(t)\varphi)(\lambda)> = <\varphi^*, \mathcal{L}_*(T_2^{**}(t)\Gamma)(\lambda)[\Delta L(\mathcal{L}(T_1(t)\varphi)(\lambda))]>.$$

Since φ^* is an arbitrary element in $\mathcal{B}^*$, we finally have that

$$\mathcal{L}(V(t)\varphi)(\lambda) = \mathcal{L}_*(T_2^{**}(t)\Gamma)(\lambda)[\Delta L(\mathcal{L}(T_1(t)\varphi)(\lambda))].$$

5.2. <u>The spectrum and the resolvent of the infinitesimal generator of solution semigroup.</u> Let $\hat{T}(t)$ and $\hat{A}$ be as in Section 5.1. For any λ in $\mathbb{C}$ and b in $\mathbb{C}^n$, we set $(\omega(\lambda)b)(\theta) = \exp(\lambda\theta)b$ for θ in $(-\infty, 0]$. If the function $\omega(\lambda)b$ belongs to $\mathcal{B}$, then $\hat{\omega}(\lambda)b$ denotes the element of $\hat{\mathcal{B}}$ which is the equivalence class of $\omega(\lambda)b$. Let $T(t,\sigma)$ be the solution operator of Equation (1.1) defined as in Section 4.2.

<u>Theorem 2.1 (Hale & Kato [1])</u>. If $\hat{A}$ is the infinitesimal

generator of $\hat{T}(t)$, then

$$P_\sigma(\hat{A}) = \{\lambda \in \mathbb{C} : \text{there exists a } b \neq 0,\ b \in \mathbb{C}^n, \text{ such that } \omega(\lambda)b \in \mathscr{B} \text{ and that } \lambda b - L(\omega(\lambda)b) = 0\}.$$

Proof. Suppose $\lambda \in P_\sigma(\hat{A})$, and let $\hat{\varphi} \in \hat{\mathscr{B}}$ be a nonzero element such that $\hat{A}\hat{\varphi} = \lambda\hat{\varphi}$. Then, $(d/dt)\hat{T}(t)\hat{\varphi} = \hat{T}(t)\hat{A}\hat{\varphi} = \lambda\hat{T}(t)\hat{\varphi}$, and hence $\hat{T}(t)\hat{\varphi} = e^{\lambda t}\hat{\varphi}$ for $t \geq 0$. Set $\hat{\xi}(t) = e^{\lambda t}\hat{\varphi}$ for $t \in \mathbb{R}$. Then $\hat{T}(t,\sigma)\hat{\xi}(\sigma) = \hat{T}(t-\sigma)(e^{\lambda\sigma}\hat{\varphi}) = e^{\lambda t}\hat{\varphi} = \hat{\xi}(t)$ for all $t \geq \sigma > -\infty$. From Theorem 4.2.9, it follows that the function $x(t) := [\hat{\xi}(t)](0) = e^{\lambda t}b$, $b = \varphi(0)$, is a solution of the equation (1.1) on $\mathbb{R}$ and $\hat{\xi}(t) = \hat{x}_t = e^{\lambda t}\hat{\omega}(\lambda)b$ for all $t \in \mathbb{R}$. Clearly, $b \neq 0$ and $\omega(\lambda)b = x_0 \in \mathscr{B}$. Moreover, we have $\lambda e^{\lambda t}b = \dot{x}(t) = L(x_t) = e^{\lambda t}L(\omega(\lambda)b)$; hence $\lambda b - L(\omega(\lambda)b) = 0$.

Conversely, suppose that $\lambda \in \mathbb{C}$ is a number such that $\omega(\lambda)b \in \mathscr{B}$ and $\lambda b - L(\omega(\lambda)b) = 0$ for some nonzero $b \in \mathbb{C}^n$. One can easily see that the function $x(t) = e^{\lambda t}b$ is a solution of Equation (1.1) for all $t \in \mathbb{R}$. Therefore, if $\varphi - x_0 = \omega(\lambda)b$, then $\hat{\varphi} \neq 0$ by Axiom (A), and $\hat{T}(t)\hat{\varphi} = \hat{x}_t = e^{\lambda t}\hat{\varphi}$ for $t \geq 0$. Thus, $(d/dt)\hat{T}(t)\hat{\varphi} = \lambda e^{\lambda t}\hat{\varphi}$ exists for $t \geq 0$; hence $\hat{\varphi} \in \mathscr{D}(\hat{A})$ and $\hat{A}\hat{\varphi} = \lambda\hat{\varphi}$, which shows $\lambda \in P_\sigma(\hat{A})$.

Let β be the number defined by Relation (4.3.3). From Theorem 4.3.5, the essential spectral radius of $\hat{T}(t)$ is given by

$$r_e(\hat{T}(t)) = e^{\beta t} \quad \text{for } t > 0.$$

Theorem 2.2 (Naito [4]). The type number $\alpha = \alpha_L$ of $T(t)$ is given by

$$\alpha_L = \max\{\beta,\ \sup\{\mathrm{Re}\,\lambda : \lambda \in P_\sigma(\hat{A})\}\}.$$

Proof. From the definition of the essential spectrum, it follows that, if λ is in $\sigma(\hat{T}(t))$ and if $|\lambda| > \exp(t\beta)$, then λ is an eigenvalue of $\hat{T}(t)$. On the other hand, for a strongly continuous semigroup $\hat{T}(t)$ with an infinitesimal generator $\hat{A}$, it is known (Hille Phillips [1]) that $P_\sigma(\hat{T}(t)) = \exp[tP_\sigma(\hat{A})]$, plus, possibly, the point $\lambda = 0$. In view of Relation (1.3), we have the result from these facts.

Consider a linear differential equation

(2.1) $$\dot{x}(t) = Cx(t),$$

where C is an $n \times n$ matrix. If we define $L(\varphi) = C\varphi(0)$ for φ in $\mathscr{B}$, L is in $\mathcal{L}(\mathscr{B},\mathbb{C}^n)$ by Axiom (A). Hence, Equation (2.1) can be considered as a special case of Equation (1.1). For this equation, λ is in $P_\sigma(\hat{A})$ if and only if there exists a $b \neq 0$, $b \in \mathbb{C}^n$, such that $\omega(\lambda)b \in \mathscr{B}$ and that $\lambda b - Cb = 0$; hence $P_\sigma(\hat{A}) \subset P_\sigma(C)$. Thus, the type number α satisfies

$$\alpha \leq \max\{\beta, \sup\{\mathrm{Re}\,\lambda : \lambda \in P_\sigma(C)\}\}.$$

In fact, we can derive the equality in the above, as the following result shows.

<u>Proposition 2.3.</u> Suppose $T(t)$ is the solution semigroup on $\mathscr{B}$ of Equation (2.1). Then we have

(2.2) $$\alpha = \max\{\beta, \sup\{\mathrm{Re}\,\lambda : \lambda \in P_\sigma(C)\}\},$$

and moreover,

(2.3) $$\mathcal{L}_*(T^{**}(t)\Gamma)(\lambda) = \hat{\omega}(\lambda)(\lambda I - C)^{-1}$$

if $\mathrm{Re}\,\lambda > \alpha$, where α is the type number of $T(t)$.

<u>Proof</u>. Suppose $\alpha < \max\{\beta, \sup\{\mathrm{Re}\,\lambda : \lambda \in P_\sigma(C)\}\}$. Then there exists a $\lambda \in P_\sigma(C)$ such that $\mathrm{Re}\,\lambda > \alpha$. Select a nonzero $a \in \mathbb{C}^n$ such that $Ca = \lambda a$, and consider the solution x of the equation $\dot{x}(t) = Cx(t) + e^{\lambda t}a$ with $x_0 = 0$. As remarked in the previous section, $e^{-\lambda t}x_t$ converges in $\mathscr{B}$ as $t \to \infty$; hence $\lim_{t\to\infty} e^{-\lambda t}x(t)$ exists, which is a contradiction, because

$$x(t) = \int_0^t e^{C(t-s)}e^{\lambda s}a\,ds = te^{\lambda t}a, \quad t \geq 0.$$

Hence, we have the equality (2.2).

Next, suppose $\mathrm{Re}\,\lambda > \alpha$. The t-segment of the solution x of the equation $\dot{x}(t) = Cx(t) + e^{\lambda t}e_j$ with $x_0 = 0$ is given by

$$x_t(\theta) = \begin{cases} (\lambda I - C)^{-1}\{(\exp \lambda(t+\theta))I - (\exp (t+\theta)C)\}e_j & \text{if } t+\theta \geq 0, \\ 0 & \text{if } t+\theta < 0. \end{cases}$$

Hence we have

$$e^{-\lambda t}x_t(\theta) = \begin{cases} (\lambda I-C)^{-1}\{(\exp \lambda\theta)I - \exp(-t(\lambda I-C) + \theta C)\}e_j & \text{if } t+\theta \geq 0, \\ 0 & \text{if } t+\theta < 0, \end{cases}$$

which converges to the function

$$\psi_j(\theta) = \{(\lambda I-C)^{-1}\exp(\lambda\theta)\}e_j = \{(\exp(\lambda\theta))(\lambda I-C)^{-1}\}e_j$$

as $t \to \infty$ compactly for $\theta \leq 0$. Since $e^{-\lambda t}\hat{x}_t$ converges to $\mathcal{L}_*(T^{**}(t)\gamma_j)(\lambda)$ as $t \to \infty$ in the norm of $\hat{\mathcal{B}}$, Axiom (C1) implies that $\hat{\psi}_j$ is in $\hat{\mathcal{B}}$, and $\hat{\psi}_j = \mathcal{L}_*(T^{**}(t)\gamma_j)(\lambda)$. Consequently, we obtain Formula (2.3).

In relation to Theorem 2.1, it is natural to ask if $\omega(\lambda)b$ belongs to $\mathcal{B}$ or not. The following result is not trivial since we know very little about elements of the space $\mathcal{B}$.

<u>Theorem 2.4</u>. If $\mathrm{Re}\,\lambda > \beta$, then $\omega(\lambda)b$ is a function in $\mathcal{B}$ for any b in $\mathbb{C}^n$. Furthermore, $\hat{\omega}(\lambda)b$ is a $\hat{\mathcal{B}}$-valued analytic function for $\mathrm{Re}\,\lambda > \beta$.

<u>Proof</u>. If we take $C = \gamma I$ for some $\gamma \leq \beta$, then the type number of $T(t)$ is equal to β; hence, Formula (2.3) becomes $\mathcal{L}_*(T^{**}(t)\Gamma)(\lambda) = \hat{\omega}(\lambda)(\lambda-\gamma)^{-1}I$ for $\mathrm{Re}\,\lambda > \beta$. This means the first statement of the theorem.

To see the second result, take two constants γ_1, γ_2, $\gamma_1 \neq \gamma_2$, γ_1, $\gamma_2 < \beta$, and consider the solution semigroups $T_i(t)$ of the equations $\dot{x}(t) = \gamma_i x(t)$, $i = 1,2$. Apply Theorem 1.3 for these semigroups $T_1(t)$ and $T_2(t)$. Denote by $x(t)$ the solution of the equation $\dot{x}(t) = \gamma_1 x(t)$ with $x_0 = \varphi$. Since $(L_1-L_2)\varphi = (\gamma_1-\gamma_2)\varphi(0)$, we have that, for $\mathrm{Re}\,\lambda > \beta$,

$$\begin{aligned} (L_1-L_2)(R(\lambda;A_1)\varphi) &= (L_1-L_2)\left(\int_0^\infty e^{-\lambda t}x_t dt\right) \\ &= \int_0^\infty (L_1-L_2)(e^{-\lambda t}x_t)dt = \int_0^\infty (\gamma_1-\gamma_2)e^{-\lambda t}x(t)dt \\ &= \int_0^\infty (\gamma_1-\gamma_2)e^{-\lambda t}e^{\gamma_1 t}\varphi(0)dt = (\gamma_1-\gamma_2)(\lambda-\gamma_1)^{-1}\varphi(0). \end{aligned}$$

Therefore, the formula in Theorem 1.3 becomes

$$R(\lambda;\hat{A}_1)\hat{\varphi} = R(\lambda;\hat{A}_2)\hat{\varphi} + \hat{\omega}(\lambda)(\lambda-\gamma_2)^{-1}(\gamma_1-\gamma_2)(\lambda-\gamma_1)^{-1}\varphi(0).$$

Since $R(\lambda;\hat{A}_i)\hat{\varphi}$ is analytic for $\mathrm{Re}\,\lambda > \alpha_{L_i}$, $i = 1,2$, and since $\alpha_{L_i} = \beta$, we have the second result of the theorem.

The derivative $\frac{d}{d\lambda}\hat{\omega}(\lambda)b$ is the equivalence class of the function $\varphi(\theta) = \frac{\partial}{\partial\lambda}(e^{\lambda\theta}b) = \theta e^{\lambda\theta}b$ for $\theta \le 0$, $\mathrm{Re}\,\lambda > \beta$. Indeed, if we set $\varphi^n(\theta) = n[\omega(\lambda+n^{-1})(\theta) - \omega(\lambda)(\theta)]b = n[\exp(\lambda+n^{-1})\theta - \exp(\lambda\theta)]b$ for $\theta \le 0$, $n = 1,2,\cdots$, then the sequence $\{\varphi^n(\theta)\}$ converges to $\varphi(\theta)$ compactly for θ in $(-\infty,0]$. On the other hand, Theorem 2.4 assures that $|\hat{\varphi}^n - \frac{d}{d\lambda}\hat{\omega}(\lambda)b| \to 0$ as $n \to \infty$. Thus, from Axiom (C1) we have that $\hat{\varphi} = \frac{d}{d\lambda}\hat{\omega}(\lambda)b$. By induction, we have generally that

$$(2.4) \qquad \frac{d^n}{d\lambda^n}\hat{\omega}(\lambda)b = \left(\frac{\partial^n}{\partial\lambda^n} e^{\lambda\theta}b\right)^{\wedge} = [\theta^n e^{\lambda\theta}b]^{\wedge}$$

for $n = 1,2,\cdots$, $\mathrm{Re}\,\lambda > \beta$.

From Theorem 2.4, we can introduce an $n \times n$ matrix $\Delta(\lambda)$, the characteristic matrix of Equation (1.1), defined by

$$\Delta(\lambda) = \lambda I - L(\omega(\lambda)I) \qquad \text{for} \quad \mathrm{Re}\,\lambda > \beta.$$

All the entries of $\Delta(\lambda)$ are analytic functions for $\mathrm{Re}\,\lambda > \beta$. Moreover, from Theorem 2.1 we know that

$$\{\lambda \in \mathbb{C} : \mathrm{Re}\,\lambda > \beta,\ \det\Delta(\lambda) = 0\} = \{\lambda \in P_\sigma(\hat{A}) : \mathrm{Re}\,\lambda > \beta\}.$$

In particular, $\det\Delta(\lambda) \ne 0$ provided that $\mathrm{Re}\,\lambda > \alpha_L$. From the proof of Theorem 2.1, one also sees that

$$(2.5) \qquad \mathcal{N}((\hat{A}-\lambda\hat{I})) = \{\hat{\omega}(\lambda)b : b \in \mathbb{C}^n \text{ with } \Delta(\lambda)b = 0\}$$

for $\lambda \in P_\sigma(\hat{A})$ with $\mathrm{Re}\,\lambda > \beta$, where $\mathcal{N}((\hat{A}-\lambda\hat{I}))$ denotes the null space of $\hat{A}-\lambda\hat{I}$. More generally, we can also characterize the space $\mathcal{N}((\hat{A}-\lambda\hat{I})^k)$ for any positive integer k if $\lambda \in P_\sigma(\hat{A})$ with $\mathrm{Re}\,\lambda > \beta$. To see this, let

$$P_{j+1} = P_{j+1}(\lambda) = (1/j!)\Delta^{(j)}(\lambda) = \frac{1}{j!}\frac{d^j\Delta(\lambda)}{d\lambda^j},$$

$$(2.6) \qquad D_j = D_j(\lambda) = \begin{bmatrix} P_1 & P_2 & \cdots & P_j \\ 0 & P_1 & \cdots & P_{j-1} \\ \vdots & & & \vdots \\ 0 & & P_1 & \vdots \\ \vdots & & & \ddots \\ 0 & \cdots & 0 & P_1 \end{bmatrix}$$

<u>Theorem 2.5 (Hale & Kato [1])</u>. If $\lambda \in P_\sigma(\hat{A})$ and $\mathrm{Re}\,\lambda > \beta$, then $\mathcal{N}((\hat{A}-\lambda\hat{I})^k)$ coincides with the space of equivalence classes $\hat{\varphi} \in \hat{\mathscr{B}}$ of the form

$$\varphi(\theta) = \sum_{j=0}^{k-1} \gamma_{j+1} \frac{\theta^j}{j!} e^{\lambda\theta} \qquad -\infty < \theta \le 0,$$

where $\gamma_{j+1} \in \mathbb{C}^n$ for each j and $\gamma = \mathrm{col}(\gamma_1, \cdots, \gamma_k)$ satisfies

$$D_k \gamma = 0.$$

<u>Proof.</u> If $k = 1$, the assertion has already been proved. Suppose $k = 2$ and $\hat{\varphi} \in \mathcal{N}((\hat{A}-\lambda\hat{I})^2)$. Since $(\hat{A}-\lambda\hat{I})\hat{\varphi}$ is in $\mathcal{N}((\hat{A}-\lambda\hat{I}))$, it follows from (2.5) that $(\hat{A}-\lambda\hat{I})\hat{\varphi} = \hat{\omega}(\lambda)b$ for some $b \in \mathbb{C}^n$ satisfying $\Delta(\lambda)b = 0$. Therefore,

$$\begin{aligned}(d/dt)\hat{T}(t)\hat{\varphi} = \hat{T}(t)\hat{A}\hat{\varphi} &= \lambda\hat{T}(t)\hat{\varphi} + \hat{T}(t)\hat{\omega}(\lambda)b \\ &= \lambda\hat{T}(t)\hat{\varphi} + e^{\lambda t}\hat{\omega}(\lambda)b\end{aligned}$$

and

$$\hat{T}(t)\hat{\varphi} = e^{\lambda t}\hat{\varphi} + te^{\lambda t}\hat{\omega}(\lambda)b \quad \text{for} \quad t \ge 0.$$

Set $\hat{\xi}(s) = e^{\lambda s}\hat{\varphi} + se^{\lambda s}\hat{\omega}(\lambda)b$ for $-\infty < s < \infty$. Then, from the above formula it follows that $\hat{T}(t-s)\hat{\xi}(s) = \hat{\xi}(t)$ for $-\infty < s < t < \infty$. Therefore, from Theorem 4.2.9, the function $x(t) = [\hat{\xi}(t)](0) = e^{\lambda t}c + te^{\lambda t}b$, where $c = \hat{\varphi}(0)$, is a solution of Equation (1.1) on $-\infty < t < \infty$ such that $\hat{x}_t = \hat{T}(t)\hat{\varphi}$ for $t \ge 0$. Then, using Formula (2.4) one sees that

$$P_1 c + P_2 b = 0, \qquad P_1 b = 0,$$

which is the assertion of the theorem for $k = 2$. The same type of calculations and an induction argument will complete the proof of the theorem for any integer k.

Let us remember that $S(t)$ is the solution semigroup of Equation (2.1) for $C = 0$. Let B be the infinitesimal generator of $S(t)$. The type number α_0 of $S(t)$ satisfies $\alpha_0 \geq 0$.

Theorem 2.6 (Naito [4]). Let $\hat{A}$ and $\hat{B}$ be the infinitesimal generators of the solution semigroups $\hat{T}(t)$ and $\hat{S}(t)$, respectively. Then, their resolvents are related to each other by the formula

$$R(\lambda;\hat{A})\hat{\varphi} = R(\lambda;\hat{B})\hat{\varphi} + \omega(\lambda)\Delta(\lambda)^{-1}\hat{L}(R(\lambda;\hat{B})\hat{\varphi}), \quad \hat{\varphi} \in \hat{\mathcal{B}}, \tag{2.7}$$

for λ in $D = \{\lambda : \text{Re}\,\lambda > \beta,\ \det\Delta(\lambda) \neq 0 \text{ and } \lambda \neq 0\}$.

Proof. Let α_L and α_0 be the type numbers of $T(t)$ and $S(t)$, respectively, and set $\alpha = \max\{\alpha_L, \alpha_0\}$. Then, from Theorem 1.3 and Proposition 2.3, we have that

$$\begin{aligned} R(\lambda;\hat{A})\hat{\varphi} &= R(\lambda;\hat{B})\hat{\varphi} + \mathcal{L}_*(S^{**}(t)\Gamma)(\lambda)\cdot L(R(\lambda;\hat{A})\hat{\varphi}) \\ &= R(\lambda;\hat{B})\hat{\varphi} + \lambda^{-1}\hat{\omega}(\lambda)L(R(\lambda;\hat{A})\hat{\varphi}) \end{aligned}$$

for $\text{Re}\,\lambda > \alpha$. The operation of $\hat{L}$ on both sides leads to the equation

$$L(R(\lambda;\hat{A})\hat{\varphi}) = L(R(\lambda;\hat{B})\hat{\varphi}) + \lambda^{-1}L(\hat{\omega}(\lambda))L(R(\lambda;\hat{A})\hat{\varphi}),$$

which can be rewritten as

$$\Delta(\lambda)\hat{L}(R(\lambda;\hat{A})\hat{\varphi}) = \lambda\hat{L}(R(\lambda;\hat{B})\hat{\varphi}).$$

Since $\det\Delta(\lambda) \neq 0$ for $\text{Re}\,\lambda > \alpha_L$, we have shown the formula (2.7) for λ such that $\text{Re}\,\lambda > \max(\alpha_L, \alpha_0)$.

From Theorem 2.1, $P_\sigma(\hat{B})$ contains one point $\lambda = 0$ at most. This implies that $P_\sigma(\hat{S}(t))$ consists of the point $\lambda = 1$ at most, plus possibly, $\lambda = 0$. On the other hand, it is well known (Hille & Phillips [1]) that if $T(t)$ is a strongly continuous semigroup on a Banach space whose infinitesimal generator is A, then $\exp[t\sigma(A)] \subset \sigma(T(t))$ for $t \geq 0$. Furthermore, from Lemma 4.3.3 it follows that $r_e(\hat{S}(t)) = \exp t\beta$ for $t > 0$. These results imply that if $\text{Re}\,\lambda > \beta$ and $\lambda \in \sigma(\hat{B})$, then $\exp t\lambda \in P_\sigma(\hat{S}(t))$, that is, $\exp t\lambda \equiv 1$ for $t > 0$. This is possible only if $\lambda = 0$. Therefore, $R(\lambda;\hat{\mathcal{B}})$ is analytic for $\text{Re}\,\lambda > \beta$ and $\lambda \neq 0$. Finally, we observe that the set D in Theorem 2.6 is a connected set in $\mathbb{C}$, since the zeros of $\det\Delta(\lambda)$

are isolated in the domain $\{\lambda : \mathrm{Re}\,\lambda > \beta\}$. These considerations lead to the result that the right hand side of (2.7) is an analytic extension of the resolvent $R(\lambda;\hat{A})$ to the domain D. Therefore, we have the conclusion of the theorem, since the resolvent cannot be continued analytically beyond the boundary of the resolvent set.

For any λ in D above, we define bounded linear operators $F(\lambda;L)$ and $G(\lambda;L)$ by

$$F(\lambda;L)\varphi = \varphi + \omega(\lambda)\Delta(\lambda)^{-1}L(\varphi)$$

$$G(\lambda;L)\varphi = \varphi - \lambda^{-1}\omega(\lambda)L(\varphi)$$

for φ in $\mathscr{B}$. An easy calculation shows that

$$F(\lambda;L)G(\lambda;L)\varphi = G(\lambda;L)F(\lambda;L)\varphi = \varphi$$

for φ in $\mathscr{B}$. Let $\hat{F}(\lambda;L)$ and $\hat{G}(\lambda;L)$ be the operators on $\hat{\mathscr{B}}$ induced by $F(\lambda;L)$ and $G(\lambda;L)$, respectively. Then Relation (2.7) can be written as

$$(2.8)\qquad R(\lambda;\hat{A}) = \hat{F}(\lambda;L)R(\lambda;\hat{B}) \quad \text{for } \lambda \in D,$$

which is also equivalent to

$$(2.9)\qquad R(\lambda;\hat{B}) = \hat{G}(\lambda;L)R(\lambda;\hat{A}) \quad \text{for } \lambda \in D.$$

Theorem 2.7 (Naito [4]). Any point λ such that $\mathrm{Re}\,\lambda > \beta$ is a normal point of $\hat{A}$; that is, λ does not lie in the essential spectrum of $\hat{A}$.

Proof. As we have already observed in the proof of Theorem 2.6, the resolvent $R(\lambda;\hat{B})$ is analytic for $\mathrm{Re}\,\lambda > \beta$ and $\lambda \neq 0$. Hence, by Relation (2.7) it is clear that, if $\lambda \neq 0$, $\mathrm{Re}\,\lambda > \beta$ and if $\det \Delta(\lambda) = 0$, then λ is a pole of $R(\lambda;\hat{A})$.

To investigate the singularity of $R(\lambda;\hat{A})$ at $\lambda = 0$, let us suppose that $\beta < 0$, and that the point 0 lies in $\sigma(\hat{A})$. Then it follows that the point 1 lies in $\sigma(\hat{T}(t))$ for $t \geq 0$, since $\exp(t\sigma(\hat{A})) \subset \sigma(\hat{T}(t))$ for $t \geq 0$. From the assumption that $\beta < 0$, we have $r_e(\hat{T}(t)) < 1$ for $t > 0$, which implies that the point 1 lies in $P_\sigma(\hat{T}(t))$ for $t \geq 0$. This is possible only if the point 0 lies in $P_\sigma(\hat{A})$, since the zeros of $\det \Delta(\lambda)$ are isolated in the domain $\{\lambda : \mathrm{Re}\,\lambda > \beta\}$. Hence, Theorem 2.1 implies that $\det L(\omega(0)I) = 0$. Consequently, for any operator L such that $\det L(\omega(0)I) \neq 0$, the point 0 lies in $\rho(\hat{A})$. Making use of this result, we can prove that $R(\lambda;\hat{B})$ has a pole at $\lambda = 0$. Indeed, we consider, for example, the operator M defined by $M\varphi = \varphi(0)$ for $\varphi \in \mathcal{B}$. Then the resolvent $R(\lambda;\hat{C})$ is analytic at $\lambda = 0$, where C is the infinitesimal generator of the solution semigroup of the equation $\dot{x}(t) = x(t)$. Applying Theorem 2.6 to this semigroup, we obtain that $R(\lambda;\hat{B}) = \hat{G}(\lambda;M)R(\lambda;\hat{C})$ (see also (2.9)). This implies our statement since $\hat{G}(\lambda;M)$ has a pole at $\lambda = 0$. Finally, returning to Relation (2.7), we see that, for the infinitesimal generator $\hat{A}$ of the semigroup $\hat{T}(t)$, the possible singularity of $R(\lambda;\hat{A})$ at $\lambda = 0$ is only a pole.

In order to apply Browder's theorem (see Appendix, Theorem 2.1) to the operator $\hat{A}$, we shall show that for $\lambda \in P_\sigma(\hat{A})$ and $\mathrm{Re}\,\lambda > \beta$, the generalized eigenspace is of finite dimension. Let m be the multiplicity of λ as a zero of $\det \Delta(\lambda)$. Then, from the analyticity of $\Delta(\lambda)$ for $\mathrm{Re}\,\lambda > \beta$, Levinger's theorem (see Appendix, Theorem 3.1) implies that the dimension of $\mathcal{N}(D_j)$ is equal to m for any $j \geq m+1$, where D_j is the matrix defined by (2.4). Hence, by Theorem 2.5 we conclude that $\mathcal{N}((\hat{A}-\lambda\hat{I})^k)$ is of dimension $\leq m$ for $k \geq m+1$. Consequently, the generalized eigenspace of λ, that is, the smallest subspace of $\hat{\mathcal{B}}$ containing all the elements of $\hat{\mathcal{B}}$ which belongs to $\mathcal{N}((\hat{A}-\lambda\hat{I})^k)$, $k = 1,2,\cdots$, coincides with $\mathcal{N}((\hat{A}-\lambda\hat{I})^k)$ for any $k \geq k_0$, where k_0 is some constant $\geq m+1$, and has a finite dimension $\leq m$. Thus, the theorem follows from the above results and Browder's theorem.

5.3. <u>The decomposition of $\hat{\mathcal{B}}$ by the spectrum of $\hat{A}$.</u> Let $\hat{T}(t)$ and $\hat{A}$ be as in the preceding sections. Suppose λ is in $\sigma(\hat{A})$ with $\mathrm{Re}\,\lambda > \beta$. Then λ is a normal eigenvalue by Theorem 2.7; hence, the generalized eigenspace $\mathcal{M}_\lambda$ for λ is finite dimensional, $\mathcal{M}_\lambda = \mathcal{N}((\hat{A}-\lambda\hat{I})^k)$ for some positive integer k, and

$$\hat{\mathcal{B}} = \mathcal{M}_\lambda \oplus \mathcal{R}((\hat{A}-\lambda\hat{I})^k).$$

Let $d = \dim \mathcal{M}_\lambda$, and let $\hat{\Phi}_\lambda = (\hat{\varphi}_1^\lambda,\cdots,\hat{\varphi}_d^\lambda)$ be a basis for $\mathcal{M}_\lambda$. Since $\mathcal{M}_\lambda \subset \mathcal{D}(\hat{A})$ and $\hat{A}(\mathcal{M}_\lambda) \subset \mathcal{M}_\lambda$, there exists a $d \times d$ constant matrix G_λ such that $\hat{A}\hat{\Phi}_\lambda = \hat{\Phi}_\lambda G_\lambda$. The eigenvalue of G_λ is only λ. In fact, for any d-vector a, we have $0 = (\hat{A}-\lambda\hat{I})^k\hat{\Phi}_\lambda a = \hat{\Phi}_\lambda(G_\lambda-\lambda I)^k a$. Hence $(G_\lambda-\lambda I)^k a = 0$ for all d-vectors a; or $(G_\lambda-\lambda I)^k = 0$, which implies that λ is the only eigenvalue of G_λ.

Also, $(d/dt)\hat{T}(t)\hat{\Phi}_\lambda = \hat{T}(t)\hat{A}\hat{\Phi}_\lambda = \hat{T}(t)\hat{\Phi}_\lambda G_\lambda$, and hence

$$\hat{T}(t)\hat{\Phi}_\lambda = \hat{\Phi} e^{tG_\lambda}$$

for all $t \geq 0$. For any d-vector a, we define a function $\hat{\xi}: \mathbb{R} \to \hat{\mathscr{X}}$ by $\hat{\xi}(t) = \hat{\Phi}_\lambda(\exp(tG_\lambda))a$ for $t \in \mathbb{R}$. Then

$$\begin{aligned}\hat{T}(t,\sigma)\hat{\xi}(\sigma) &= \hat{T}(t-\sigma)\hat{\Phi}_\lambda e^{\sigma G_\lambda} a \\ &= \hat{\Phi}_\lambda e^{(t-\sigma)G_\lambda} e^{\sigma G_\lambda} a \\ &= \hat{\Phi}_\lambda e^{tG_\lambda} a \\ &= \hat{\xi}(t)\end{aligned}$$

for all $t \geq \sigma > -\infty$; hence Theorem 4.2.9 implies that the function $x(t) := [\hat{\xi}(t)](0) = \hat{\Phi}_\lambda(0)e^{tG_\lambda}a$ is a solution of Equation (1.1) on $\mathbb{R}$ and $x_t \in \hat{\xi}(t)$ for all $t \in \mathbb{R}$, and so

$$\hat{\Phi}_\lambda a = \hat{\xi}(0) = \hat{x}_0 = \hat{\Phi}_\lambda(0)\hat{\omega}(G_\lambda)a$$

for all d-vectors a, where $[\omega(G_\lambda)](\theta) = \exp \theta G_\lambda$ for $\theta \leq 0$. This shows that $\hat{T}(t)$ can be extended to $(-\infty,\infty)$ on $\mathscr{K}_\lambda$ to be a group of solution operators; more precisely, for all d-vectors a, $\hat{T}(t)\hat{\Phi}_\lambda a$ with initial value $\hat{\Phi}_\lambda a$ at $t = 0$ may be defined on $(-\infty,\infty)$ by the relation

$$\hat{T}(t)\hat{\Phi}_\lambda a = \hat{\Phi}_\lambda e^{tG_\lambda} a$$

$$\hat{\Phi}_\lambda(\theta) = \hat{\Phi}_\lambda(0)e^{\theta G_\lambda}, \qquad -\infty < \theta \leq 0.$$

On the other hand, since $\hat{T}(t)\hat{A}\hat{\varphi} = \hat{A}\hat{T}(t)\hat{\varphi}$ for $\hat{\varphi} \in \mathscr{D}(\hat{A})$, the complementary subspace $\mathscr{R}((\hat{A}-\lambda\hat{I}))^k)$ is clearly invariant under $\hat{T}(t)$. Moreover, the restriction of $\hat{T}(t)$ to $\mathscr{R}((\hat{A}-\lambda\hat{I})^k)$ is a strongly continuous semigroup whose infinitesimal generator is the restriction of $\hat{A}$ to $\mathscr{D}(\hat{A}) \cap \mathscr{R}((\hat{A}-\lambda\hat{I})^k)$, and another normal eigenvalue of $\hat{A}$ is also a normal eigenvalue of the restricted infinitesimal generator.

By a repeated application of the above process, we obtain the following result (Naito [4, Theorem 5.1]).

Theorem 3.1. Suppose Λ is a finite set $\{\lambda_1,\cdots,\lambda_p\}$ of eigenvalues of $\hat{A}$ such that $\mathrm{Re}\,\lambda_j > \beta$, $j = 1,\cdots,p$, and let $\hat{\Phi}_\Lambda = (\hat{\Phi}_{\lambda_1},\cdots,\hat{\Phi}_{\lambda_p})$, $G_\Lambda = \mathrm{diag}(G_{\lambda_1},\cdots,G_{\lambda_p})$, where $\hat{\Phi}_{\lambda_j}$ is a basis for the generalized eigenspace of λ_j and G_{λ_j} is the matrix defined by $\hat{A}\hat{\Phi}_{\lambda_j} = \hat{\Phi}_{\lambda_j} G_{\lambda_j}$, $j = 1,\cdots,p$. Then the eigenvalue of G_{λ_j} is only λ_j, and for any vector a of the same dimension as $\hat{\Phi}_\Lambda$, $\hat{T}(t)\hat{\Phi}_\Lambda a$ with initial value $\hat{\Phi}_\Lambda a$ at $t = 0$ may be defined on $(-\infty,\infty)$ by the relation

$$(3.1)\qquad \begin{aligned} &\hat{T}(t)\hat{\Phi}_\Lambda a = \hat{\Phi}_\Lambda e^{tG_\Lambda} a \\ &\hat{\Phi}_\Lambda(\theta) = \hat{\Phi}_\Lambda(0)e^{\theta G_\Lambda} \qquad -\infty < \theta \le 0. \end{aligned}$$

Moreover, there exists a subspace $\hat{Q}_\Lambda$ of $\hat{\mathscr{B}}$ such that $\hat{T}(t)\hat{Q}_\Lambda \subset \hat{Q}_\Lambda$ for all $t \ge 0$ and

$$\hat{\mathscr{B}} = \hat{P}_\Lambda \oplus \hat{Q}_\Lambda,$$

where $\hat{P}_\Lambda = \{\hat{\varphi}$ in $\hat{\mathscr{B}} : \hat{\varphi} = \hat{\Phi}_\Lambda a$ for some vector $a\}$. Furthermore, $\sigma(\hat{A}|_{\hat{P}_\Lambda}) = \Lambda$ and $\sigma(\hat{A}|_{\hat{Q}_\Lambda\cap\mathscr{D}(\hat{A})}) = \sigma(\hat{A}) \setminus \Lambda$.

Let us fix a real number γ such that $\gamma > \beta$, and set

$$\Lambda(\gamma) = \{\lambda \in \sigma(\hat{A}) : \mathrm{Re}\,\lambda \ge \gamma\}.$$

Then each number in the set $\Lambda(\gamma)$ is a normal eigenvalue of $\hat{A}$.

Proposition 3.2. $\Lambda(\gamma)$ is a finite set.

Proof. Let $\lambda \in \Lambda(\gamma)$. First, observe that $\mathscr{N}((\hat{A}-\lambda\hat{I})^k) \subset \mathscr{N}((\hat{T}(t)-e^{\lambda t}\hat{I})^k)$ for all positive integers k. Indeed, if $k = 1$, then the assertion is already assured in the proof of Theorem 2.1. Also, we can prove by induction that the assertion holds for all $k = 1,2,\cdots$.

Now, suppose $\Lambda(\gamma)$ is not a finite set. Then, each space which contains generalized eigenspaces of $\hat{A}$ for all λ in $\Lambda(\gamma)$ must be of infinite dimension. In particular, the linear extension of the generalized eigenspaces $\mathscr{M}_{e^{\lambda t}}$, $\lambda \in \Lambda(\gamma)$, of $\hat{T}(t)$ is of infinite dimension. This is a contradiction. In fact, each point $e^{\lambda t}$, $\lambda \in \Lambda(\gamma)$, is a normal eigenvalue of $\hat{T}(t)$, and any normal eigenvalue of $\hat{T}(t)$ does not accumulate in the domain $\{\lambda \in \mathbb{C} : |\lambda| \geq e^{\gamma t}\}$ since $e^{\gamma t} > r_e(\hat{T}(t))$. Thus, the set $\{e^{\lambda t} : \lambda \in \Lambda(\gamma)\}$ is a finite set, and the linear extension of the generalized eigenspaces $\mathcal{M}_{e^{\lambda t}}$ is of finite dimension.

By Proposition 3.2, one can apply Theorem 3.1 as $\Lambda = \Lambda(\gamma)$ to obtain the decomposition of $\hat{\mathcal{B}}$ by $\Lambda(\gamma)$. Since by Proposition 3.2 one can easily see that $\sup\{\mathrm{Re}\,\lambda : \lambda \in \sigma(\hat{A})\setminus\Lambda(\gamma)\} < \gamma$, from (1.4) and Theorem 2.2 it follows that for sufficiently small $\varepsilon > 0$ there exists a constant $c(\varepsilon) > 0$ such that

$$|\hat{T}(t)\hat{\varphi}| \leq c(\varepsilon)e^{(\gamma-\varepsilon)t}|\hat{\varphi}|, \quad t \geq 0, \ \hat{\varphi} \in \hat{Q}_{\Lambda(\gamma)}.$$

Combining this fact with the behavior of $\hat{T}(t)$ on $\hat{P}_{\Lambda(\gamma)}$ shown in Theorem 3.1, we obtain the following result.

<u>Theorem 3.3 (Naito [4])</u>. For any real $\gamma > \beta$, let $\Lambda = \Lambda(\gamma) = \{\lambda \text{ in } \sigma(\hat{A}) : \mathrm{Re}\,\lambda \geq \gamma\}$. Then $\hat{\mathcal{B}}$ is decomposed by Λ as in Theorem 3.1;

$$\hat{\mathcal{B}} = \hat{P}_\Lambda \oplus \hat{Q}_\Lambda.$$

Furthermore, for sufficiently small $\varepsilon > 0$ there exists a $c(\varepsilon) > 0$ such that

$$|\hat{T}(t)\hat{\varphi}| \leq c(\varepsilon)e^{(\gamma-\varepsilon)t}|\hat{\varphi}| \quad \text{for } t \leq 0, \ \hat{\varphi} \in \hat{P}_\Lambda$$

$$|\hat{T}(t)\hat{\varphi}| \leq c(\varepsilon)e^{(\gamma-\varepsilon)t}|\hat{\varphi}| \quad \text{for } t \geq 0, \ \hat{\varphi} \in \hat{Q}_\Lambda.$$

In particular, if the set $\{\lambda \text{ in } \sigma(\hat{A}) : \mathrm{Re}\,\lambda = \gamma\}$ is empty, then the estimate on $\hat{P}_\Lambda$ can be strengthened:

$$|\hat{T}(t)\hat{\varphi}| \leq c(\varepsilon)e^{(\gamma+\varepsilon)t}|\hat{\varphi}| \quad \text{for } t \leq 0, \ \hat{\varphi} \in \hat{P}_\Lambda.$$

<u>Corollary 3.4.</u> Suppose $\beta < 0$ and the characteristic equation $\det \Delta(\lambda) = 0$ has no roots on the imaginary axis. Then Equation (1.1) has a saddle point property; that is, $\hat{\mathcal{B}}$ can be decomposed as $\hat{\mathcal{B}} = \hat{P} \oplus \hat{Q}$, where $\hat{P}$ is finite dimensional and the semigroup $\hat{T}(t)$ can be defined on $\hat{P}$ for all $t \in (-\infty, \infty)$ and satisfies the relation

$$|\hat{T}(t)\hat{\varphi}| \le ce^{\nu t}|\hat{\varphi}|, \quad t \le 0, \ \hat{\varphi} \in \hat{P},$$

$$|\hat{T}(t)\hat{\varphi}| \le ce^{-\nu t}|\hat{\varphi}|, \quad t \ge 0, \ \hat{\varphi} \in \hat{Q},$$

for some constants $c > 0$ and $\nu > 0$.

Corollary 3.5. Suppose $\beta < 0$ and the equation $\det \Delta(\lambda) = 0$ has no roots with $\mathrm{Re}\,\lambda \ge 0$. Then

$$|T(t)\varphi| \le ce^{-\nu t}|\varphi|, \quad t \ge 0, \ \varphi \in \mathcal{X},$$

for some constants $c > 0$ and $\nu > 0$.

Next, in order to present a representation of the projection associated with the decomposition guaranteed in the preceding theorems, we introduce some results on the general situation.

Suppose a Banach space X is a direct sum of closed subspaces Y and Z,

$$X = Y \oplus Z,$$

and let $\pi = \pi_Y$ be the projection on Y along this direct sum: $\pi(x) = y$ if $x = y + z$, $y \in Y$, $z \in Z$. Since X is a Banach space, π is a bounded linear operator on X; see Rudin [2, Theorem 5.16]. Hence, π^* is also a bounded linear operator on X^*, π^*X^* is closed and $\pi^*\pi^* = \pi^*$. Repeating this argument, we have also that π^{**} has the same property. Hence

$$(3.2) \qquad X^* = \pi^*X^* \oplus (I-\pi^*)X^* \qquad X^{**} = \pi^{**}X^{**} \oplus (I-\pi^{**})X^{**}.$$

For $x \in X$, $x^* \in X^*$ and $x^{**} \in X^{**}$, we set $\langle x^*, x\rangle = x^*(x)$, $\langle x^*, x^{**}\rangle = x^{**}(x^*)$; X is regarded as a subspace of X^{**} as usual.

For any subspace Y of X, the annihilator $Y^\perp$ of Y is defined as

$$Y^\perp = \{x^* \in X^* : \langle x^*, x\rangle = 0 \ \text{ for all } \ x \in Y\}.$$

Clearly, Y is contained in $(Y^\perp)^\perp$.

Lemma 3.6. Let X, Y, Z and π be as above. Then $\pi^*X^* = Z^\perp$, $(I-\pi^*)X^* = Y^\perp$, $\pi^{**}X^{**} = (Y^\perp)^\perp$, $(I-\pi^{**})X^{**} = (Z^\perp)^\perp$, and hence

(3.3) $X^* = Z^\perp \oplus Y^\perp$ and $X^{**} = (Y^\perp)^\perp \oplus (Z^\perp)^\perp$.

Suppose that $\dim Y = d < \infty$. Then, we have that $\dim Z^\perp = \dim(Y^\perp)^\perp = d$; in particular, $Y = (Y^\perp)^\perp$ and $X^{**} = Y \oplus (Z^\perp)^\perp$. Furthermore, for any basis $\{\varphi_1, \cdots, \varphi_d\}$ of Y, there exists uniquely a basis $\{\psi_1, \cdots, \psi_d\}$ of $Z^\perp$ such that

$$\langle \psi_i, \varphi_j \rangle = \delta_{ij}. \tag{3.4}$$

In this case, we have

$$\pi x = \sum_{i=1}^{d} \langle \psi_i, x \rangle \varphi_i \qquad x \in X \tag{3.5}$$

$$\pi^* x^* = \sum_{i=1}^{d} \langle x^*, \varphi_i \rangle \psi_i \qquad x^* \in X^* \tag{3.6}$$

$$\pi^{**} x^{**} = \sum_{i=1}^{d} \langle \psi_i, x^{**} \rangle \varphi_i \qquad x^{**} \in X^{**}. \tag{3.7}$$

<u>Proof</u>. It is obvious that $\pi^* X^* \subset Z^\perp$. Suppose x^* is in $Z^\perp$. Then $\langle x^*, x \rangle = \langle x^*, \pi x + (I-\pi)x \rangle = \langle x^*, \pi x \rangle = \langle \pi^* x^*, x \rangle$ for all $x \in X$; hence $x^* = \pi^* x^*$. Thus, we have $\pi^* X^* = Z^\perp$, and Relation (3.3) follows from Relation (3.2).

Suppose that $\dim Y = d < \infty$, and let $\{\varphi_1, \cdots, \varphi_d\}$ be a basis of Y. Then, any element x in X is represented as

$$x = c_1 \varphi_1 + \cdots + c_d \varphi_d + z,$$

where $c_1, \cdots, c_d \in \mathbb{C}$, $z \in Z$. If we set $\psi_i(x) = c_i$ for such an x, then it is easy to see that $\{\psi_1, \cdots, \psi_d\}$ is a basis of $Z^\perp$ such that $\langle \psi_i, \varphi_j \rangle = \delta_{ij}$. If $\{\tilde{\psi}_1, \cdots, \tilde{\psi}_d\}$ is another basis of $Z^\perp$ satisfying Relation (3.4), then $\psi_i - \tilde{\psi}_i \in Z^\perp$ and $\langle \psi_i - \tilde{\psi}_i, \varphi_j \rangle = 0$ for all $i, j = 1, \cdots, d$; consequently, $\psi_i - \tilde{\psi}_i$ vanishes on the entire space X, and hence $\{\psi_1, \cdots, \psi_d\} = \{\tilde{\psi}_1, \cdots, \tilde{\psi}_d\}$. Thus, a basis $\{\psi_1, \cdots, \psi_d\}$ of $Z^\perp$ satisfying Relation (3.4) is unique. Relation (3.5) follows immediately since $\pi x = c_1 \varphi_1 + \cdots + c_d \varphi_d$ for x in the above. This implies that

$$\langle \pi^* x^*, x \rangle = \langle x^*, \pi x \rangle = \langle x^*, \sum_{i=1}^{d} \langle \psi_i, x \rangle \varphi_i \rangle$$

$$= \sum_{i=1}^{d} \langle \psi_i, x \rangle \langle x^*, \varphi_i \rangle = \langle \sum_{i=1}^{d} \langle x^*, \varphi_i \rangle \psi_i, x \rangle$$

for every x in X; hence, we have (3.6). Similarly, we have (3.7) since

$$<x^*,\pi^{**}x^{**}> = <\pi^*x^*,x^{**}> = <\sum_{i=1}^{d} <x^*,\varphi_i>\psi_i,x^{**}>$$

$$= \sum_{i=1}^{d} <x^*,\varphi_i><\psi_i,x^{**}> = <x^*, \sum_{i=1}^{d} <\psi_i,x^{**}>\varphi_i>$$

for every x^* in X^*.

For the bases in Lemma 3.6, we set

$$\Phi = (\varphi_1,\cdots,\varphi_d), \qquad \Psi = \mathrm{col}(\psi_1,\cdots,\psi_d),$$

and call Φ the basis vector of Y, and Ψ the basis vector of $Z^\perp$ associated with Φ. We usually write

$$<x^*,\Phi> = (<x^*,\varphi_1>,\cdots,<x^*,\varphi_d>),$$

$$<\Psi,x> = \mathrm{col}(<\psi_1,x>,\cdots,<\psi_d,x>)$$

for x in X and x^* in X^*. Moreover, we denote by $<\Psi,\Phi>$ the $d\times d$ matrix whose (i,j)-component is $<\psi_i,\varphi_j>$. Then, Relations (3.4), (3.5), (3.6) and (3.7) can be rewritten, respectively, as follows:

(3.4′) $\qquad <\Psi,\Phi> = I_d$ (= the $d\times d$ unit matrix);

(3.5′) $\qquad \pi x = \Phi<\Psi,x>$ for $x\in X$;

(3.6′) $\qquad \pi^*x^* = <x^*,\Phi>\Psi$ for $x^*\in X^*$;

(3.7′) $\qquad \pi^{**}x^{**} = \Phi<\Psi,x^{**}>$ for $x^{**}\in X^{**}$.

Now, consider a linear functional differential equation

$$\dot{x}(t) = L(x_t) + h(t), \tag{3.8}$$

where L is in $\mathcal{L}(\mathcal{B},\mathbb{C}^n)$ and $h:\mathbb{R}\to\mathbb{C}^n$ is locally integrable. Let $\hat{\mathcal{B}} = \hat{P}_\Lambda \oplus \hat{Q}_\Lambda$ be the decomposition of $\hat{\mathcal{B}}$ ensured in Theorem 3.1 for a set Λ. For simplicity, we shall abbreviate the subscript Λ as $\hat{P}_\Lambda = \hat{P}$, $\hat{Q}_\Lambda = \hat{Q}$, $\hat{\Phi}_\Lambda = \hat{\Phi}$ and so on. By Lemma 3.6, there exists a basis vector Ψ of $\hat{Q}^{\perp} \subset \hat{\mathscr{B}}^{*} = \mathscr{B}^{*}$ associated with $\hat{\Phi}$. Let π_P and π_Q be the projections on $\hat{P}$ and $\hat{Q}$, respectively. Since π_P^{**} is given by $\pi_P^{**}\varphi^{**} = \hat{\Phi}\langle\Psi,\varphi^{**}\rangle$ from Formula (3.7′), Relation (3.1) yields that

$$
\begin{aligned}
T^{**}(t,s)\pi_P^{**}\Gamma &= \hat{T}(t-s)\hat{\Phi}\langle\Psi,\Gamma\rangle \\
&= \hat{\Phi}e^{(t-s)G}\langle\Psi,\Gamma\rangle \\
&= \hat{\Phi}e^{(t-s)G}(-\tilde{\Psi}(0^-))
\end{aligned}
$$

for $t \geq s$. Then, from the above fact and Theorem 4.2.8, we obtain the following theorem.

<u>Theorem 3.7.</u> Let $\hat{\mathscr{B}} = \hat{P} \oplus \hat{Q}$, $\hat{\Phi}$ and G be the ones ensured in Theorem 3.1 for a set Λ, and let Ψ be the basis vector of $\hat{Q}^{\perp}$ associated with $\hat{\Phi}$. Then the $\hat{P}$-component $\xi_P(t)$ and $\hat{Q}$-component $\xi_Q(t)$ of the t-segment $\hat{x}_t(\sigma,\varphi,h)$ of the solution of Equation (3.8) satisfy

$$
\xi_P(t) = \hat{\Phi}e^{(t-\sigma)G}\langle\Psi,\xi_P(t)\rangle - \int_\sigma^t \hat{\Phi}e^{(t-s)G}\tilde{\Psi}(0^-)h(s)ds
$$

and

$$
\xi_Q(t) = \hat{T}(t-\sigma)\xi_Q(t) + {}_*\!\!\int_\sigma^t T^{**}(t-s)\pi_Q^{**}\Gamma h(s)ds
$$

for $t \geq s$, respectively.

The following corollary is an immediate consequence of Theorem 3.7. It tells us that the behavior on the space P of solutions of Equation (3.8) is reflected by that of solutions of an ordinary differential equation.

<u>Corollary 3.8.</u> The coordinate $u(t) = \langle\Psi,\xi_P(t)\rangle$ of $\xi_P(t)$ with respect to $\hat{\Phi}$ satisfies the ordinary differential equation

$$
\dot{u}(t) = Gu(t) - \tilde{\Psi}(0^-)h(t)
$$

for a.e. $t \in \mathbb{R}$.

5.4. The infinitesimal generator of solution semigroup. In the preceding sections, we have obtained some results on the infinitesimal generator A of the solution semigroup $T(t)$ of Equation (1.1), which is defined on the space $\mathscr{B}$ satisfying Axioms (A), (A1), (B) and (C1). Nevertheless, we have not succeeded in determining the form of A. In this section, we first treat the case $\mathscr{B} = UC_g$, C_g, C_g^0 or $C([-r,0]) \times L^p(g)$. Characterizing the infinitesimal generator A in these particular cases of $\mathscr{B}$, we know well the role played by Axiom (A1).

Denote by $x(t,\varphi)$ the solution of Equation (1.1) such that $x_0 = \varphi$.

a. The case $\mathscr{B} = UC_g$, C_g or C_g^0.

Theorem 4.1. Let $\mathscr{B} = UC_g$, C_g or C_g^0. Then

$$\mathscr{D}(A) = \{\varphi \in \mathscr{B} : \varphi \text{ is continuously differentiable, } \dot{\varphi}(0) = L(\varphi) \text{ and } \dot{\varphi} \in \mathscr{B}\}$$

and

$$A\varphi = \dot{\varphi} \qquad \text{for } \varphi \text{ in } \mathscr{D}(A).$$

Proof. Suppose that φ is in $\mathscr{D}(A)$, and set $A\varphi = \psi$. Then, we have

$$\left|\frac{1}{t}[x(t+\theta,\varphi) - \varphi(\theta)] - \psi(\theta)\right| \le g(\theta)\left|\frac{1}{t}[x_t(\varphi) - \varphi] - \psi\right|_{\mathscr{B}}$$

for each $\theta \in (-\infty,0]$ and for each $t > 0$. Taking the limit as $t \to 0^+$, we have that $\psi(0) = \dot{x}(0) = L(\varphi)$, and that $D^+\varphi(\theta) = \psi(\theta)$ for $\theta < 0$, where $D^+\varphi$ denotes the right derivative of φ. Since ψ is continuous on $(-\infty,0]$, the function φ is really differentiable on $(-\infty,0]$ and $\dot{\varphi}(\theta) = \psi(\theta)$ for $\theta \le 0$.

Conversely, suppose that φ is continuously differentiable, $\dot{\varphi}(0) = L(\varphi)$ and $\dot{\varphi} \in \mathscr{B}$. Then, the solution $x(t,\varphi)$ is also continuously differentiable for t in $(-\infty,\infty)$. Hence, for each $t > 0$ and for each $\theta \le 0$, there exists a σ, $0 < \sigma < 1$, such that

$$\left|\frac{1}{t}[x(t+\theta,\varphi) - x(\theta,\varphi)] - \dot{x}(\theta,\varphi)\right| \le |\dot{x}(\sigma t+\theta,\varphi) - \dot{x}(\theta,\varphi)|.$$

If we set $y(t) = \dot{x}(t,\varphi)$ for t in $(-\infty,\infty)$, we can write

$$\frac{1}{g(\theta)}\left|\frac{1}{t}[x(t+\theta,\varphi) - \varphi(\theta)] - \dot{\varphi}(\theta)\right| \le \frac{1}{g(\theta)}|y_{\sigma t}(\theta) - y_0(\theta)|.$$

Since $y_0 = \dot{\varphi}$ belongs to $\mathscr{B}$ and $y(t)$ is continuous for t in $(-\infty,\infty)$, Axiom (A1) implies that y_t is in $\mathscr{B}$ for $t \ge 0$ and $|y_t - y_0|_{\mathscr{B}} \to 0$ as $t \to 0^+$. Thus, we have

$$\frac{1}{g(\theta)}|y_{\sigma t}(\theta) - y_0(\theta)| \le |y_{\sigma t} - y_0|_{\mathscr{B}} \le \sup\{|y_s - y_0|_{\mathscr{B}} : 0 < s < t\}$$

for each $t > 0$ and $\theta \le 0$. Consequently, it follows that

$$\left|\frac{1}{t}[x_t(\varphi) - \varphi] - \dot{\varphi}\right|_{\mathscr{B}} \le \sup\{|y_s - y_0|_{\mathscr{B}} : 0 < s < t\}$$

for $t > 0$. Hence, φ is in $\mathscr{D}(A)$ and $A\varphi = \dot{\varphi}$.

b. <u>The case $\mathscr{B} = C([-r,0]) \times L^p(g)$</u>. We assume that g satisfies Conditions (g-5) and (g-6) in Section 1.3. Then, the space $\mathscr{B}$ satisfies Axioms (A) and (A1) by Theorem 1.3.8. Furthermore, from Theorem 1.4.10, we can assume that

$$g(\theta) = 0 \quad \text{a.e. in} \quad (-\infty,-R]$$

$$g(\theta) > 0 \quad \text{a.e. in} \quad (-R,-r]$$

for some R, $r \le R \le \infty$. For any $u \le 0$, set

$$G_1(u) = \inf\{k \ge 0 : g(u+\theta) \le kg(\theta) \quad \text{a.e. } \theta \in [-R,-r]\}.$$

Clearly, we have

$$(4.1) \qquad G_1(u+v) \le G_1(u)G_1(v) \quad \text{for} \quad u \le 0 \quad \text{and} \quad v \le 0$$

and

$$(4.2) \qquad g(u+\theta) \le G_1(u)g(\theta) \quad \text{for} \quad u \le 0 \quad \text{and a.e. } \theta \in (-\infty,-r].$$

Since $G_1(u) \le G(u)$ for all $u \le 0$ by (g-6), (4.1) and (g-6) imply that the function $G_1(u)$ is locally bounded on $(-\infty,0]$. Hence, by (4.2), we can easily derive that

$$(4.3) \qquad \int_{-t}^{-r}|\varphi(\theta)|d\theta \le c(t)\left(\int_{-\infty}^{-r}|\varphi(\theta)|^p g(\theta)d\theta\right)^{1/p}$$

for a.e. $t \in [-R,-r]$ and every Lebesgue measurable function φ, where $c(t)$ is a constant. In fact, $c(t)$ may be taken as

$$c(t) = \left(\sup_{\tau\in[-t,0]} G_1(\tau)/g(-t)\right)^{1/p}\cdot(t-r)^{1-\frac{1}{p}}.$$

In particular, if φ is in $\mathscr{B}$, then φ is locally (Lebesgue) integrable on $(-R,-r)$ by (4.3).

We say that a function φ in $\mathscr{B}$ is regular, and define $D_L\varphi$ for such a function φ if φ satisfies the following conditions:

(i) in case $r = 0$, φ is locally absolutely continuous on $(-R,0]$, and

$$(D_L\varphi)(\theta) = \begin{cases} L(\varphi) & \theta = 0 \\ \dot{\varphi}(\theta) & \text{a.e. in } (-R,0]; \end{cases}$$

(ii) in case $r > 0$, $\varphi|_{[-r,0]}$ is continuously differentiable, φ is locally absolutely continuous on $(-R,-r]$, and

$$(D_L\varphi)(\theta) = \begin{cases} L(\varphi) & \theta = 0 \\ \dot{\varphi}(\theta) & -r \le \theta < 0 \\ \dot{\varphi}(\theta) & \text{a.e. in } (-R,-r]. \end{cases}$$

If φ and ψ in $\mathscr{B}$ are regular and $|\varphi - \psi|_{\mathscr{B}} = 0$, then $\varphi(\theta) = \psi(\theta)$ for $-R < \theta \le 0$ by Inequality (4.3); hence $(D_L\varphi)(\theta) = (D_L\psi)(\theta)$ for all θ in $[-r,0]$ and a.e. in $(-R,-r]$. Thus, if $D_L\varphi$ is in $\mathscr{B}$, then $D_L\psi$ is in $\mathscr{B}$ and $|D_L\varphi - D_L\psi|_{\mathscr{B}} = 0$. The equivalence class $\hat{\varphi}$ of φ in $\mathscr{B}$ is said to be regular if $\hat{\varphi}$ contains a regular function η in $\mathscr{B}$. If $D_L\eta$ is in $\mathscr{B}$, we define $\hat{D}_L\hat{\varphi} = (D_L\eta)^{\wedge}$.

<u>Theorem 4.2.</u> Suppose g satisfies Conditions (g-5) and (g-6). Then

$$\mathscr{D}(\hat{A}) = \{\hat{\varphi} \in \hat{\mathscr{B}} : \hat{\varphi} \text{ is regular and } \hat{D}_L\hat{\varphi} \in \hat{\mathscr{B}}\}$$

and

$$\hat{A}\hat{\varphi} = \hat{D}_L\hat{\varphi}.$$

<u>Proof</u>. Suppose φ is in $\mathscr{D}(A)$ and $A\varphi = \psi$. It is easy to see that $\psi(0) = L(\varphi)$. If $r > 0$, as in the proof of Theorem 4.1, we see immediately that $\varphi|_{[-r,0]}$ is continuously differentiable and $\dot{\varphi}(\theta)$

$= \psi(\theta)$ for $-r \le \theta \le 0$.

Since φ is locally integrable on $(-R,-r]$ by Inequality (4.3), we can define a function $F_N(\theta)$ by

$$F_N(\theta) = N \int_{\theta}^{\theta+\frac{1}{N}} x(s,\varphi)ds$$

for $N > 0$ and $-R < \theta \le -r$. Then, F_N is locally absolutely continuous on $(-R,-r]$,

$$(d/d\theta)F_N(\theta) = N\{[T(1/N)\varphi](\theta) - \varphi(\theta)\} \quad \text{a.e. in } (-R,-r]$$

and

$$\lim_{N\to\infty} F_N(\theta) = \varphi(\theta) \quad \text{a.e. in } (-R,-r].$$

Since ψ is also locally integrable on $(-R,-r]$, we have

$$|F_N(\theta) - \varphi(-r) - \int_{-r}^{\theta} \psi(s)ds| \le |F_N(-r) - \varphi(-r)|$$

$$+ \int_{\theta}^{-r} |F_N'(s) - \psi(s)|ds.$$

Since $x(s,\varphi) = \varphi(s)$ is right continuous at $s = -r$, we have $F_N(-r) \to \varphi(-r)$ as $N \to \infty$. Inequality (4.3) implies that

$$\int_{\theta}^{-r} |F_N'(s) - \psi(s)|ds \le c(-\theta)\{\int_{\theta}^{-r} |F_N'(s) - \psi(s)|^p g(s)ds\}^{1/p}$$

$$\le c(-\theta)|N[T(1/N)\varphi - \varphi] - \psi|_{\mathcal{B}}$$

a.e. $\theta \in (-R,-r]$. These results shows that

$$\varphi(\theta) = \varphi(-r) + \int_{-r}^{\theta} \psi(s)ds \quad \text{a.e. in } (-R,-r].$$

Hence, $\hat{\varphi}$ is regular and $\hat{D}_L\hat{\varphi} = \hat{\psi}$.

Conversely, assume that φ is regular and $D_L\varphi$ is in $\mathcal{B}$. Suppose $0 < t_n$ and $t_n \to 0$ as $n \to \infty$. Put $\varphi_n = (1/t_n)\{T(t_n)\varphi - \varphi\}$. Then, as in the proof of Theorem 4.1, we see that $\sup\{|\varphi_n(\theta) - \dot{\varphi}(\theta)| : -r \le \theta \le 0\}$ as $n \to \infty$. To observe the integral part of $|\varphi_n - D_L\varphi|_{\mathcal{B}}$, we write

$$\varphi_n(\theta) - D_L\varphi(\theta) = (1/t_n)\int_0^{t_n} \{\dot{\varphi}(\theta+u) - \dot{\varphi}(\theta)\}du$$

for a.e. $\theta \in (-R,-r-t_n]$. Then, using Hölder's inequality, we have

$$\int_{-R}^{-r-t_n} |\varphi_n(\theta) - D_L\varphi(\theta)|^p g(\theta)d\theta$$

$$\le (1/t_n)\int_0^{t_n}\{\int_{-R}^{-r-t_n} |\dot{\varphi}(\theta+u) - \dot{\varphi}(\theta)|^p g(\theta)d\theta\}du.$$

However, the definition of the norm in $\mathscr{B}$ yields that

$$\int_{-R}^{-r-t_n} |\dot{\varphi}(\theta+u) - \dot{\varphi}(\theta)|^p g(\theta)d\theta \le |S(u)(D_L\varphi) - (D_L\varphi)|^p \quad \text{for} \quad u \ge 0,$$

which implies that

$$\int_{-R}^{-r} |\varphi_n(\theta) - (D_L\varphi)(\theta)|^p g(\theta)d\theta \to 0 \quad \text{as} \quad n \to \infty.$$

Therefore, $|\varphi_n - (D_L\varphi)|_{\mathscr{B}} \to 0$ as $n \to \infty$; that is, φ is in $\mathscr{D}(A)$.

For any $\delta \in (-\infty,\infty)$, set

$$\mathbb{C}_\delta = \{\lambda \in \mathbb{C} : \mathrm{Re}\,\lambda > \delta\}$$

and

$$\bar{\mathbb{C}}_\delta = \{\lambda \in \mathbb{C} : \mathrm{Re}\,\lambda \ge \delta\}.$$

We define a constant β_0, $-\infty \le \beta_0 < \infty$, by

$$\beta_0 = \inf\{\mathrm{Re}\,\lambda \in \mathbb{C} : \omega(\lambda)a \in \mathscr{B} \text{ for all } a \in \mathbb{C}^n\}.$$

From Theorem 2.4, it follows that $\beta_0 \le \beta$, where β is the number defined by Relation (4.3.3). In particular, if $\mathscr{B} = C_\gamma$, then $\beta_0 = \beta = -\gamma$ by Example 4.3.2.

Now, recall that $P_\sigma(A)$ and $\rho(A)$ denote the point spectrum and the resolvent set of A, respectively. We shall give a result on the distribution of $P_\sigma(A)$ and $\rho(A)$ in the case where $\mathscr{B} = C([-r,0]) \times L^p(g)$ or $\mathscr{B} = C_g$. The following theorems are essentially due to Naito [3].

<u>Theorem 4.3.</u> Let $\mathscr{B} = C([-r,0]) \times L^p(g)$. Then

(4.4) $$\mathbb{C}_\beta \subset P_\sigma(A) \cup \rho(A) \subset \bar{\mathbb{C}}_{\beta_0}$$

and

(4.5) $$\mathbb{C}_\beta \setminus P_\sigma(A) = \rho(A).$$

In particular, if g satisfies the condition

$$\text{"}g(u+v) \le g(u)g(v) \quad \text{for a.e. } u, v \in (-\infty,-r)\text{"},$$

then

(4.6) $$\mathbb{C}_\beta \subset P_\sigma(A) \cup \rho(A) \subset \bar{\mathbb{C}}_\beta.$$

<u>Theorem 4.4.</u> Let $\mathscr{B} = C_g$. Then Relations (4.4) and (4.5) hold. In particular, if g satisfies the condition

$$\text{"}g(u+v) \ge g(u)g(v) \quad \text{for } u, v \le 0\text{"},$$

then Relation (4.6) holds.

We shall prove only Theorem 4.3, because the same argument is applicable to the proof of Theorem 4.4. Now, the first part of Relation (4.4) follows from Theorem 2.7. Also, from the definition of β_0, it is clear that $P_\sigma(A) \subset \bar{\mathbb{C}}_{\beta_0}$. So, if we can prove that

(4.7) $$\rho(A) \subset \mathbb{C}_\beta,$$

then Relations (4.4) and (4.5) follow from the fact $\beta_0 \le \beta$. To verify (4.7), we let $\operatorname{Re}\lambda < \beta < \operatorname{Re}\mu$, and consider the equation

(4.8) $$(\lambda I - A)\varphi = \omega_\mu a,$$

where $a \in \mathbb{C}^n$. By Theorem 4.2, a solution of (4.7) must satisfy

(4.9) $$\lambda\varphi(\theta) - \dot{\varphi}(\theta) = e^{\mu\theta}a \quad \text{for a.e. } \theta \in (-\infty,0].$$

and

(4.10) $$\lambda\varphi(0) - L(\varphi) = a.$$

The solution of (4.9) which belongs to $\mathcal{D}(A)$ is given by $\varphi = (\lambda-\mu)^{-1}\omega_\mu a$, which satisfies (4.10) if and only if $\Delta(\mu)a = 0$. Hence, if $\mu \bar{\in} P_\sigma(A)$ and $a \neq 0$, Equation (4.8) has no solution in $\mathcal{D}(A)$, because $\det \Delta(\mu) \neq 0$; consequently, $\mathcal{D}((\lambda I-A)^{-1})$ does not coincide with the entire space $\mathcal{B}$, which means that λ is not in $\rho(A)$. Thus, we must have $\rho(A) \subset \bar{\mathbb{C}}_\beta$. On the one hand, $\rho(A)$ is an open set. Therefore, any boundary point of $\mathbb{C}_\beta$ can not belong to the set $\rho(A)$. Thus, we have $\rho(A) \subset \mathbb{C}_\beta$.

Next, under the additional condition on g, we shall verify that $\beta \le \beta_0$. Then, Relation (4.6) follows from Relation (4.4). Notice that

$$\beta \le \frac{1}{t}\log[\operatorname*{ess\,sup}_{s\in(-R,-r)} \frac{g(s-t)}{g(s)}]^{1/p} \le \frac{1}{pt}\log g(-t)$$

or

$$e^{p\beta t} \le g(-t)$$

for a.e. $t > r$ (cf. Example 4.3.2). Hence, if $\operatorname{Re}\lambda > \beta_0$, then $\operatorname{Re}\lambda > \beta$, because

$$\int_{-\infty}^{-r} |e^{\lambda\theta}a|^p e^{-p\beta\theta}d\theta \le \int_{-\infty}^{-r} |e^{\lambda\theta}a|^p g(\theta)d\theta \le |\omega_\lambda a|^p < \infty$$

for all $a \in \mathbb{C}^n$. Thus, we must have $\beta_0 \ge \beta$.

<u>5.5. Adjoint theory of solution semigroup.</u> We first recall the general theory of dual semigroups, cf. Butzer & Berens [1], Hille & Phillips [1], Yosida [1], etc. . Let X be a Banach space, X^* the dual space of X, and X^{**} the dual space of X^*. Let $\iota : X \to X^{**}$ be the map defined by $\iota(x)(x^*) = \langle x^*, x\rangle$ for $x \in X$ and $x^* \in X^*$. From the Hahn-Banach theorem, the map ι is isometric; hence X is regarded as a subspace of X^{**}. If $X = X^{**}$, then X is said to be reflexive.

The weak*-topology of X^* is the weakest topology on X^* that makes every $\iota(x)$, $x \in X$, continuous. A sequence $\{x_n^*\}$ of X^* converges to x_0^* in the weak*-topology if and only if $\langle x_n^*, x\rangle$ converges to $\langle x_0^*, x\rangle$ for every x in X.

Suppose $T(t)$, $t \ge 0$, is a strongly continuous semigroup of bounded linear operators on X, and that A is the infinitesimal generator of $T(t)$. The adjoint operator $T^*(t)$ has the semigroup property, but it is not necessarily strongly continuous, and the domain of A^* is not necessarily dense in X^*. However, if X^* is equipped with the weak*-topology, then $T^*(t)$ is strongly continuous and A^* is the infinitesimal generator of $T^*(t)$. The fundamental facts about the dual semigroups are summarized as follows, cf. Clément et al. [1].

<u>Theorem 5.1.</u> Let $T(t)$ and A be as cited above. Then the following statements hold:

(i) For any x^* in X^*, the map $t \to T^*(t)x^*$ from $[0,\infty)$ into X^* equipped with the weak*-topology is continuous.

(ii) A^* is the weak* generator of $T^*(t)$, i.e., x^* belongs to $\mathcal{D}(A^*)$ if and only if $t^{-1}[T^*(t)x^* - x^*]$ converges in the weak*-topology as $t \to 0^+$, and whenever there is convergence, the limit equals A^*x^*.

(iii) If x^* belongs to $\mathcal{D}(A^*)$, so does $T^*(t)x^*$ for any $t \geq 0$ and $A^*T^*(t)x^* = T^*(t)A^*x^*$.

<u>Definition 5.2.</u> $X^{\odot} = \{x^* \in X^* : \lim_{t\to 0^+} |T^*(t)x^* - x^*| = 0\}$.

Clearly, $X^{\odot}$ is an originally closed subspace of X^*, and is an invariant subspace of $T^*(t)$. Denote by $T^{\odot}(t)$ the restriction of $T^*(t)$ on $X^{\odot}$. Then, $T^{\odot}(t)$ is a strongly continuous semigroup on $X^{\odot}$. Let $A^{\odot}$ denote its generator.

<u>Theorem 5.3.</u>

(i) $X^{\odot} = \overline{\mathcal{D}(A^*)}$,

(ii) $\mathcal{D}(A^{\odot}) = \{x^* \in \mathcal{D}(A^*) : A^*x^* \in X^{\odot}\}$,

(iii) $\mathcal{D}(A^{\odot})$ is weak* dense in X.

Obviously, an element x in X is regarded as an element of $(X^{\odot})^* = X^{\odot *}$: $x(x^{\odot}) = \langle x^{\odot}, x\rangle$. The norm $|x^{\odot *}|_{\odot *}$ in $X^{\odot *}$ is defined by

$$|x^{\odot *}|_{\odot *} = \sup\{|\langle x^{\odot}, x^{\odot *}\rangle| : |x^{\odot}| \leq 1\}.$$

This norm is equivalent in X to the original norm as follows.

<u>Theorem 5.4.</u>

(i) $|x|_{\odot *} \leq |x|_X \leq M|x|_{\odot *}$ for every x in X, where $M = \liminf_{\lambda\to\infty} |\lambda(\lambda I - A)^{-1}|$.

(ii) If we equip with the norm $|\cdot|_{\odot *}$ on X, the norm $X^{\odot}$ remains unchanged, i.e.,

$$|x^{\odot}| := \sup\{|\langle x^{\odot}, x\rangle| : x \in X, |x| \le 1\}$$

$$= \sup\{|\langle x^{\odot}, x\rangle| : x \in X, |x|_{\odot *} \le 1\}.$$

Consider the dual space $X^{\odot *}$ of $X^{\odot}$, and dual operators $T^{\odot *}(t)$ and $A^{\odot *}$. If x is in X, we have that, for every $x^{\odot}$ in $X^{\odot}$,

$$\langle x^{\odot}, T^{\odot *}(t)x\rangle = \langle T^{\odot}(t)x^{\odot}, x\rangle = \langle T^{*}(t)x^{\odot}, x\rangle = \langle x^{\odot}, T(t)x\rangle,$$

which means that $|T^{\odot *}(t)x - T(t)x|_{\odot *} = 0$; hence, $|T^{\odot *}(t)x - T(t)x| = 0$ by Theorem 5.4. Therefore, $T^{\odot *}(t)$ is an extension of $T(t)$ to an operator on $X^{\odot *}$. Set

$$X^{\odot\odot} = \{x^{\odot *} \in X^{\odot *} : \lim_{t \to 0} |T^{\odot *}(t)x^{\odot *} - x^{\odot *}|_{\odot *} = 0\},$$

and let $T^{\odot\odot}(t)$ be the restriction of $T^{\odot *}(t)$ on $X^{\odot\odot}$. Then, from the above remark it follows that $X \subset X^{\odot\odot}$ and

$$T^{\odot\odot}(t)x = T(t)x \quad \text{for } x \in X.$$

Thus, $T^{\odot\odot}(t)$ is an extension of $T(t)$ to a strongly continuous semigroup on $X^{\odot\odot}$. Let $A^{\odot\odot}$ denote its infinitesimal generator. Then $Ax = A^{\odot\odot}x$ for x in $\mathcal{D}(A)$; consequently if x is in $\mathcal{D}(A^{\odot\odot}) \cap X$ and $A^{\odot\odot}x \in X$, then x is in $\mathcal{D}(A)$. Thus,

$$\mathcal{D}(A) = \{x \in \mathcal{D}(A^{\odot\odot}) \cap X : A^{\odot\odot}x \in X\}.$$

However, more careful discussion leads to the following result.

Proposition 5.5.

$$\mathcal{D}(A) = \{x \in \mathcal{D}(A^{\odot *}) \cap X : A^{\odot *}x \in X\}. \tag{5.1}$$

Proof. Since $A^{\odot\odot}$ is a restriction of $A^{\odot *}$, it is clear that $\mathcal{D}(A)$ is contained by the set in the right hand side of Relation (5.1). The converse relation of inclusion follows from the fact that $X \subset X^{\odot\odot}$ and that

$$\mathcal{D}(A^{\odot\odot}) = \{x^{\odot *} \in \mathcal{D}(A^{\odot *}) : A^{\odot *}x^{\odot *} \in X^{\odot\odot}\}.$$

Now, consider the solution semigroup $T(t)$ on $\mathscr{B}$ of Equation (1.1). Let $S(t)$ be the solution semigroup on $\mathscr{B}$ of the equation $\dot{x}(t) = 0$, and B its generator. We define an operator $U(t)$, $t \geq 0$, by

$$(5.2) \qquad T(t)\varphi = S(t)\varphi + U(t)\varphi \quad \text{for } \varphi \text{ in } \mathscr{B}.$$

Clearly, it is a bounded linear operator on $\mathscr{B}$, and is given explicitly by the relation

$$(5.3) \qquad [U(t)\varphi](\theta) = \begin{cases} x(t+\theta,\varphi) - \varphi(0) & \text{for } t+\theta \geq 0 \\ 0 & \text{for } t+\theta < 0, \end{cases}$$

where $x(t,\varphi)$ is the solution of Equation (1.1) such that $x_0 = \varphi$. The first observation follows directly from this relation and Theorem 5.1. Denote by $L_i(\varphi)$ the i-th component of $L(\varphi)$, $i = 1,\cdots,n$; then, $L_i \in \mathscr{B}^*$. The element γ_i in $\mathscr{B}^{**}$ is defined as in Chapter 4:

$$\gamma_i(\psi) = -\tilde{\psi}_i(0^-) \quad \text{for } \psi \text{ in } \mathscr{B}^*.$$

Set $\Gamma = (\gamma_1,\cdots,\gamma_n)$.

<u>Theorem 5.6</u>. $\mathscr{D}(A^*) = \mathscr{D}(B^*)$ and

$$(5.4) \qquad A^*\varphi^* = B^*\varphi^* + \Gamma(\varphi^*)L \quad \text{for } \varphi^* \in \mathscr{D}(A^*),$$

where $\Gamma(\varphi^*)L = \sum_{i=1}^{n} \gamma_i(\varphi^*)L_i$.

<u>Proof</u>. From Equation (5.2), we have

$$(5.5) \qquad \frac{1}{t}[T^*(t)\varphi^* - \varphi^*] = \frac{1}{t}[S^*(t)\varphi^* - \varphi^*] + \frac{1}{t}U^*(t)\varphi^*$$

for $t > 0$ and $\varphi^* \in \mathscr{B}^*$. Since $U(t)\varphi$ is in C_{00} for every φ in $\mathscr{B}$, it follows that

$$<\frac{1}{t}U^*(t)\varphi^*,\varphi> = \frac{1}{t}<\varphi^*,U(t)\varphi> = \frac{1}{t}\int_{-t}^{0} d\tilde{\varphi}^*(\theta)[x(t+\theta,\varphi) - x(0)],$$

which converges to $-\tilde{\varphi}^*(0^-)L(\varphi)$ as $t \to 0^+$. Since we can write

$$-\tilde{\varphi}^*(0^-)L(\varphi) = \sum_{i=1}^{n} \gamma_i(\varphi^*)L_i(\varphi) = <\sum_{i=1}^{n} \gamma_i(\varphi^*)L_i,\varphi>,$$

we have that

(5.6) $$\lim_{t\to 0^+} \langle \frac{1}{t} U^*(t)\varphi^*, \varphi\rangle = \lim_{t\to 0^+} \langle \varphi^*, \frac{1}{t} U(t)\varphi\rangle = \langle \Gamma(\varphi^*)L, \varphi\rangle$$

for every φ in $\mathscr{B}$ and φ^* in $\mathscr{B}^*$. In other words, $\frac{1}{t} U^*(t)\varphi^*$ converges to $\Gamma(\varphi^*)L$ as $t \to 0^+$ in the weak*-topology of $\mathscr{B}^*$. Thus, from Equation (5.5), $\frac{1}{t}[T^*(t)\varphi^* - \varphi^*]$ and $\frac{1}{t}[S^*(t)\varphi^* - \varphi^*]$ converges as $t \to 0^+$ simultaneously in the weak*-topology of $\mathscr{B}^*$, that is, $\mathscr{D}(A^*) = \mathscr{D}(B^*)$ by Theorem 5.1; and if $\varphi^* \in \mathscr{D}(A^*) = \mathscr{D}(B^*)$, then Relation (5.4) holds.

Thus, in view of Theorem 5.3, we know that

$$\mathscr{B}^\odot = \overline{\mathscr{D}(A^*)} = \overline{\mathscr{D}(B^*)},$$

which is independent of the solution semigroup $T(t)$ of (1.1); $T^\odot(t)$ is defined on this fixed space $\mathscr{B}^\odot$.

Corollary 5.7.

$$\mathscr{D}(A^\odot) = \{\varphi^* \in \mathscr{D}(B^*) : B^*\varphi^* + \Gamma(\varphi^*)L \in \mathscr{B}^\odot\}$$

and

$$A^\odot\varphi^\odot = B^*\varphi^\odot + \Gamma(\varphi^\odot)L \quad \text{for} \quad \varphi^\odot \in \mathscr{D}(A^\odot).$$

The independence property of $\mathscr{B}^\odot$ is also derived from the following simple observation.

Proposition 5.8. $|U(t)| \to 0$ as $t \to 0$.

Proof. Since $U(t)\varphi$ is in C_{00}, Axiom (A) implies

$$|U(t)\varphi| \le K(t)\cdot\sup\{|x(s,\varphi) - \varphi(0)| : 0 \le s \le t\}$$

for every φ in $\mathscr{B}$. However, we have $|x(s,\varphi) - \varphi(0)| = |x(s,\varphi) - x(0,\varphi)| \le |\dot{x}(u,\varphi)|s$ for some u in $(0,s)$, and $\dot{x}(u,\varphi) = L(x_u(\varphi)) = L(T_u(\varphi))$. Thus, it follows that

$$|U(t)\varphi| \le tK(t)|L|\cdot\sup\{|T(s)| : 0 \le s \le t\}|\varphi|,$$

which implies $|U(t)| \to 0$ as $t \to 0$.

This proposition and Definition 5.2 also show that $\mathscr{X}^{\odot}$ is independent of $T(t)$; however, they imply a further result about the second dual space.

<u>Proposition 5.9</u>.

$$\mathscr{X}^{\odot\odot} := \{\varphi^{\odot *} \in \mathscr{X}^{\odot *} : \lim_{t\to 0} |S^{\odot *}(t)\varphi^{\odot *} - \varphi^{\odot *}| = 0\}$$

$$= \{\varphi^{\odot *} \in \mathscr{X}^{\odot *} : \lim_{t\to 0} |T^{\odot *}(t)\varphi^{\odot *} - \varphi^{\odot *}| = 0\}.$$

<u>Proof.</u> Let $U^{\odot}(t)$ be the restriction of $U^{*}(t)$ on $\mathscr{X}^{\odot}$. Since $T^{*}(t) = S^{*}(t) + U^{*}(t)$, and since $\mathscr{X}^{\odot}$ is an invariant set of $T^{\odot}(t)$ and $S^{\odot}(t)$, it follows that $\mathscr{X}^{\odot}$ is an invariant set of $U^{\odot}(t)$ and

$$T^{\odot}(t) = S^{\odot}(t) + U^{\odot}(t).$$

Hence, we have

$$T^{\odot *}(t)\varphi^{\odot *} = S^{\odot *}(t)\varphi^{\odot *} + U^{\odot *}(t)\varphi^{\odot *} \quad \text{for } \varphi^{\odot *} \in \mathscr{X}^{\odot *}. \tag{5.7}$$

However, the definition of operator norm implies

$$|U^{\odot *}(t)| = |U^{\odot}(t)| \le |U^{*}(t)| = |U(t)|.$$

In particular, $|U^{\odot *}(t)\varphi^{\odot *}| \to 0$ as $t \to 0^{+}$ for every $\varphi^{\odot *}$ in $\mathscr{X}^{\odot *}$. Therefore, we have the relation in Proposition 5.9.

Repeating the argument in the above proof, we also have from Relation (5.7) that

$$T^{\odot\odot}(t)\varphi^{\odot\odot} = S^{\odot\odot}(t)\varphi^{\odot\odot} + U^{\odot\odot}(t)\varphi^{\odot\odot} \quad \text{for } \varphi^{\odot\odot} \in \mathscr{X}^{\odot\odot},$$

where $U^{\odot\odot}(t)$ is the restriction of $U^{\odot *}(t)$ on $\mathscr{X}^{\odot\odot}$. This proposition and Theorem 5.3 imply that

$$\mathscr{X}^{\odot\odot} = \overline{\mathscr{D}(A^{\odot *})} = \overline{\mathscr{D}(B^{\odot *})}.$$

Since γ_i is in $\mathscr{X}^{**}$, it is naturally regarded as an element of $\mathscr{X}^{\odot *}$; that is, the restriction of γ_i on $\mathscr{X}^{\odot}$ is in $\mathscr{X}^{\odot *}$. Recall that $\hat{\mathscr{X}} = \mathscr{X}/|\cdot|_{\mathscr{X}}$ is the quotient space, and $\hat{T}(t)$ and $\hat{A}$ are operators on $\hat{\mathscr{X}}$ induced by $T(t)$ and A, respectively. $\hat{\mathscr{X}}$ is a Banach space, and $\hat{\mathscr{X}}$ is a subspace of $\mathscr{X}^{\odot\odot}$.

Theorem 5.10.

$$\mathscr{D}(\hat{A}) = \{\hat{\varphi} \in \mathscr{D}(B^{\odot *}) \cap \hat{\mathscr{X}} : B^{\odot *}\hat{\varphi} + \Gamma L(\varphi) \in \hat{\mathscr{X}}\}$$

and

$$\hat{A}\hat{\varphi} = B^{\odot *}\hat{\varphi} + \Gamma L(\varphi) \quad \text{for} \quad \hat{\varphi} \quad \text{in} \quad \mathscr{D}(\hat{A}),$$

where $\Gamma L(\varphi) := \sum_{i=1}^{n} \gamma_i L_i(\varphi)$ is generally an element of $\mathscr{X}^{\odot *}$.

Proof. Since Proposition 5.5 says that

$$\mathscr{D}(\hat{A}) = \{\hat{\varphi} \in \mathscr{D}(A^{\odot *}) \cap \hat{\mathscr{X}} : A^{\odot *}\hat{\varphi} \in \hat{\mathscr{X}}\},$$

we consider the condition that $\hat{\varphi}$ in $\hat{\mathscr{X}}$ is in $\mathscr{D}(A^{\odot *})$. Since $A^{\odot *}$ is the weak*-infinitesimal generator of $T^{\odot *}(t)$ as in Theorem 5.1, an element $\hat{\varphi}$ in $\hat{\mathscr{X}}$ is in $\mathscr{D}(A^{\odot *})$ if and only if $t^{-1}[T^{\odot *}(t)\hat{\varphi} - \hat{\varphi}]$ converges as $t \to 0^+$ in the weak*-topology of $\mathscr{X}^{\odot *}$. Thus, we consider the convergence of $\langle\varphi^{\odot}, t^{-1}[T^{\odot *}(t)\hat{\varphi} - \hat{\varphi}]\rangle$ as $t \to 0^+$ for $\varphi^{\odot}$ in $\mathscr{X}^{\odot}$. However, since $T^{\odot *}(t)\hat{\varphi} = T(t)\hat{\varphi}$ for $\hat{\varphi}$ in $\hat{\mathscr{X}}$, we can write

$$\langle\varphi^{\odot}, \frac{1}{t}[T^{\odot *}(t)\varphi - \varphi]\rangle = \langle\varphi^{\odot}, \frac{1}{t}[\hat{T}(t)\hat{\varphi} - \hat{\varphi}]\rangle$$

for $t > 0$ and $\hat{\varphi}$ in $\hat{\mathscr{X}}$.

From the definition of $U(t)$, we have

$$\frac{1}{t}[\hat{T}(t)\hat{\varphi} - \hat{\varphi}] = \frac{1}{t}[\hat{S}(t)\hat{\varphi} - \hat{\varphi}] + \frac{1}{t}\,\hat{U}(t)\hat{\varphi}$$

for $t > 0$ and $\hat{\varphi}$ in $\hat{\mathscr{X}}$. Since Relation (5.6) holds for every $\hat{\varphi}$ in $\hat{\mathscr{X}}$ and for every φ^* in $\mathscr{X}^*$, we can replace φ^* by $\varphi^{\odot}$ in $\mathscr{X}^{\odot}$, that is,

$$\lim_{t\to 0} \langle\varphi^{\odot}, \frac{1}{t}\,\hat{U}(t)\hat{\varphi}\rangle = \lim_{t\to 0} \langle\varphi^{\odot}, \frac{1}{t}\,\hat{U}(t)\hat{\varphi}\rangle = \langle\Gamma(\varphi^{\odot})L, \hat{\varphi}\rangle$$

for every $\hat{\varphi}$ in $\hat{\mathscr{X}}$ and $\varphi^{\odot}$ in $\mathscr{X}^{\odot}$. Since we can write

$$\langle\Gamma(\varphi^{\odot})L,\hat{\varphi}\rangle = \langle \sum_{i=1}^{n} \gamma_i(\varphi^{\odot})L_i, \hat{\varphi}\rangle = \sum_{i=1}^{n} \gamma_i(\varphi^{\odot})L_i(\varphi)$$

$$= \sum_{i=1}^{n} \langle\varphi^{\odot},\gamma_i\rangle L_i(\varphi) = \langle\varphi^{\odot}, \sum_{i=1}^{n} \gamma_i L_i(\varphi)\rangle,$$

we can say that, for every $\hat{\varphi}$ in $\hat{\mathcal{B}}$, $t^{-1}[\hat{U}(t)\hat{\varphi}]$ converges to $\Gamma L(\varphi)$ as $t \to 0^+$ in the weak*-topology of $\mathcal{B}^{\odot *}$. Hence, $t^{-1}[\hat{T}(t)\hat{\varphi} - \hat{\varphi}]$ and $t^{-1}[\hat{S}(t)\hat{\varphi} - \hat{\varphi}]$ converges simultaneously as $t \to 0^+$ in the weak*-topology of $\mathcal{B}^{\odot *}$; that is,

$$\mathcal{D}(A^{\odot *}) \cap \hat{\mathcal{B}} = \mathcal{D}(B^{\odot *}) \cap \hat{\mathcal{B}},$$

and if $\hat{\varphi}$ is in this set, then

$$A^{\odot *}\hat{\varphi} = B^{\odot *}\hat{\varphi} + \Gamma L(\varphi).$$

Consequently, from Proposition 5.5, we have the relations in the theorem.

If $\mathcal{B}^{\odot\odot} = \mathcal{B}$, we say that $\mathcal{B}$ is $\odot$-reflexive (with respect to functional differential equations). For example, if $\mathcal{B} = \mathbb{C}^n \times L^p(g)$, $1 < p < \infty$, it is reflexive; hence, it is $\odot$-reflexive. The space $C([-r,0],\mathbb{C}^n)$, $0 < r < \infty$, is usually taken as the phase space for finite delay equations. Diekmann [1] shows that this space is $\odot$-reflexive.

The space $C([-r,0],\mathbb{C}^n)$ is clearly isometric to the space

$$C_\gamma = \{\varphi \in C((-\infty,0],\mathbb{C}^n) : \alpha_\gamma(\varphi) := \lim_{\theta\to-\infty} e^{\gamma\theta}\varphi(\theta) \quad \text{exists}\},$$

with the norm $|\varphi| = \sup\{|e^{\gamma\theta}\varphi(\theta)| : \theta \le 0\}$, see Section 1.3.,(c). However, we have the following theorem. Set

$$UC_\gamma = \{\varphi \in C((-\infty,0],\mathbb{C}^n) : e^{\gamma\theta}\varphi(\theta) \text{ is bounded, uniformly continuous for } \theta \le 0\}.$$

and define the same norm in this space as in the above.

<u>Theorem 5.11.</u> C_γ is not $\odot$-reflexive; $C_\gamma^{\odot\odot}$ is isometric to the space $\mathbb{C}^n \times UC_\gamma$ with the norm $|(\alpha,\varphi)| = \max\{|\alpha|, |\varphi|\}$.

For simplicity, write $\mathbb{C}^n = \mathbb{E}$. A function $f : (-\infty,0] \to \mathbb{E}^*$ is said to be of bounded variation (on $(-\infty,0]$), if

$$\mathrm{Var}(f,(-\infty,0]) := \sup_{r\geq 0}\{\mathrm{Var}(f,[-r,0]) : r \geq 0\}$$

is finite, and it is said to be normalized if $f(0) = 0$ and f is left continuous on $(-\infty,0)$. Let BV be the set of functions f which are normalized, and of bounded variation on $(-\infty,0]$. As was shown in Section 3.4, the dual space of C_0 is identified with the product space $\mathbb{E}^* \times BV$ with respect to the duality relation

$$<(a,f),\xi> = a\xi(-\infty) + \int_{-\infty}^{0} df(\theta)\xi(\theta)$$

for (a,f) in $\mathbb{E}^* \times BV$ and for ξ in C_0, where the integral in the right hand side is an improper integral on $(-\infty,0]$. The norm in $\mathbb{E}^* \times BV$ is given by $|(a,f)| = |a| + \mathrm{Var}(f,(-\infty,0])$.

Clearly, the space C_γ and C_0 are isometric; an isomorphism $j : C_\gamma \to C_0$ is given by

$$j(\varphi)(\theta) = e^{\gamma\theta}\varphi(\theta) \quad \text{for} \quad \varphi \in C_\gamma.$$

Thus, the isomorphism $j^* : \mathbb{E}^* \times BV \to C_\gamma{}^*$ is induced as

$$<j^*(a,f),\varphi> = <(a,f),j(\varphi)> = a\alpha_\gamma(\varphi) + \int_{-\infty}^{0} df(\theta)e^{\gamma\theta}\varphi(\theta)$$

for $(a,f) \in \mathbb{E}^* \times BV$, $\varphi \in C_\gamma$. If $j^*(a,f) = \varphi^*$, we call (a,f) the coordinate of $\varphi^* \in C_\gamma{}^*$.

From these remarks, a continuous linear operator $L : C_\gamma \to \mathbb{E}$ is represented as

$$L(\varphi) = P\alpha_\gamma(\varphi) + \int_{-\infty}^{0} dQ(\theta)e^{\gamma\theta}\varphi(\theta) \quad \text{for} \quad \varphi \in C_\gamma, \tag{5.8}$$

where P and $Q(\theta)$ are $n \times n$ matrices, and $Q(\theta)$ is of bounded variation on $(-\infty,0]$. From this representation, a continuous linear operator M on a product space $\mathbb{E} \times UC_\gamma$ is naturally induced as

$$M(u,\varphi) = Pu + \int_{-\infty}^{0} dQ(\theta)e^{\gamma\theta}\varphi(\theta) \quad \text{for} \quad (u,\varphi) \in \mathbb{E} \times UC_\gamma.$$

<u>Theorem 5.12.</u> Let L be represented as in (5.8), and $T(t)$ be the solution semigroup of the equation

$$\dot{x}(t) = L(x_t). \tag{5.9}$$

Then the semigroup $T^{\odot\odot}(t)$ on $C_\gamma^{\odot\odot} = \mathbb{E} \times UC_\gamma$ is a solution semigroup of the equation

$$(5.10) \qquad \begin{cases} \dot{u}(t) = -\gamma u(t) \\ \dot{x}(t) = M(u(t), x_t). \end{cases}$$

We prove these two theorems in the rest of this section. It is easy to see the first proposition.

Proposition 5.13. If $j(\varphi) = \xi$, then

$$j(S(t)\varphi)(\theta) = \begin{cases} e^{\gamma\theta}\xi(0) & \text{for } -t \le \theta \le 0 \\ e^{-\gamma t}\xi(t+\theta) & \text{for } \theta < -t. \end{cases}$$

Proposition 5.14. A function φ is in $\mathcal{D}(B)$ if and only if $\xi = j(\varphi)$ is represented as

$$(5.11) \qquad \xi(\theta) = \alpha + \int_{-\infty}^{\theta} \zeta(s)ds \qquad \text{for } \theta \le 0,$$

for some α in $\mathbb{E}$ and for some ζ in C_0 such that the improper integral in (5.11) exists, and that

$$(5.12) \qquad \zeta(0) = \gamma[\alpha + \int_{-\infty}^{0} \zeta(s)ds].$$

For such a function φ, we have

$$j(B\varphi) = -\gamma\xi + \zeta \; (= -\gamma\xi + \dot{\xi}).$$

Proof. A function φ in C_γ is continuously differentiable if and only if ξ is so, and $\dot{\varphi}(\theta) = e^{-\gamma\theta}\{-\gamma\xi(\theta) + \dot{\xi}(\theta)\}$. Thus $\alpha_\gamma(\dot{\varphi})$ exists if and only if $\lim_{\theta\to-\infty} \{-\gamma\xi(\theta) + \dot{\xi}(\theta)\}$ exists; and $\dot{\varphi}(0) = 0$ if and only if $-\gamma\xi(0) + \dot{\xi}(0) = 0$. Set $\zeta = \dot{\xi}$. Then ξ is in C_0 if and only if the improper integral of ζ on $(-\infty, 0]$ exists. Hence, we have the representation in (5.11), where $\alpha = \xi(-\infty)$. The condition $-\gamma\xi(0) + \dot{\xi}(0) = 0$ is then written as condition (5.12). If $\xi(\theta)$ and $-\gamma\xi(\theta) + \dot{\xi}(\theta)$ converges as $\theta \to -\infty$, then $\zeta(\theta)$ also converges as $\theta \to -\infty$.

Proposition 5.15. For (a, f) in $\mathbb{E}^* \times BV$, the coordinate of $S^*(t)j^*(a, f)$ is given by $(e^{-\gamma t}a, v)$, where v is defined by

$$v(\theta) = \begin{cases} 0 & \text{for } \theta = 0 \\ \gamma\int_{-t}^{0} e^{\gamma s} f(s)ds + e^{-\gamma t} f_{-t}(\theta) & \text{for } \theta < 0. \end{cases}$$

Proof. For $(a,f) \in E^* \times BV$, $\varphi \in C_\gamma$, we have that

$$<(a,f), j(S(t)\varphi)> = ae^{-\gamma t}\alpha_\gamma(\varphi)$$
$$+ \int_{-\infty}^{-t} df(\theta)e^{\gamma\theta}\varphi(t+\theta) + \int_{-t}^{0} df(\theta)e^{\gamma\theta}\varphi(0).$$

The first integral is written as

$$\int_{-\infty}^{-t} df(\theta)e^{\gamma\theta}\varphi(t+\theta) = e^{-\gamma t}\int_{-\infty}^{0} d_\theta f(-t+\theta)e^{\gamma\theta}\varphi(\theta).$$

Since $f(0) = 0$, the second integral becomes

$$\int_{-t}^{0} df(\theta)e^{\gamma\theta} = -e^{-\gamma t} f(-t) - \gamma\int_{-t}^{0} e^{\gamma s} f(s)ds.$$

Thus, if v is defined as in the proposition, we have that $<j^*(a,f), S(t)\varphi> = <(e^{-\gamma t}a, v), j(\varphi)>$.

In the following proof, we sometimes use the property that, if $\xi(-\infty) := \lim_{s\to-\infty} \xi(s)$ exists, and if the improper integral of ξ on $(-\infty, 0]$ exists, then $\xi(-\infty) = 0$.

Proposition 5.16. $\mathcal{D}(B^*)$ is the set of $j^*(a,f)$ such that a is free in E^* and f is represented as

$$(5.13) \qquad f(\theta) = \begin{cases} 0 & \theta = 0 \\ b + \int_{\theta}^{0} g(s)ds & \theta < 0 \end{cases}$$

for some b in E^* and for some function g in $BV \cap L^1((-\infty,0])$. For such an (a,f), $j^*(p,u) := B^*(j^*(a,f))$ is given by $p = -\gamma a$ and

$$(5.14) \qquad u(\theta) = \begin{cases} 0 & \theta = 0 \\ -\gamma\int_{\theta}^{0} g(s)ds + g(\theta) & \theta < 0. \end{cases}$$

Proof. The condition $B^*(\jmath^*(a,f)) = \jmath^*(p,u)$ means, by definition, that

$$(5.15)\qquad \langle \jmath^*(a,f), B\varphi\rangle = \langle \jmath^*(p,u),\varphi\rangle \quad \text{for every} \quad \varphi \in \mathcal{D}(B),$$

which is rewritten as

$$\langle (a,f), -\gamma\xi+\dot{\xi}\rangle = \langle (p,u),\xi\rangle \quad \text{for every} \quad \xi = \jmath(\varphi),\ \varphi \in \mathcal{D}(B).$$

Suppose that ξ is represented as in Proposition 5.14. The condition $f,\ u \in BV$ implies that $f(0) = u(0) = 0$; and the condition $\varphi \in \mathcal{D}(B)$ implies that $\dot{\xi}(-\infty) = 0$. From integration by parts, we then have that

$$\langle (a,f), -\gamma\xi+\dot{\xi}\rangle = -[a-f(-\infty)]\gamma\alpha + \int_{-\infty}^{0} [\gamma f(\theta)d\theta + df(\theta)]\zeta(\theta)$$

$$\langle (p,u),\xi\rangle = [p-u(-\infty)]\alpha - \int_{-\infty}^{0} u(\theta)\zeta(\theta)d\theta.$$

Thus Condition (5.15) and the following one are equivalent:

$$(5.16)\qquad \int_{-\infty}^{0} [df(\theta)+\gamma f(\theta)d\theta+u(\theta)d\theta]\zeta(\theta) = \{\gamma[a-f(-\infty)]+p-u(-\infty)\}\alpha$$

for every α and ζ satisfying the condition in Proposition 5.14.

At first, consider the case that $\gamma = 0$. Then the condition for α and ζ in Proposition 5.14 reduces to the one that $\zeta \in D$, where

$$D = \{\zeta \in C_0 : \zeta(0) = 0 \text{ and } \lim_{r\to\infty} \int_{-r}^{0} \zeta(s)ds \text{ exists}\}.$$

There is no restriction on α. Setting $\zeta \equiv 0$ in (5.16), we have $[p-u(-\infty)]\alpha = 0$ for every α in E^*; hence $p = u(-\infty)$, and

$$(5.17)\qquad \int_{-\infty}^{0} [df(\theta)+u(\theta)d\theta]\zeta(\theta) = 0 \quad \text{for} \quad \zeta \in D.$$

Suppose ζ is in C_{00}, the set of continuous functions on $(-\infty,0]$ with compact support. Define ζ^n by $\zeta^n(\theta) = \zeta(\theta)$ for $\theta \le -1/n$ and $\zeta^n(\theta) = \zeta(\theta)-(n\theta+1)\zeta(0)$ for $-1/n < \theta \le 0$, $n = 1,2,\dots$. Then ζ^n is in D. Setting $\zeta = \zeta^n$ in (5.17), we take the limit as $n \to \infty$. Then we obtain the relation

$$\int_{-\infty}^{0} [df(\theta)+u(\theta)d\theta]\xi(\theta) + f(0-)\xi(0) = 0 \quad \text{for} \quad \xi \in C_{00}.$$

From the uniqueness of the kernel function in BV, f is given by Relation (5.13) for $b = f(0-)$ and $g = u$. Such an f is clearly locally absolutely continuous on $(-\infty,0)$. Hence, the total variation of f is given by

$$\text{(5.18)} \qquad \mathrm{Var}(f,(-\infty,0]) = |b| + \int_{-\infty}^{0} |g(\theta)|d\theta,$$

which implies that $g \in BV \cap L^1((-\infty,0])$; so $g(-\infty)$ exists, and is equal to zero. Thus, we have that $p = u(-\infty) = g(-\infty) = 0$.

Conversely, if f has the form in (5.13), then, for every $\alpha \in \mathbb{E}$ and $\xi \in D$, Condition (5.16) with $\gamma = 0$ holds for $u = g$ and $p = 0$. Therefore, $\jmath^*(a,f) \in \mathcal{D}(B^*)$ for every $a \in \mathbb{E}^*$, and $(0,g)$ is the coordinates of $B^*(\jmath^*(a,f))$.

For the case that $\gamma \neq 0$, we start again with Condition (5.16). Set $c = -\{\gamma[a-f(-\infty)]+p-u(-\infty)\}$. From Relation (5.12), this condition reduces to the one that the equation

$$\text{(5.19)} \qquad \int_{-\infty}^{0} [df(\theta)+\gamma f(\theta)d\theta+u(\theta)d\theta]\xi(\theta) = -\frac{c}{\gamma}\xi(0) + \int_{-\infty}^{0} c\xi(\theta)d\theta$$

holds for every $\xi \in E$, where

$$E := \{\xi \in C_0 : \lim_{r\to\infty}\int_{-r}^{0} \xi(s)ds \ \text{exists}\}.$$

Since $E \supset C_{00}$, it follows immediately that

$$\text{(5.20)} \qquad f(\theta) + \gamma\int_0^{\theta} f(s)ds + \int_0^{\theta} u(s)ds = h(\theta) + c\theta \quad \text{for} \quad \theta \le 0,$$

where $h(0) = 0$ and $h(\theta) = c/\gamma$ for $\theta < 0$. Hence f is locally absolutely continuous on $(-\infty,0)$. Since $f \in BV$, it then follows that the derivative $\dot{f}$ exists a.e. in $(-\infty,0)$, and is integrable on $(-\infty,0)$. Set

$$g(\theta) = \begin{cases} 0 & \theta = 0 \\ \gamma f(\theta)+u(\theta)-c & \theta < 0. \end{cases}$$

Then $g \in BV$ and $\dot{f} = -g$ a.e. in $(-\infty,0)$. Hence, g lies in $BV \cap L^1((-\infty,0])$, and we have Representation (5.13). Since $g(-\infty) = 0$

for $g \in BV \cap L^1((-\infty,0])$, it follows that $\gamma f(-\infty)+u(-\infty)-c = 0$. Comparing this relation with definition of c, we have that $\gamma a+p = 0$, or $p = -\gamma a$. Since $f(0-) = h(0-) = c/\gamma$ from Equation (5.20), it holds that, for $\theta < 0$, $u(\theta) = -\gamma f(\theta)+c+g(\theta)$ $= -\gamma[f(\theta)-f(0-)]+g(\theta)$. Hence u has Representation (5.14).

Conversely, set $p = -\gamma a$, and suppose that f and u are represented as in the above. Take a ξ in E. Then, for any $r > 0$, we have

$$(5.21)\qquad \int_{-r}^{0} [df(\theta)+\gamma f(\theta)d\theta+u(\theta)d\theta]\xi(\theta) = -b\xi(0) + \gamma b\int_{-r}^{0} \xi(\theta)d\theta.$$

Since $f \in BV$ and ξ is bounded, we see that $df(\theta)\xi(\theta)$ is integrable on $(-\infty,0]$. Take integration by parts:

$$\int_{-r}^{-s} f(\theta)\xi(\theta)d\theta = f(\theta)\int_{-\infty}^{\theta} \xi(s)ds\Big|_{-r}^{-s} - \int_{-r}^{-s} df(\theta)\int_{-\infty}^{\theta} \xi(s)ds.$$

Since $f \in BV$ and ξ is in E, the right hand side converges to zero as $r, s \to \infty$. Thus, taking the limit in (5.21), we see that

$$(5.22)\qquad \int_{-\infty}^{0} [df(\theta)+\gamma f(\theta)d\theta+u(\theta)d\theta]\xi(\theta) = -b\xi(0) + \gamma b\int_{-\infty}^{0} \xi(\theta)d\theta.$$

Since $p = -\gamma a$, and $g(-\infty) = 0$ for $g \in BV \cap L^1((-\infty,0])$, we have that $c = u(-\infty)+\gamma f(-\infty) = \gamma b + g(-\infty) = \gamma b$. Therefore, Relation (5.22) is Relation (5.19) itself.

Suppose that $\jmath^*(a,f)$ in $\mathcal{D}(B^*)$ is represented as in Proposition 5.16. Since $Var(f,(-\infty,0])$ is given as in (5.18), we have that

$$(5.23)\qquad |(a,f)| = |a| + |b| + \int_{-\infty}^{0} |g(s)|ds.$$

Let $\mathbb{L}^1$ denote the Banach space of the equivalence classes of functions in $L^1((-\infty,0),E^*)$; but we do not distinguish between them for the sake of simplicity of language. The right hand side of Relation (5.23) defines a norm in $(E^*)^2\times \mathbb{L}^1$; the set $\mathcal{D}(B^*)$ is isometrically imbedded onto $(E^*)^2 \times \{g \in \mathbb{L}^1 : g \in BV\}$. Thus, the completion $C_\gamma^{\odot} = \overline{\mathcal{D}(B^*)}$ is identified with the space $(E^*)^2 \times \mathbb{L}^1$. Define a map ι: $(E^*)^2 \times \mathbb{L}^1 \to C_\gamma^{\odot}$ by

$$\iota(a,b,g) = j^*(a,f) \text{ for } (a,b,g) \in (\mathbb{E}^*)^2 \times \mathbb{L}^1,$$

where f is given by Formula (5.13). Then, it is a metric preserving isomorphism, and (a,b,g) is regarded as the coordinate of $j^*(a,f) \in C_\gamma^\odot$, cf. Diekmann [1].

Let $S^\odot(t)$ be the restriction of $S^*(t)$ on $C_\gamma^\odot$, and set $S^\odot(t) = \iota^{-1}S^\odot(t)\iota$. Then, $S^\odot(t)$ is a strongly continuous semigroup on $(\mathbb{E}^*)^2 \times \mathbb{L}^1$.

<u>Proposition 5.17.</u> $S^\odot(t)$ is given by

$$S^\odot(t)(a,b,g) = (e^{-\gamma t}a, b + \int_{-t}^{0} e^{\gamma s}g(s)ds, e^{-\gamma t}g_{-t}).$$

<u>Proof.</u> In fact, if f is written as in (5.13), the function v in Proposition 5.15 has the form

$$v(\theta) = b + \int_{-t}^{0} e^{\gamma s}g(s)ds + \int_{\theta}^{0} e^{-\gamma t}g(-t+s)ds \quad \text{for } \theta < 0.$$

Let A be the infinitesimal generator of the solution semigroup $T(t)$ of Equation (5.9).

<u>Proposition 5.18.</u> The domain $A^\odot$ consists of $\iota(a,b,g)$ such that a and b are free in E^*, and g is given by

$$(5.24) \qquad g(\theta) = \begin{cases} 0 & \theta = 0 \\ b[Q(\theta)-Q(-\infty)] - \int_{-\infty}^{\theta} h(s)ds & \theta < 0 \end{cases}$$

for some $h \in \mathbb{L}^1$ which is required to satisfy the condition

$$(5.25) \qquad \int_{-\infty}^{0} \left| b[Q(\theta)-Q(-\infty)] - \int_{-\infty}^{\theta} h(s)ds \right| d\theta < \infty.$$

If $\iota(a,b,g)$ is given in this manner, then

$$(5.26) \qquad \iota^{-1} A^\odot \iota(a,b,g) = (-\gamma a - bP, -bQ(-\infty) - \int_{-\infty}^{0} h(s)ds, h - \gamma g).$$

<u>Proof.</u> From Corollary 5.7, $j^*(a,f)$ is in $\mathcal{D}(A^\odot)$ if and only if $j^*(a,f) \in \mathcal{D}(B^*)$, and

(5.27) $$B^*(j^*(a,f))+\Gamma(j^*(a,f))L \in C_\gamma^{\odot}.$$

If $j^*(a,f) \in \mathcal{D}(B^*)$, then f is represented as in (5.13) : that is, we can write $j^*(a,f) = \iota(a,b,g)$. Furthermore, $B^*(j^*(a,f)) = j^*(-\gamma a,u)$, where u is given in (5.14). Since $\Gamma(j^*(a,f)) = b$, and since $L = j^*(P,Q)$, cf. (5.8), we have that

(5.28) $$B^*(j^*(a,f))+\Gamma(j^*(a,f))L = j^*(-\gamma a-bP,u-bQ).$$

Thus Condition (5.27) implies that

(5.29) $$-\gamma\int_\theta^0 g(s)ds+g(\theta)-bQ(\theta) = d+\int_\theta^0 k(s)ds \quad \text{for} \quad \theta < 0$$

and for some $d \in \mathbb{E}^*$, $k \in \mathbb{L}^1$. From this relation, it follows that

$$-\gamma\int_{-\infty}^0 g(s)ds-bQ(-\infty) = d+\int_{-\infty}^0 k(s)ds.$$

Eliminating d, we have that

$$g(\theta) = b[Q(\theta)-Q(-\infty)]-\int_{-\infty}^\theta [\gamma g(s)+k(s)]ds \quad \theta < 0.$$

Since $g,k \in \mathbb{L}^1$, Formula (5.24) is valid for $h := \gamma g+k \in \mathbb{L}^1$, and Condition (5.25) holds.

Conversely, let $b \in \mathbb{E}^*$, $h \in \mathbb{L}^1$, and Condition (5.25) holds. Then the function g defined as in (5.24) is in $BV \cap \mathbb{L}^1$; so $\iota(a,b,g)$ is in $\mathcal{D}(B^*)$ for any $a \in \mathbb{E}^*$. Since we can write

$$-\gamma\int_\theta^0 g(s)ds+g(\theta)-bQ(\theta) = -bQ(-\infty)-\int_{-\infty}^0 h(s)ds+\int_\theta^0 [h(s)-\gamma g(s)]ds,$$

we have Relation (5.29) with $k = h-\gamma g \in \mathbb{L}^1$. This implies Condition (5.27); hence $\iota(a,b,g) \in \mathcal{D}(A^{\odot})$ and Relation (5.26) is valid.

Finally, notice that $\mathcal{D}(A^{\odot})$ is dense in $C_\gamma^{\odot}$ as a domain of an infinitesimal generator of the strongly continuous semigroup $T^{\odot}(t)$. Thus the set $\{(a,b,g) : \iota(a,b,g) \in \mathcal{D}(A^{\odot})\}$ is dense in $(\mathbb{E}^*)^2 \times \mathbb{L}^1$, which implies that a and b attain arbitrary value in $\mathbb{E}^*$.

The dual space of $(\mathbb{E}^*)^2 \times \mathbb{L}^1$ is isomorphic to the space $\mathbb{E}^2 \times \mathbb{L}^\infty$, where $\mathbb{L}^\infty$ denotes the equivalence classes of functions in $L^\infty((-\infty,0),\mathbb{E})$. We define the duality between them by

$$\langle (b,a,g),(\alpha,\beta,\xi)\rangle = a\alpha - b\beta - \int_{-\infty}^{0} g(s)\xi(s)ds$$

for (a,b,g) in $(\mathbb{E}^*)^2 \times \mathbb{L}^1$ and for (α,β,ξ) in $(\mathbb{E})^2 \times \mathbb{L}^\infty$. Since the norm in $(\mathbb{E}^*)^2 \times \mathbb{L}^1$ is now given by Relation (5.23), the norm in $\mathbb{E}^2 \times \mathbb{L}^\infty$ is induced as

$$|(\alpha,\beta,\xi)| = \max\{|\alpha|,|\beta|,\text{ess sup}\{|\xi(s)| : s \le 0\}\}.$$

The dual map $\iota^* : C_0^{\odot *} \to (\mathbb{E})^2 \times \mathbb{L}^\infty$ is induced by the map ι, and $\iota^*(\varphi^{\odot *})$ is regarded as the coordinate of $\varphi^{\odot *}$. On the other hand, if φ is in C_γ, we have that

$$[\iota(a,b,g)](\varphi) = a\alpha_\gamma(\varphi) - b\varphi(0) - \int_{-\infty}^{0} g(\theta)e^{\gamma\theta}\varphi(\theta)d\theta$$

$$= \langle (a,b,g),(\alpha_\gamma(\varphi),\varphi(0),j(\varphi))\rangle.$$

Since $\sup\{|j(\varphi)(\theta)| : -\infty < \theta \le 0\} = |(\alpha_\gamma(\varphi),\varphi(0),j(\varphi))|$, it follows that C_γ is imbedded isometrically into the space $C^{\odot *}$.

Consider the dual semigroup $S^{\odot *}(t)$ on $(\mathbb{E})^2 \times \mathbb{L}^\infty$. It is defined by the relation

$$\langle (a,b,g),S^{\odot *}(t)(\alpha,\beta,\xi)\rangle = \langle S^{\odot}(t)(a,b,g),(\alpha,\beta,\xi)\rangle$$

for $(a,b,g) \in (\mathbb{E}^*)^2 \times \mathbb{L}^1$, $(\alpha,\beta,\xi) \in (\mathbb{E})^2 \times \mathbb{L}^\infty$. Computing the right hand side by using the representation of $S^{\odot}(t)$, we have the following proposition.

<u>Proposition 5.19.</u> $S^{\odot *}(t)$ is given by

$$S^{\odot *}(t)(\alpha,\beta,\xi) = (e^{-\gamma t}\alpha,\beta,\zeta) \quad \text{for} \quad (\alpha,\beta,\xi) \in (\mathbb{E})^2 \times \mathbb{L}^\infty,$$

where ζ is given by

$$\zeta(\theta) = \begin{cases} e^{\gamma\theta}\beta & -t \le \theta \le 0 \\ e^{-\gamma t}\xi(t+\theta) & \theta < -t. \end{cases}$$

<u>Proposition 5.20.</u> The domain of $A^{\odot *}$ consists of $\varphi^{\odot *} \in C_\gamma^{\odot *}$ such that the coordinate $\iota^*(\varphi^{\odot *}) = (\alpha,\beta,\xi)$ satisfies the condition that ξ is bounded, globally Lipschitz continuous on $(-\infty,0]$, and

$\beta = \xi(0)$. For such a $\varphi^{\odot *}$, the coordinate $(\sigma,\tau,\zeta) := \iota^*(A^{\odot *}\varphi^{\odot *})$ is given by

$$\sigma = -\gamma\alpha,\ \tau = P\alpha + \int_{-\infty}^{0} dQ(\theta)\xi(\theta),\ \zeta = \dot{\xi} - \gamma\xi.$$

<u>Proof.</u> Suppose that $\varphi^{\odot *}$ is in $\mathcal{D}(A^{\odot *})$ and $A^{\odot *}\varphi^{\odot *} = \psi^{\odot *}$. Set $\iota^*(\varphi^{\odot *}) = (\alpha,\beta,\xi)$, $\iota^*(\psi^{\odot *}) = (\sigma,\tau,\zeta)$. Let $\varphi^{\odot} := \iota(a,b,g)$ be in $\mathcal{D}(A^{\odot})$, where (a,b,g) is given as in Proposition 5.18. Then the relation $\langle A^{\odot}\varphi^{\odot},\varphi^{\odot *}\rangle = \langle\varphi^{\odot},\psi^{\odot *}\rangle$ is written as

$$\langle\iota^{-1}A^{\odot}\iota(a,b,g),(\alpha,\beta,\xi)\rangle = \langle(a,b,g),(\sigma,\tau,\zeta)\rangle.$$

Using Relation (5.26), we have that

$$(-\gamma a-bP)\alpha + \{bQ(-\infty) + \int_{-\infty}^{0} h(s)ds\}\beta - \int_{-\infty}^{0} [h(s)-\gamma g(s)]\xi(s)ds$$
$$= a\sigma - b\tau - \int_{-\infty}^{0} g(s)\zeta(s)ds.$$

Arranging the terms together with the substitution g as in Formula (5.24), we have that

$$(5.30)\quad \int_{-\infty}^{0} h(s)[\beta-\xi(s)]ds + \int_{-\infty}^{0}\{b[Q(s)-Q(-\infty)] - \int_{-\infty}^{s} h(t)dt\}(\gamma\xi(s)+\zeta(s))ds$$
$$= a(\gamma\alpha+\sigma) - b[\tau - P\alpha + Q(-\infty)\beta].$$

At first, take $b = 0$. Then Condition (5.25) reduces to the condition

$$(5.31)\quad \int_{-\infty}^{0}\left|\int_{-\infty}^{\theta} h(s)ds\right|d\theta < \infty.$$

Hence we can set $h = 0$ in (5.30). Then we have $a(\gamma\alpha+\sigma) = 0$; since a is arbitrary in E, $\sigma = -\gamma\alpha$. If h satisfies Condition (5.31), we obtain

$$\int_{-\infty}^{0}\int_{-\infty}^{s} h(t)dt(\gamma\xi(s)+\zeta(s))ds = \int_{-\infty}^{0} h(t)\int_{t}^{0}(\gamma\xi(s)+\zeta(s))dsdt.$$

Condition (5.30) now becomes

$$\int_{-\infty}^{0} h(s)[\beta-\xi(s)-\int_{s}^{0}(\gamma\xi(t)+\zeta(t))dt]ds = 0.$$

Since every $h \in C_{00}$ satisfies condition (5.31), it follows that

$$\xi(s) = \beta-\int_{s}^{0}(\gamma\xi(t)+\zeta(t))dt \quad \text{a.e. in} \quad (-\infty,0].$$

From this we can assume that ξ is bounded, globally Lipschitz continuous on $(-\infty,0]$, $\xi(0) = \beta$ and $\dot{\xi} = \gamma\xi+\zeta$, or $\zeta = -\gamma\xi+\dot{\xi}$.

Using these results, we can write Condition (5.30) as

$$(5.32) \qquad \int_{-\infty}^{0} h(s)[\beta-\xi(s)]ds+\int_{-\infty}^{0} \{b[Q(s)-Q(-\infty)]-\int_{-\infty}^{s} h(t)dt\}\dot{\xi}(s)ds$$

$$= -b(\tau-P\alpha+Q(-\infty)\beta).$$

Let us compute the integrals in the left hand side. Since $h \in \mathbb{L}^1$ and $\dot{\xi} \in \mathbb{L}^\infty$, from the change of the order of integration we have that, for $r > 0$,

$$\int_{-r}^{0}\int_{-\infty}^{s} h(t)dt\dot{\xi}(s)ds = \int_{-\infty}^{-r} h(t)[\beta-\xi(-r)]dt+\int_{-r}^{0} h(t)[\beta-\xi(t)]dt.$$

Since the integrals in the right hand side converge as $r \to \infty$, we obtain

$$\int_{-\infty}^{0}\int_{-\infty}^{s} h(t)dt\dot{\xi}(s)ds = \int_{-\infty}^{0} h(t)[\beta-\xi(t)]dt.$$

From integration by parts we get

$$\int_{-\infty}^{0} b[Q(s)-Q(-\infty)]\dot{\xi}(s)ds = -bQ(-\infty)\beta-\int_{-\infty}^{0} bdQ(s)\xi(s).$$

Consequently, we can write Relation (5.32) as

$$b[\tau-P\alpha-\int_{-\infty}^{0} dQ(s)\xi(s)] = 0.$$

Since b is arbitrary in E^*, we obtain

$$\tau = P\alpha+\int_{-\infty}^{0} dQ(s)\xi(s).$$

It is not difficult to prove the inverse direction.

Let BC denote the space of bounded, continuous functions on $(-\infty,0]$, $BU = \{\xi \in BC : \xi$ is uniformly continuous on $(-\infty,0]\}$ and $BL = \{\xi \in BC : \xi$ is Lipschitz continuous on $(-\infty,0]\}$. The space C_γ has been identified with the subset $\{(\alpha_\gamma(\varphi),\varphi(0),\jmath(\varphi)) : \varphi \in C_\gamma\}$ of the space $\mathbb{E}^2 \times \mathbb{L}^\infty$. It is a proper subset of $\{(\alpha,\xi(0),\xi) : \alpha \in \mathbb{E},\ \xi \in BU\}$, which is isomorphic to the space $\mathbb{E} \times BU$ with the norm

$$|(\alpha,\xi)| = \max\ [|\alpha|,\ \sup\{\ |\xi(\theta)| : -\infty < \theta \leq 0\}].$$

Proof of Theorem 5.11. From Proposition 5.20 we have that

$$\imath^*(C_\gamma^{\odot\odot}) = \text{the closure of } \{(\alpha,\xi(0),\xi) : \alpha \in \mathbb{E},\ \xi \in BL\}$$

on the space of $\mathbb{E}^2 \times \mathbb{L}^\infty$, which implies that $\imath^*(C_\gamma^{\odot\odot})$ is contained, at least, by $\{(\alpha,\xi(0),\xi) : \alpha \in \mathbb{E},\ \xi \in BU\}$. However, for all $\alpha \in \mathbb{E}$, and for all $\xi \in BU$, it holds that

$$|\imath^* S^{\odot *}(t)(\imath^*)^{-1}(\alpha,\xi(0),\xi) - (\alpha,\xi(0),\xi)| \to 0 \quad \text{as} \quad t \to 0^+.$$

In view of Definition 5.2, we therefore have that

$$\imath^*(C_\gamma^{\odot\odot}) = \{(\alpha,\xi(0),\xi) : \alpha \in \mathbb{E},\ \xi \in BU\}.$$

For $\imath(a,b,g) \in C_\gamma^{\odot}$, and for $(\alpha,\varphi) \in \mathbb{E} \times UC_\gamma$, define

$$\langle \imath(a,b,g),(\alpha,\varphi)\rangle = a\alpha - b\varphi(0) - \int_{-\infty}^{0} g(\theta)e^{\gamma\theta}\varphi(\theta)d\theta.$$

Then, we can regard that that (α,φ) is an element of $C_\gamma^{\odot *}$. If we set $\jmath(\varphi)(\theta) = e^{\gamma\theta}\varphi(\theta)$, then the right hand side of this relation is written as $\langle (a,b,g),(\alpha,\varphi(0),\jmath(\varphi))\rangle$. Since (a,b,g) is taken in $(\mathbb{E}^*)^2 \times \mathbb{L}^1$ arbitrarily, we see that $\imath^*(\alpha,\varphi) = (\alpha,\varphi(0),\jmath(\varphi))$ for $(\alpha,\varphi) \in \mathbb{E} \times UC_\gamma$. Furthermore, it is clear that $\imath^*(\mathbb{E} \times UC_\gamma) = \{(\alpha,\xi(0),\xi) : \alpha \in \mathbb{E},\ \xi \in BU\}$. Thus we can regard that $C_\gamma^{\odot\odot} = \mathbb{E} \times UC_\gamma$.

Proof of Theorem 5.12. From Proposition 5.20, $\imath^*(\alpha,\beta,\xi)$ is in $\mathcal{D}(A^{\odot\odot})$ if and only if ξ and $\dot{\xi}$ are in BU, $\beta = \xi(0)$, and

$$\dot{\xi}(0)-\gamma\xi(0) = P\alpha+\int_{-\infty}^{0} dQ(\theta)\xi(\theta).$$

From the identification of $C_{\gamma}^{\odot\odot}$ with $\mathbb{E} \times UC_{\gamma}$, we read this condition as follows. An element $(\alpha,\varphi) \in \mathbb{E} \times UC_{\gamma}$ is in $\mathcal{D}(A^{\odot\odot})$ if and only if φ is continuously differentiable, $\dot{\varphi}$ is in UC_{γ}, and

$$\dot{\varphi}(0) = P\alpha+\int_{-\infty}^{0} dQ(\theta)e^{\gamma\theta}\varphi(\theta).$$

Furthermore, $A^{\odot\odot}(\alpha,\varphi) = (-\gamma\alpha,\dot{\varphi})$ for $\varphi \in \mathcal{D}(A^{\odot\odot})$. Hence $A^{\odot\odot}$ coincides with the infinitesimal generator of the solution semigroup of Equation (5.10).

In the case $\mathcal{B} = C_{\gamma}$ or $C([-r,0]) \times L^{p}(g)$, we can construct the formal adjoint theory through the well defined bilinear form defined on the product space of an appropriate space and the space $\mathcal{B}$. As a consequence of this consideration, we can more explicitly represent the projection operator associated with the decomposition of the space $\mathcal{B}$ discussed in Section 5.3. See Naito [2,3] for the detail.

CHAPTER 6.

LINEAR PERIODIC SYSTEMS

6.1. Periodic families of bounded linear operators. Let X be a Banach space, and let $T(t,s): X \to X$, $t \ge s$, be a family of bounded linear operators satisfying the following properties:

(i) $T(s,s) = I$, the identity operator on X, for all $s \in \mathbb{R}$;

(ii) $T(t,u)T(u,s) = T(t,s)$ for all $t \ge u \ge s$;

(iii) there exists a constant $\omega > 0$ such that for any $t \ge s$,

$$T(t+\omega, s+\omega) = T(t,s);$$

(iv) there exists a constant $\theta > 0$ such that $|T(t,s)| \le \theta$ for all $0 \le s \le \omega$ and $s \le t \le s+\omega$.

The above family $T(t,s)$, $t \ge s$, is called an ω-periodic family of bounded linear operators on X, or an ω-periodic family on X, cf. Stech [1].

For any $s \in \mathbb{R}$, set

$$P(s) = T(s+\omega, s).$$

$P(s)$ is called the period map. From Properties (ii) and (iii), it follows that $P^k(s) = T(s+k\omega, s)$, and hence

$$T(t,s)P^k(s) = P^k(t)T(t,s) \tag{1.1}$$

for all $t \ge s$ and $k = 1,2,\cdots$. Moreover, $P(s)$ is ω-periodic; $P(s+\omega) = P(s)$ for all $s \in \mathbb{R}$.

Now, let $\mu \ne 0$ be an eigenvalue of $P(s)$, and φ be an eigenvector associated with μ; $P(s)\varphi = \mu\varphi$. Then, (1.1) implies that $[P(t)-\mu I]T(t,s)\varphi = T(t,s)[P(s)-\mu I]\varphi = 0$ for $t \ge s$. Note that $T(t,s)\varphi \ne 0$ for $t \ge s$. Indeed, if $T(t,s)\varphi = 0$, then we may choose a nonnegative integer m so that $s + m\omega \ge t$, and show that $0 = T(s+m\omega,t)T(t,s)\varphi = T(s+m\omega,s)\varphi = P^m(s)\varphi = \mu^m\varphi$. This contradicts the fact that both μ and φ are nonzero. Thus, if $t \ge s$, $T(t,s)$ maps eigenvectors of $P(s)$ into eigenvectors of $P(t)$, and any nonzero eigenvalue of $P(s)$ is also an eigenvalue of $P(t)$. Conversely, if

μ is a nonzero eigenvalue of $P(t)$, then it is also an eigenvalue of $P(s)$, because μ is an eigenvalue of $P(s+m\omega) = P(s)$ for all m such that $s + m\omega > t$ as shown above. Thus, the nonzero point spectrum of $P(s)$ is independent of $s \in \mathbb{R}$, and the null spaces $\mathcal{N}(P(s)-\mu I)$, $s \in \mathbb{R}$, are all of the same (perhaps infinite) dimension.

A complex number $\mu \neq 0$ is said to be a characteristic multiplier of the ω-periodic family of $T(t,s)$, if it is a normal eigenvalue of $P(s)$ for all s. A complex number λ for which $e^{\lambda\omega}$ is a characteristic multiplier of $T(t,s)$ is called a characteristic exponent of $T(t,s)$.

Let μ be a characteristic multiplier of $T(t,s)$, and $\{\varphi_1, \cdots, \varphi_d\}$ be a basis for the generalized eigenspace $\mathcal{N}((P(0)-\mu I)^k)$. Define the basis vector $\Phi = (\varphi_1, \cdots, \varphi_d)$.

<u>Theorem 1.1 (Stech [1])</u>. Let $T(t,s)$, μ, k and Φ be as above. Then, there exist a $d \times d$ constant matrix G and a d-row vector $\Phi(t)$ in X such that $\sigma(e^{G\omega}) = \{\mu\}$, $\Phi(0) = \Phi$ and $\Phi(t+\omega) = \Phi(t)$ for all $t \in \mathbb{R}$, and

$$(1.2) \qquad T(t,s)\Phi(s) = \Phi(t)e^{G(t-s)} \quad \text{for } t \geq s.$$

If b is any d-vector, then $T(t,s)\Phi(s)b$ can be defined for all $t \in \mathbb{R}$ by the relation

$$T(t,s)\Phi(s)b = \Phi(t)e^{G(t-s)}b.$$

Moreover, the generalized eigenspace of $P(t)$ for μ is of the form $\mathcal{N}((P(t)-\mu I)^k)$ for the common number k, its dimension is independent of $t \in \mathbb{R}$, and $\Phi(t)$ defines a basis of $\mathcal{N}((P(t)-\mu I)^k)$.

<u>Proof.</u> Since μ is a normal eigenvalue of $P(0)$, we can obtain the decomposition of X as $X = \mathcal{N} \oplus \mathcal{R}$ with $\sigma(P(0)|_{\mathcal{N}}) = \{\mu\}$ and $\sigma(P(0)|_{\mathcal{R}}) = \sigma(P(0)) \setminus \{\mu\}$ by Corollary 2.2 in Appendix, where $\mathcal{N} = \mathcal{N}((P(0)-\mu I)^k)$ and $\mathcal{R} = \mathcal{R}((P(0)-\mu I)^k)$. Since $P(0)\mathcal{N} \subset \mathcal{N}$, there is a $d \times d$ matrix M such that $P(0)\Phi = \Phi M$. The spectrum of M is exactly $\{\mu\}$, because $\sigma(P(0)|_{\mathcal{N}}) = \{\mu\}$. Thus, there exists a $d \times d$ matrix G such that $M = e^{G\omega}$. Set $\Phi(t) = T(t,0)\Phi e^{-Gt}$ for $t \geq 0$. Then

$$\Phi(t+\omega) = T(t+\omega,0)\Phi e^{-G(t+\omega)}$$

$$= T(t,0)T(\omega,0)\Phi e^{-G\omega}\, e^{-Gt}$$

$$= T(t,0)\Phi e^{G\omega} e^{-G\omega} e^{-Gt}$$

$$= \Phi(t)$$

for $t \geq 0$. Therefore, we can extend $\Phi(t)$ for $t < 0$ by defining $\Phi(t) = \Phi(t+m\omega)$ for any m such that $m\omega + t \geq 0$. Clearly, $\Phi(t)$ is ω-periodic in $t \in \mathbb{R}$. Let $t \geq s$, and choose an integer m so that $m\omega + s \geq 0$. Then

$$T(t,s)\Phi(s) = T(t+m\omega, s+m\omega)\Phi(s+m\omega)$$

$$= T(t+m\omega, s+m\omega)T(s+m\omega, 0)\Phi e^{-G(s+m\omega)}$$

$$= T(t+m\omega, 0)\Phi e^{-G(t+m\omega)} e^{G(t-s)}$$

$$= \Phi(t+m\omega)e^{G(t-s)}$$

$$= \Phi(t)e^{G(t-s)},$$

which proves Relation (1.2).

Notice that

$$(1.3) \qquad T(t,s)\mathcal{N}((P(s)-\mu I)^{j}) = \mathcal{N}((P(t)-\mu I)^{j}), \quad t \geq s,$$

for $j = 1, 2, \cdots$. Indeed, the above relation for $j = 1$ has already been proved. The general case can easily be proved by the induction on j. Relation (1.3) shows that $T(t,s)$, $t \geq s$, maps the generalized eigenspace of $P(s)$ for μ onto the generalized eigenspace of $P(t)$ for μ. We claim that the restriction of $T(t,s)$, $t \geq s$, to the generalized eigenspace of $P(s)$ for μ is injective. Indeed, if φ is an element of the generalized eigenspace of $P(s)$ for μ and $T(t,s)\varphi = 0$ for some $t \geq s$ with $t \leq s + m\omega$ for an integer m, then $(P(s)-\mu I)^{k}\varphi = 0$ for some integer k, and moreover $P^{m}(s)\varphi = T(s+m\omega, s)\varphi = T(s+m\omega, t)T(t,s)\varphi = 0$. Since the polynomials $(x-\mu)^{k}$ and x^{m} are prime to each other, there exist two polynomials $g(x)$ and $h(x)$ such that $g(x)(x-\mu)^{k} + h(x)x^{m} = 1$; hence $\varphi = g(P(s))(P(s)-\mu I)^{k}\varphi + h(P(s))P^{m}(s)\varphi = 0$, which proves the claim. Combining this claim with Relation (1.3), we see that the generalized eigenspace of $P(t)$ for μ is equal to $\mathcal{N}((P(t)-\mu I)^{k})$ with the same number k as the one for $P(0)$, and that $\Phi(t)$ defines a basis of $\mathcal{N}((P(t)-\mu I)^{k})$.

Let μ be a characteristic multiplier of $T(t,s)$, $t \geq s$. Since $\mathcal{R}((P(t)-\mu I)^k)^\perp = \mathcal{N}((P^*(t)-\mu I)^k)$ (see Rudin [2, Theorem 4.12]), by Lemma 5.3.6 there exists uniquely a basis vector $\Psi(t) = \mathrm{col}(\psi_1(t),\cdots,\psi_d(t))$, $t \in \mathbb{R}$, for $\mathcal{N}((P^*(t)-\mu I)^k)$ such that $\langle\Psi(t),\Phi(t)\rangle = I_d$, the $d \times d$ unit matrix.

<u>Theorem 1.2.</u> Let $\Phi(t)$, $\Psi(t)$ and G be as above. Then $\Psi(t)$ is ω-periodic in t, and

$$(1.4) \qquad T^*(t,s)\Psi(t) = e^{G(t-s)}\Psi(s) \quad \text{for } t \geq s.$$

<u>Proof</u>. Since $\langle\Psi(t),\Phi(t+\omega)\rangle = \langle\Psi(t),\Phi(t)\rangle = I_d$ and $\Psi(t) \subset \mathcal{N}((P^*(t)-\mu I)^k) = \mathcal{N}((P^*(t+\omega)-\mu I)^k)$, the uniqueness of $\Psi(t+\omega)$ implies that $\Psi(t) = \Psi(t+\omega)$. Thus, $\Psi(t)$ is ω-periodic.

Now, by Relation (1.2) we have

$$I_d = \langle\Psi(t),\Phi(t)\rangle = \langle\Psi(t),T(t,s)\Phi(s)e^{-G(t-s)}\rangle$$

$$= \langle T^*(t,s)\Psi(t),\Phi(s)\rangle e^{-G(t-s)},$$

and hence

$$I_d = e^{-G(t-s)}\langle T^*(t,s)\Psi(t),\Phi(s)\rangle$$

$$= \langle e^{-G(t-s)}T^*(t,s)\Psi(t),\Phi(s)\rangle \quad \text{for } t \geq s.$$

Furthermore, Relation (1.1) implies that $T^*(t,s)(P^*(t)-\mu I)^k = (P^*(s)-\mu I)^k T^*(t,s)$; hence $T^*(t,s)\Psi(t) \subset \mathcal{N}((P^*(s)-\mu I)^k)$. Thus, the uniqueness of $\Psi(s)$ implies that $e^{-G(t-s)}T^*(t,s)\Psi(t) = \Psi(s)$ for $t \geq s$, which proves Relation (1.4).

Let $\Lambda = \{\mu_1,\cdots,\mu_m\}$ be a set such that each μ_i is a characteristic multiplier of $T(t,s)$, $t \geq s$. From the preceding argument, we may decompose the space X as

$$X = E(s) \oplus F(s), \qquad s \in \mathbb{R},$$

where $E(s) = \sum_{i=1}^{m} \oplus\, \mathcal{N}((P(s)-\mu_i I)^{k_i})$, and $\mathcal{N}((P(s)-\mu_i I)^{k_i})$ is the generalized eigenspace of $P(s)$ for μ_i. Let $\Phi_i(s)$ and $\Psi_i(s)$ denote the basis vectors associated with $\mathcal{N}((P(s)-\mu_i I)^{k_i})$ and $\mathcal{N}((P^*(s)-\mu_i I)^{k_i})$, respectively, and set $\Phi(s) = (\Phi_1(s),\cdots,\Phi_m(s))$

and $\Psi(s) = \mathrm{col}(\Psi_1(s), \cdots, \Psi_m(s))$. Then, the projections $\pi_E(s)$, $\pi_F(s)$ of X onto $E(s)$, $F(s)$, respectively, are given by

$$\pi_E(s)\varphi = \Phi(s)\langle\Psi(s),\varphi\rangle$$

$$\pi_F(s)\varphi = \varphi - \Phi(s)\langle\Psi(s),\varphi\rangle, \qquad \varphi \in X,$$

by Lemma 5.3.6. In particular, an element $\varphi \in X$ belongs to the space $F(s)$ if and only if $\langle\Psi(s),\varphi\rangle = 0$. Therefore, in virtue of (1.4), we see that

$$T(t,s)F(s) \subset F(t), \qquad t \geq s.$$

Theorem 1.3 (Stech [1]). Let $\gamma \in \mathbb{R}$, and suppose that a finite number of characteristic multipliers, $\{\mu_1, \cdots, \mu_m\}$, satisfy $|\mu_i| \geq e^{\gamma\omega}$ and $\sup\left(|\mu| : \mu \in \sigma(P(0))\setminus\{\mu_1, \cdots, \mu_m\}\right) < e^{\gamma\omega}$. Then, for sufficiently small $\varepsilon > 0$ there exists a constant $c(\varepsilon) > 0$ such that

$$|T(t,s)\varphi| \leq c(\varepsilon)e^{(\gamma-\varepsilon)(t-s)}|\varphi|$$

for any $\varphi \in F(s)$ and all $t \geq s$.

Proof. For simplicity, we write $\tilde{P}(s) = P(s)|_{F(s)}$. Since $\sigma(\tilde{P}(0)) = \sigma(P(0)) \setminus \{\mu_1, \cdots, \mu_m\}$, we can assume that the number ε is so small that $r_\sigma(\tilde{P}(0)) < e^{(\gamma-\varepsilon)\omega}$, where $r_\sigma(\tilde{P}(0))$ denotes the spectral radius of $\tilde{P}(0)$. Then Gelfand formula (see, e.g., Dunford & Schwartz [1]) implies that

$$\lim_{n\to\infty} |\tilde{P}(0)^n|^{1/n} e^{-(\gamma-\varepsilon)\omega} < \beta$$

for some number β, $0 < \beta < 1$, from which it follows that $|\tilde{P}(s)^n|e^{(\varepsilon-\gamma)n\omega} \to 0$ as $n \to \infty$ uniformly for $s \in \mathbb{R}$, because $|\tilde{P}(s)^n| \leq \theta^2|\tilde{P}(0)^{n-1}|$, $s \in \mathbb{R}$, by Properties (ii) and (iv). In particular, combining this fact with Property (iv), we obtain that

$$\sup\{|\tilde{P}(s)^n|e^{(\varepsilon-\gamma)n\omega} : n \geq 0,\ s \in \mathbb{R}\} =: b(\varepsilon) < \infty.$$

Let $s \leq t$, and choose an integer j so that $j\omega + s \leq t < (j+1)\omega + s$. Then, $T(t,s) = T(t,j\omega+s)T(j\omega+s,s) = T(t,j\omega+s)P^j(s) = T(t-j\omega,s)P^j(s)$. By Property (iv) and the preceding remark, we have

$$(1.5) \qquad |T(t,s)\varphi| \le \theta|\tilde{P}^j(s)||\varphi|$$

$$\le \theta e^{(\gamma-\varepsilon)j\omega} \max_{n\ge 0}\{|\tilde{P}(s)^n|e^{(\varepsilon-\gamma)n\omega}\}|\varphi|$$

$$\le \theta b(\varepsilon)e^{(\gamma-\varepsilon)j\omega}|\varphi|$$

for $\varphi \in F(s)$. If $\gamma-\varepsilon > 0$, then $\omega j \le t-s$ means $(\gamma-\varepsilon)j\omega \le (\gamma-\varepsilon)(t-s)$. If $\gamma-\varepsilon \le 0$, then $t-s \le \omega(j+1)$ means $(\gamma-\varepsilon)(j+1)\omega \le (\gamma-\varepsilon)(t-s)$. In either case, it follows that

$$(1.6) \qquad e^{(\gamma-\varepsilon)j\omega} \le e^{|\gamma-\varepsilon|\omega} e^{(\gamma-\varepsilon)(t-s)}.$$

From (1.5) and (1.6), we have

$$|T(t,s)\varphi| \le c(\varepsilon)e^{(\gamma-\varepsilon)(t-s)}|\varphi|$$

for $t \ge s$ and $\varphi \in F(s)$, where $c(\varepsilon) = \theta b(\varepsilon)e^{|\gamma-\varepsilon|\omega}$.

<u>6.2. The decomposition of $\hat{\mathscr{B}}$ by normal eigenvalues of the solution operator for linear periodic systems.</u> In this section, we assume the same conditions on $\mathscr{B}$ as in Chapter 5, and consider a linear periodic functional differential equation

$$(2.1) \qquad \dot{x}(t) = L(t,x_t),$$

where $L : \mathbb{R} \times \mathscr{B} \to \mathbb{C}^n$ is a continuous function such that $L(t,\varphi)$ is linear in φ and that $L(t+\omega,\varphi) = L(t,\varphi)$ for all $(t,\varphi) \in \mathbb{R} \times \mathscr{B}$ and for some constant $\omega > 0$. In the following, we shall extend the results in Section 5.3 to the periodic case.

Let $T(t,s)$, $t \ge s$, be the solution operator defined by Equation (2.1). Since $L(t,\varphi)$ is ω-periodic in t, we can select a constant as a function $n(t)$ arising in Lemma 4.1.1; hence, Theorem 4.2.1 and Properties (a), (b) and (d) in Section 4.3 imply that the family $\hat{T}(t,s)$, $t \ge s$, is an ω-periodic family of bounded linear operators on the quotient space $\hat{\mathscr{B}}$. Moreover, from Theorem 4.3.5 it follows that

$$r_e(\hat{P}(s)) = e^{\beta\omega} \quad \text{for all } s \in \mathbb{R},$$

where $\hat{P}(s) = \hat{T}(s+\omega,s)$ and β is the number defined by Relation (3.3) in Section 4.3. Hence, the following theorem follows immediately from the results in the previous section.

Theorem 2.1 (Murakami [4]). For any $\gamma > \beta$, let $\Lambda = \Lambda(\gamma) = \{\mu \in \sigma(\hat{P}(0)) : |\mu| \geq e^{\gamma\omega}\}$. Then there exist closed subspaces $\hat{E}(s)$ and $\hat{F}(s)$ of $\hat{\mathcal{B}}$, basis vectors $\hat{\Phi}(s)$ for $\hat{E}(s)$, $\Psi(s)$ associated with $\hat{\Phi}(s)$ and a $d \times d$ matrix G such that:

(i) $\hat{\mathcal{B}} = \hat{E}(s) \oplus \hat{F}(s) \qquad s \in \mathbb{R},$

and

$$\hat{T}(t,s)\hat{E}(s) = \hat{E}(t)$$

$$\hat{T}(t,s)\hat{F}(s) \subset \hat{F}(t) \qquad \text{for} \quad t \geq s;$$

(ii) $\dim \hat{E}(s) = d$ (independent of $s \in \mathbb{R}$), $\hat{E}(s+\omega) = \hat{E}(s)$, $\hat{F}(s+\omega) = \hat{F}(s)$ and the projections $\pi_E(s) : \hat{\mathcal{B}} \to \hat{E}(s)$ and $\pi_F(s) : \hat{\mathcal{B}} \to \hat{F}(s)$ are given by

$$\pi_E(s)\hat{\varphi} = \hat{\Phi}(s)\langle\Psi(s),\hat{\varphi}\rangle$$

$$\pi_F(s)\hat{\varphi} = \hat{\varphi} - \hat{\Phi}(s)\langle\Psi(s),\hat{\varphi}\rangle$$

for $\hat{\varphi} \in \hat{\mathcal{B}}$, respectively;

(iii) $\hat{T}(t,s)\hat{\Phi}(s) = \hat{\Phi}(t)e^{G(t-s)}$

$$T^*(t,s)\Psi(t) = e^{G(t-s)}\Psi(s)$$

for all $t \geq s$, and

$$\sigma(e^{G\omega}) = \Lambda(\gamma);$$

(iv) for any d-vector a, $\hat{T}(t,s)\hat{\Phi}(s)a$ with the initial value $\hat{\Phi}(s)a$ at $t = s$ may be defined on $\mathbb{R}$ by the relation

$$\hat{T}(t,s)\hat{\Phi}(s)a = \hat{\Phi}(t)e^{G(t-s)}a;$$

(v) for sufficiently small $\varepsilon > 0$, there exists a constant $c(\varepsilon) > 0$ such that

$$|\hat{T}(t,s)\hat{\varphi}| \leq c(\varepsilon)e^{(\gamma-\varepsilon)(t-s)}|\hat{\varphi}| \qquad \text{for} \quad t \leq s,\ \hat{\varphi} \in \hat{E}(s),$$

$$|\hat{T}(t,s)\hat{\varphi}| \leq c(\varepsilon)e^{(\gamma-\varepsilon)(t-s)}|\hat{\varphi}| \qquad \text{for} \quad t \geq s,\ \hat{\varphi} \in \hat{F}(s).$$

In particular, if the set $\{\mu \in \sigma(\hat{P}(0)) : |\mu| = e^{\gamma\omega}\}$ is empty, the estimate on $\hat{E}(s)$ can be strengthened;

$$|\hat{T}(t,s)\hat{\varphi}| \le c(\varepsilon)e^{(\gamma+\varepsilon)(t-s)}|\hat{\varphi}| \quad \text{for } t \le s,\ \hat{\varphi} \in \hat{E}(s).$$

The following result will be needed in the next chapter.

Proposition 2.2. Let $\hat{\Phi}(t)$, $\Psi(t)$, $\pi_E(t)$ and $\pi_F(t)$ be as in Theorem 2.1. Then $\hat{\Phi}(t)$ is continuous in $t \in \mathbb{R}$ and $\Psi(t)$ is w^*-continuous in $t \in \mathbb{R}$. In particular, $\sup_{t\in\mathbb{R}} |\pi_E(t)| < \infty$ and $\sup_{t\in\mathbb{R}} |\pi_F(t)| < \infty$.

Proof. The continuity of $\hat{\Phi}(t)$ follows immediately from the ω-periodicity of $\hat{\Phi}(t)$ and (iii) in Theorem 2.1. On the other hand, if $t \to s$, $t \in [-\omega,0]$, then from (iii) in Theorem 2.1 and the continuous dependence on initial values of solutions of Equation (2.1), it follows that $\langle\Psi(t),\hat{\varphi}\rangle = \langle e^{Gt}T^*(0,t)\Psi(0),\hat{\varphi}\rangle = e^{Gt}\langle\Psi(0),T(0,t)\varphi\rangle \to e^{Gs}\langle\Psi(0),T(0,s)\varphi\rangle = \langle e^{Gs}T^*(0,s)\Psi(0),\hat{\varphi}\rangle = \langle\Psi(s),\hat{\varphi}\rangle$. Thus $\langle\Psi(t),\hat{\varphi}\rangle$ is continuous in $t \in \mathbb{R}$ for each $\hat{\varphi} \in \hat{\mathscr{B}}$, since $\langle\Psi(t),\hat{\varphi}\rangle$ is ω-periodic. This proves the w^*-continuity of $\Psi(t)$ in $t \in \mathbb{R}$. The remainder of the assertions follows from (ii) in Theorem 2.1 and the Banach-Steinhaus theorem.

Combining Theorem 4.2.9 with (ii) and (iv) in Theorem 2.1, we obtain a Floquet type theorem.

Corollary 2.3. Suppose $\operatorname{Re}\lambda > \beta$. Then $e^{\lambda\omega}$ is in $\sigma(\hat{P}(0))$ if and only if there exists a nonzero solution of Equation (2.1) defined on $\mathbb{R}$ of the form $x(t) = p(t)e^{\lambda t}$, $t \in \mathbb{R}$, where $p(t+\omega) = p(t)$ on $\mathbb{R}$.

Let $\beta < 0$. Equation (2.1) is said to be nonsingular if $\{\lambda \in \sigma(\hat{P}(0)) : |\lambda| = 1\}$ is empty. The following results follow immediately from Theorem 2.1 with $\gamma = 0$.

Corollary 2.4. Suppose $\beta < 0$ and that Equation (2.1) is nonsingular. Then there exist some positive constants ℓ and ν such that

$$|\hat{T}(t,s)\hat{\varphi}| \le \ell e^{\nu(t-s)}|\hat{\varphi}|, \quad t \le s,\ \hat{\varphi} \in \hat{E}(s),$$

$$|\hat{T}(t,s)\hat{\varphi}| \le \ell e^{-\nu(t-s)}|\hat{\varphi}|, \quad t \ge s,\ \hat{\varphi} \in \hat{F}(s).$$

In other words, Equation (2.1) has an exponential dichotomy.

Corollary 2.5. Suppose $\beta < 0$ and that the set $\{\lambda \in \sigma(\hat{P}(0)) : |\lambda| \geq 1\}$ is empty. Then there exist some positive constants ℓ and ν such that

$$|T(t,s)\varphi| \leq \ell e^{-\nu(t-s)}|\varphi|$$

for $t \geq s$ and $\varphi \in \mathscr{B}$.

Now, consider a linear functional differential equation

$$\dot{x}(t) = L(t,x_t) + h(t), \tag{2.2}$$

where $L : \mathbb{R} \times \mathscr{B} \to \mathbb{C}^n$ is an ω-periodic continuous function such that $L(t,\varphi)$ is linear in φ, and $h : \mathbb{R} \to \mathbb{C}^n$ is locally integrable. Let $\hat{\mathscr{B}} = \hat{E}(s) \oplus \hat{F}(s)$, $s \in \mathbb{R}$, be the decomposition of $\hat{\mathscr{B}}$ ensured in Theorem 2.1 for a set $\Lambda = \Lambda(\gamma)$. The projection $\pi_E(s) : \hat{\mathscr{B}} \to \hat{E}(s)$ is given by $\pi_E(s)\hat{\varphi} = \hat{\Phi}(s)\langle\Psi(s),\hat{\varphi}\rangle$ for $\hat{\varphi} \in \hat{\mathscr{B}}$, and it satisfies

$$\pi_E^{**}(s)\varphi^{**} = \hat{\Phi}(s)\langle\Psi(s),\varphi^{**}\rangle \quad \text{for} \quad \varphi^{**} \in \mathscr{B}^{**};$$

see Relations (3.5′) and (3.7′) in Section 5.3. Then,

$$\begin{aligned} T^{**}(t,s)\pi_E^{**}(s)\Gamma &= \hat{T}(t,s)\hat{\Phi}(s)\langle\Psi(s),\Gamma\rangle \\ &= -\hat{\Phi}(t)e^{G(t-s)}[\tilde{\Psi}(s)](0^-), \end{aligned} \tag{2.3}$$

where Γ is the one introduced in Section 4.2. Moreover, we have

$$e^{G(t-s)}[\tilde{\Psi}(s)](0^-) = y(s,t,\tilde{\Psi}(t)) \quad \text{for} \quad s < t. \tag{2.4}$$

Indeed, if $s < t$, then

$$\begin{aligned} y(s,t,\tilde{\Psi}(t)) = [T^*(t,s)\Psi(t)]^\sim(0^-) &= [e^{G(t-s)}\Psi(s)]^\sim(0^-) \\ &= e^{G(t-s)}[\tilde{\Psi}(s)](0^-) \end{aligned}$$

by Theorem 4.2.2 and (iii) in Theorem 2.1. Since $y(s,t,\tilde{\Psi}(t))$ is locally of bounded variation for s on $(-\infty,t]$, the function $\hat{\Phi}(t)e^{G(t-s)}[\Psi(s)]^\sim(0^-)h(s) = \hat{\Phi}(t)y(s,t,\tilde{\Psi}(t))h(s)$ is locally (Bochner) integrable for s on $(-\infty,t]$. Thus, Lemma 4.2.6 and

Theorem 4.2.8 lead to the following results.

Theorem 2.6. Let $\hat{\mathscr{B}} = \hat{E}(t) \oplus \hat{F}(t)$, $\hat{\Phi}(t)$, $\Psi(t)$ and G be the ones ensured in Theorem 2.1 for a set $\Lambda = \Lambda(\gamma)$. Then, $\hat{E}(t)$-component $\xi_E(t)$ and $\hat{F}(t)$-component $\xi_F(t)$ of $\hat{x}_t(\sigma,\varphi,h)$, respectively, satisfy the equations

$$(2.5) \qquad \xi_E(t) = \hat{\Phi}(t)e^{G(t-\sigma)}\langle\Psi(\sigma),\xi_E(\sigma)\rangle - \int_\sigma^t \hat{\Phi}(t)e^{G(t-s)}[\tilde{\Psi}(s)](0^-)h(s)ds$$

and

$$(2.6) \qquad \xi_F(t) = \hat{T}(t,\sigma)\xi_F(\sigma) + {}_*\!\int_\sigma^t T^{**}(t,s)\pi_F^{**}(s)\Gamma h(s)ds$$

for $t \geq \sigma$.

Corollary 2.7. The coordinate $u(t) = \langle\Psi(t),\xi_E(t)\rangle$ of $\xi_E(t)$ with respect to $\hat{\Phi}(t)$ satisfies the ordinary differential equation

$$(2.7) \qquad \dot{u}(t) = Gu(t) - [\tilde{\Psi}(t)](0^-)h(t) \quad \text{a.e. on } \mathbb{R}.$$

CHAPTER 7

FADING MEMORY SPACES AND FUNCTIONAL DIFFERENTIAL EQUATIONS

7.1. Fading memory spaces. In this section, we introduce the concept of fading memory spaces. The concept will be needed when we discuss the stability and the existence of periodic solutions and almost periodic solutions, later. We will start the explanation of the space by considering the following axiom on $\mathcal{B}$. Recall that C_{00}, the space of all continuous functions on $(-\infty,0]$ into E with compact support, is a subset of the space $\mathcal{B}$ (see Proposition 1.2.1).

(C2) If a uniformly bounded sequence $\{\varphi^k(\theta)\}$ in C_{00} converges to a function $\varphi(\theta)$ compactly on $(-\infty,0]$, then φ is in $\mathcal{B}$ and $|\varphi^k - \varphi|_{\mathcal{B}} \to 0$ as $k \to \infty$.

The following result shows that if $\mathcal{B}$ satisfies Axiom (C2), then the Banach space BC is contained in $\mathcal{B}$ and it is continuously imbedded in $\mathcal{B}$; hence, for any function x arising in Axiom (A) with $x_\sigma \in BC$, the function $x_t \in BC$ is continuous in t with respect to the seminorm $|\cdot|_{\mathcal{B}}$, whereas it is not necessarily continuous in t with respect to the norm $|\cdot|_{BC}$.

Proposition 1.1. Suppose the space $\mathcal{B}$ satisfies Axiom (C2). Then the following hold:

(i) $BC \subset \overline{C_{00}}$, where $\overline{C_{00}}$ denotes the closure of C_{00} in $\mathcal{B}$.

(ii) If a uniformly bounded sequence $\{\varphi^k(\theta)\}$ in BC converges to a function $\varphi(\theta)$ compactly on $(-\infty,0]$, then φ is in $\mathcal{B}$ and $|\varphi^k - \varphi|_{\mathcal{B}} \to 0$ as $k \to \infty$.

(iii) $|\varphi|_{\mathcal{B}} \le L\cdot|\varphi|_{BC}$, $\varphi \in BC$

for some constant $L > 0$.

Proof. Let $\varphi \in BC$. For $m = 1,2,\cdots$, we define a $\varphi^m \in C_{00}$ by

$$(1.1)\qquad \varphi^m(\theta) = \begin{cases} \varphi(\theta) & \text{if } -m \le \theta \le 0, \\ 0 & \text{if } \theta < -m-1, \\ \text{linear} & \text{if } -m-1 \le \theta \le -m. \end{cases}$$

Clearly, the sequence $\{\varphi^m(\theta)\}$ converges to the function $\varphi(\theta)$ as $m \to \infty$ compactly on $(-\infty,0]$, and $|\varphi^m|_{BC_m} = \sup_{-m\le\theta\le 0} |\varphi(\theta)| \le |\varphi|_{BC}$ for $m = 1,2,\cdots$. Hence, $\varphi \in \mathscr{B}$ and $|\varphi^m - \varphi|_{\mathscr{B}} \to 0$ as $m \to \infty$ by Axiom (C2). This proves the assertion (i).

Next, we prove the assertion (ii). For each φ^k and $m = 1,2,\cdots$, consider the function $\varphi^{k,m} \in C_{00}$ defined by (1.1) with $\varphi = \varphi^k$. Since $|\varphi^{k,m} - \varphi^k|_{\mathscr{B}} \to 0$ as $m \to \infty$ as shown in the proof of (i), we can choose an integer $m(k)$, $m(k) \ge k$, so that

$$(1.2)\qquad |\varphi^{k,m(k)} - \varphi^k|_{\mathscr{B}} < 1/k$$

for $k = 1,2,\cdots$. Clearly, the sequence $\{\varphi^{k,m(k)}(\theta)\}$ converges to the function $\varphi(\theta)$ as $k \to \infty$ compactly on $(-\infty,0]$, and $\sup_k|\varphi^{k,m(k)}|_{BC} \le \sup_k|\varphi^k|_{BC} < \infty$. Then, Axiom (C2) implies that $\varphi \in \mathscr{B}$ and

$$(1.3)\qquad |\varphi^{k,m(k)} - \varphi|_{\mathscr{B}} \to 0 \quad \text{as} \quad k \to \infty.$$

Hence, we have that $|\varphi^k - \varphi|_{\mathscr{B}} \to 0$ as $k \to \infty$ by (1.2) and (1.3), which is the desired result.

Finally, we shall prove the assertion (iii). First, we claim that there exists a constant $L > 0$ such that $|\varphi|_{\mathscr{B}} \le L$ whenever $\varphi \in BC$ and $|\varphi|_{BC} = 1$. If the claim is true, we have $|\varphi|_{\mathscr{B}} = |(1/|\varphi|_{BC})\varphi|_{\mathscr{B}}|\varphi|_{BC} \le L|\varphi|_{BC}$ for $\varphi \in BC$, which proves (iii). Now, to prove the claim by a reduction to absurdity, we suppose that there exists a sequence $\{\varphi^k\}$ in BC such that $|\varphi^k|_{BC} = 1$ and $|\varphi^k|_{\mathscr{B}} \ge k$ for $k = 1,2,\cdots$. Since $|(1/k)\varphi^k|_{BC} = 1/k$, from (ii) of this proposition it follows that $|(1/k)\varphi^k|_{\mathscr{B}} \to 0$ as $k \to \infty$, whereas $|(1/k)\varphi^k|_{\mathscr{B}} \ge 1$ for $k = 1,2,\cdots$. This is a contradiction.

In what follows, we always assume that $\mathscr{B}$ satisfies Axiom (C2). By Proposition 1.1, $\mathscr{B}$ possesses the following property: if $x : (-\infty,\infty) \to \mathbb{E}$ is a bounded continuous function such that $x(t) \to 0$ as $t \to \infty$, then $|x_t|_{\mathscr{B}} \to 0$ as $t \to \infty$. Does this property hold in case that x is not bounded on $(-\infty,0]$? To consider this problem, we introduce the operator $S_0(t)$ as the restriction of the operator $S(t)$ to the closed subspace

$$\mathscr{B}_0 = \{\varphi \in \mathscr{B} : \varphi(0) = 0\}.$$

Clearly, the family $S_0(t)$, $t \ge 0$, is a strongly continuous semigroup of bounded linear operators on $\mathscr{B}_0$; it is given explicitly as

$$[S_0(t)\varphi](\theta) = \begin{cases} 0 & \text{if } -t \le \theta \le 0, \\ \varphi(t+\theta) & \text{if } \theta < -t \end{cases}$$

for φ in $\mathscr{B}_0$.

Definition 1.2. The space $\mathscr{B}$ is called a fading memory space if it satisfies Axioms (B) and (C2) and if

$$|S_0(t)\varphi| \to 0 \text{ as } t \to \infty \text{ for each } \varphi \in \mathscr{B}_0.$$

For any function $x : (-\infty,\infty) \to E$ such that $x_\sigma \in \mathscr{B}$ for some σ and that x is continuous on $[\sigma,\infty)$, we have the following decomposition of x_t;

$$x_t = y_t + S_0(t-\sigma)[x_\sigma - x(\sigma)\chi], \tag{1.4}$$

where

$$y(s) = \begin{cases} x(s) & s \ge \sigma \\ x(\sigma) & s < \sigma \end{cases}$$

and

$$\chi(\theta) = 1 \quad \text{for all } \theta \le 0.$$

The following result is an immediate consequence of Proposition 1.1 and the decomposition given above.

Proposition 1.3. Suppose $\mathscr{B}$ satisfies Axiom (B). Then the space $\mathscr{B}$ is a fading memory space if and only if $|x_t|_{\mathscr{B}} \to 0$ as $t \to \infty$ whenever $x : (-\infty,\infty) \to E$ satisfies the condition $x_\sigma \in \mathscr{B}$ for some σ, x is continuous on $[\sigma,\infty)$ and $x(t) \to 0$ as $t \to \infty$.

Notice that a fading memory space means that the operator $S_0(t)$ tends to the zero as $t \to \infty$ in the strong topology. Now, we shall consider the case that the operator $S_0(t)$ tends to the zero as $t \to \infty$ in the operator norm topology.

Definition 1.4. The space $\mathscr{B}$ is called a uniform fading memory space if it satisfies Axioms (B) and (C2) and if

$$|S_0(t)| \to 0 \text{ as } t \to \infty.$$

Obviously, a uniform fading memory space implies a fading memory space. Observe that $|S_0(t)| \le M(t)$ for $t \ge 0$, where $M(t)$ is the function in Axiom (A-iii). Hence, if $M(t) \to 0$ as $t \to \infty$, then $|S_0(t)| \to 0$ as $t \to \infty$. The following result shows that the converse holds in a sense. Its proof is easy by the decomposition (1.4), Proposition 1.1 and the Banach-Steinhaus theorem; so we omit it.

<u>Proposition 1.5.</u> The following statements hold:

(i) If $\mathcal{B}$ satisfies Axiom (C2), then

$$|x_t| \le L\cdot\sup_{\sigma\le s\le t}|x(s)| + (1 + LH)|S_0(t-\sigma)||x_\sigma|_{\mathcal{B}}$$

for any function x arising in Axiom (A), where L is the constant insured in (iii) of Proposition 1.1. In particular, if $\mathcal{B}$ is a fading memory space, one can choose the functions $K(\cdot)$ and $M(\cdot)$ in Axiom (A-iii) so that $K(\cdot) = K$ and $M(\cdot) = M$, constants.

(ii) If $\mathcal{B}$ satisfies Axioms (B) and (C2), then it is a uniform fading memory space if and only if one can choose functions $K(\cdot)$ and $M(\cdot)$ in Axiom (A-iii) so that $K(t) = K$, a constant, and $M(t) \to 0$ as $t \to \infty$.

<u>Example 1.6.</u> Let g be a positive continuous function on $(-\infty,0]$ satisfying Conditions (g2) and (g3) in Section 1.3, and consider the space C_g^∞ (see Section 1.3). The space C_g^∞ satisfies Axioms (B) and (C2); see Theorem 1.3.3. Moreover, for $t \ge 0$ and $\varphi \in C_g^\infty$ with $\varphi(0) = 0$, we have $|S_0(t)\varphi|_{g;\infty} = |\tilde{\varphi}(-\infty)|a_t$, where $\tilde{\varphi}(-\infty) = \lim_{s\to-\infty}\varphi(s)/g(s)$ and $a_t = \lim_{s\to-\infty} g(t+s)/g(s)$. In particular, $|S_0(t)| = a_t$ for $t \ge 0$. Therefore, if $g(s) \equiv e^{-s}$, then the space C_g^∞ is a uniform fading memory space. On the other hand, if $g(s) \equiv 1 + |s|^k$ for some $k > 0$, then the space C_g^∞ is not a fading memory space.

<u>Example 1.7.</u> Let g be a positive continuous function on $(-\infty,0]$ satisfying Conditions (g1) (with G bounded) and (g2) in Section 1.3. Then the space C_g^0 is a fading memory space. Moreover, it is a uniform fading memory space if and only if $\sup\{g(\theta+t)/g(\theta) : -\infty < \theta \le -t\} \to 0$ as $t \to \infty$ (cf. Theorem 1.4.2). In particular, the space C_γ is a uniform fading memory space if and only if $\gamma > 0$.

<u>Example 1.8.</u> Suppose that $1 \le p < \infty$, $0 \le r < \infty$ and g is a

nonnegative Borel measurable function on $(-\infty,-r)$ satisfying Conditions (g-5), (g-6) and (g-7) in Section 1.3. Then the space $C \times L^p(g)$ is a fading memory space. Moreover, it is a uniform fading memory space if and only if $\text{ess sup}\{g(\theta-t)/g(\theta): \theta \le -r\} \to 0$ as $t \to \infty$ (cf. Theorem 1.4.11).

From Example 1.7, we know that if $g(s) = 1 + |s|^k$ for some $k > 0$, then the space C_g^0 is a fading memory space, but not a uniform fading memory space. As the following result shows, however, the space UC_g (cf. Section 1.3) has no such an aspect.

<u>Proposition 1.9.</u> Let g be a positive continuous function such that $g(s) \to \infty$ as $s \to -\infty$. Then the space UC_g is a fading memory space if and only if it is a uniform fading memory space.

<u>Proof.</u> The "if" part is obvious. We shall verify the "only if" part. Suppose the space UC_g is a fading memory space. Fix a z in $\mathbb{E}$ with $|z| = 1$, and set $\varphi(\theta) = (g(\theta) - g(0))z$ for $\theta \le 0$. Then $\varphi \in UC_g$ and $\varphi(0) = 0$; hence $|S_0(t)\varphi|_g \to 0$ as $t \to \infty$ by the assumption. Thus, Theorem 1.4.2 implies that

$$\begin{aligned}\overline{\lim_{t\to\infty}}\, |S_0(t)| &= \overline{\lim_{t\to\infty}}\, [\sup_{s\le 0} g(s)/g(s-t)] \\ &\le \overline{\lim_{t\to\infty}}\, [\sup_{s\le 0} (g(s)-g(0))/g(s-t)] + \overline{\lim_{t\to\infty}}\, [\sup_{s\le 0} g(0)/g(s-t)] \\ &\le \overline{\lim_{t\to\infty}}\, |S_0(t)\varphi|_g + \overline{\lim_{t\to\infty}}\, [\sup_{s\le 0} g(0)/g(s-t)] \\ &= 0,\end{aligned}$$

which shows that the space UC_g is a uniform fading memory space.

<u>7.2. Fading memory spaces and stabilities.</u> In the previous section, we introduced fading memory spaces and uniform fading memory spaces. In this section, we shall present some characterizations for fading memory spaces or uniform fading memory spaces in connection with the stabilities in functional differential equations.

Let $\mathbb{R}^+ = [0,\infty)$, and consider a system of FDEs

$$(2.1) \qquad \dot{x}(t) = F(t, x_t),$$

where $F : \mathbb{R}^+ \times \mathcal{B} \to \mathbb{E}$ is continuous. Let u be a solution of System (2.1) on $\mathbb{R}^+$. We shall define the stabilities of the solution u. Recall that $x(\cdot,\sigma,\varphi,F)$ denotes any solution of System (2.1) through $(\sigma,\varphi) \in \mathbb{R}^+ \times \mathcal{B}$.

Definition 2.1. The solution u of System (2.1) is said to be stable in $\mathcal{B}$, if for any $\varepsilon > 0$ and $\sigma \geq 0$ there exists a $\delta := \delta(\varepsilon,\sigma) > 0$ such that $|u_\sigma - \varphi|_{\mathcal{B}} < \delta$ implies

$$(2.2) \qquad |u_t - x_t(\sigma,\varphi,F)|_{\mathcal{B}} < \varepsilon$$

for all $t \geq \sigma$. In the above, if one can choose the number δ independent of σ, then the solution u is said to be uniformly stable (US, in short) in $\mathcal{B}$.

Definition 2.2. The solution u of System (2.1) is said to be asymptotically stable (AS, in short) in $\mathcal{B}$, if it is stable in $\mathcal{B}$ and moreover, for any $\sigma \geq 0$ there exists a $\delta_0 := \delta_0(\sigma) > 0$ such that

$$(2.3) \qquad |u_t - x_t(\sigma,\varphi,F)|_{\mathcal{B}} \to 0 \quad \text{as} \quad t \to \infty$$

whenever $|u_\sigma - \varphi|_{\mathcal{B}} < \delta_0$.

Definition 2.3. The solution u of System (2.1) is said to be weakly uniformly asymptotically stable (WUAS, in short) in $\mathcal{B}$, if it is US in $\mathcal{B}$ and moreover, there exists a $\delta_0 > 0$ such that

$$(2.4) \qquad |u_t - x_t(\sigma,\varphi,F)|_{\mathcal{B}} \to 0 \quad \text{as} \quad t \to \infty$$

whenever $\sigma \geq 0$ and $|u_\sigma - \varphi|_{\mathcal{B}} < \delta_0$.

Definition 2.4. The solution u of System (2.1) is said to be uniformly asymptotically stable (UAS, in short) in $\mathcal{B}$, if it is US in $\mathcal{B}$ and moreover, there exists a $\delta_0 > 0$ with the property that for any $\varepsilon > 0$ there exists a $T = T(\varepsilon) > 0$ such that

$$(2.5) \qquad |u_t - x_t(\sigma,\varphi,F)|_{\mathcal{B}} < \varepsilon$$

for all $t \geq \sigma + T(\varepsilon)$, whenever $\sigma \geq 0$ and $|u_\sigma - \varphi|_{\mathcal{B}} < \delta_0$.

In the above definitions, if the $\mathcal{B}$-seminorm $|u_t - x_t(\sigma,\varphi,F)|_{\mathcal{B}}$

in Relations (2.2), (2.3), (2.4) and (2.5) is replaced by the $\mathbb{E}$-norm $|u(t) - x(t,\sigma,\varphi,F)|_{\mathbb{E}}$, we get the stabilities "in $\mathbb{E}$". The practical phenomena involved in the FDEs with infinite delay intimates an advantage of the adoption of the concept of the stability "in $\mathbb{E}$" rather than "in $\mathscr{B}$". Moreover, in some concrete problems, the verification of the stability in $\mathbb{E}$ is comparatively easier than that in $\mathscr{B}$. The following result is useful when one tries to establish the stability in $\mathscr{B}$.

Proposition 2.5 (Hale & Kato [1]). The following statements hold:

(i) If the solution u of System (2.1) is stable (uniformly stable) (asymptotically stable) (weakly uniformly asymptotically stable) (uniformly asymptotically stable) in $\mathscr{B}$, then so is it in $\mathbb{E}$.

(ii) Let $\mathscr{B}$ be a fading memory space. If the solution u of System (2.1) is stable (uniformly stable) (asymptotically stable) (weakly uniformly asymptotically stable) in $\mathbb{E}$, then so is it in $\mathscr{B}$.

(iii) Let $\mathscr{B}$ be a uniform fading memory space. If the solution u of System (2.1) is uniformly asymptotically stable in $\mathbb{E}$, then so is it in $\mathscr{B}$.

Proof. The claim (i) is obvious by Axiom (A-ii). Also, the claim (ii) is a direct consequence of Propositions 1.3 and 1.5. We shall verify the claim (iii). Let $\mathscr{B}$ be a uniform fading memory space, and suppose the solution u is UAS in $\mathbb{E}$. By (ii), the solution u is US in $\mathscr{B}$. Now, let $\varepsilon > 0$ be given, and let $|u_\sigma - \varphi|_{\mathscr{B}} < \min(\delta_0, \delta(1))$. Since u is UAS in $\mathbb{E}$, we have

$$|u(t) - x(t,\sigma,\varphi,F)| < \varepsilon \quad \text{for all} \quad t \ge \sigma + T(\varepsilon),$$

and moreover, Proposition 1.5 implies

$$|u_t - x_t(\sigma,\varphi,F)|_{\mathscr{B}} \le K + M\delta_0 =: C \quad \text{for all} \quad t \ge \sigma.$$

Choose a $\tau > 0$ so that $|S_0(\tau)| < \varepsilon/C$, which is possible because $\mathscr{B}$ is a uniform fading memory space. Then, by Proposition 1.5 we have that, for all $t \ge \sigma+\tau+T(\varepsilon)$,

$$|u_t - x_t(\sigma,\varphi,F)|_{\mathscr{B}} \le L \cdot \sup_{t-\tau \le s \le t} |u(s) - x(s,\sigma,\varphi,F)|$$
$$+ (1+HL)|S_0(\tau)|\,|u_{t-\tau} - x_{t-\tau}(\sigma,\varphi,F)|_{\mathscr{B}}$$

$$\leq (1+L+HL)\varepsilon.$$

This proves the claim (iii).

In the statement (ii) (resp. (iii)) of the above proposition, we cannot generally remove the assumption that $\mathcal{B}$ is a fading memory space (resp. a uniform fading memory space). The following result shows the fact, and moreover, it provides necessary and sufficient conditions for the space $\mathcal{B}$ to be a fading memory space or a uniform fading memory space.

Theorem 2.6 (Murakami & Naito [1,2]). Suppose the space $\mathcal{B}$ satisfies Axioms (B) and (C2). Then the following statements hold:

(i) (Characterizations for uniform fading memory spaces) The following conditions are equivalent:

(a) $\mathcal{B}$ is a uniform fading memory space.

(b) Whenever a solution of a system of FDEs is UAS in $\mathbb{E}$, so is it in $\mathcal{B}$.

(c) There exists a system of FDEs which has a uniformly asymptotically stable solution in $\mathcal{B}$.

(ii) (Characterizations for fading memory spaces) The following conditions are equivalent:

(d) $\mathcal{B}$ is a fading memory space.

(e) Whenever a solution of a system of FDEs is AS in $\mathbb{E}$, so is it in $\mathcal{B}$.

(f) There exists a system of FDEs which has an asymptotically stable solution in $\mathcal{B}$.

Proof. That (a) implies (b) (or (d) implies (e)) is already shown in (iii) (or (ii)) of Proposition 2.5. Also, that (b) implies (c) (or (e) implies (f)) is obvious, because the zero solution of $\dot{x}(t) = -x(t)$ is UAS in $\mathbb{E}$. We shall prove that (c) implies (a). Suppose a solution u of System (2.1) is UAS in $\mathcal{B}$ with the triple $(\delta_0, \delta(\cdot), T(\cdot))$. Since there is no special assumption for F, we may consider $u(t) \equiv 0$. Now, suppose (a) is false. Then, there exist an $\varepsilon > 0$, sequences $\{t_k\}$, $t_k \to \infty$ as $k \to \infty$, and $\{\varphi^k\} \subset \mathcal{B}_0$, $|\varphi^k|_{\mathcal{B}} < \min(\delta_0, \delta(1))$, such that

$$(2.6) \qquad |S_0(t_k)\varphi^k|_{\mathcal{B}} \geq \varepsilon$$

for all $k = 1, 2, \cdots$. Put $x^k(t) = x(t, 0, \varphi^k, F)$ for $t \in \mathbb{R}$, and let

$$x^k_t = y^k_t + S_0(t)\varphi^k, \qquad t \ge 0,$$

be the decomposition of $(x^k)_t$ corresponding to (1.4); thus, $y^k : (-\infty,\infty) \to E$ is a continuous function defined by

$$y^k(t) = \begin{cases} x^k(t) & \text{if } t \ge 0, \\ 0 & \text{if } t < 0. \end{cases}$$

Since the zero solution of System (2.1) is UAS in $\mathcal{B}$ by the assumption, $|\varphi^k|_{\mathcal{B}} < \min(\delta_0, \delta(1))$ yields that $|x^k_t|_{\mathcal{B}} \le 1$ for $t \ge 0$ and for $k = 1,2,\cdots$, and that $|x^k_t|_{\mathcal{B}} \to 0$ as $t \to \infty$ uniformly for k. Hence the sequence $\{y^k\}$ is uniformly bounded on $\mathbb{R}$, and $y^k(t) \to 0$ as $t \to \infty$ uniformly for k. Since $t_k \to \infty$ as $k \to \infty$, Proposition 1.1 implies that $|y^k_{t_k}|_{\mathcal{B}} \to 0$ as $k \to \infty$; hence, $|S_0(t_k)\varphi^k|_{\mathcal{B}} \to 0$ as $k \to \infty$. This contradicts (2.6). This contradiction shows the desired implication.

By the same argument as above, we can prove that $|S_0(t)\varphi|_{\mathcal{B}} \to 0$ as $t \to \infty$ for each $\varphi \in \mathcal{B}_0$ under the assumption that (f) holds. In fact, we may set $\varphi^k = \varphi$, $x^k = x$ and $y^k = y$ in the proof given above. Hence, (f) implies (d).

As an application, we discuss the asymptotic stability of the zero solution of the linear autonomous system

$$(2.7) \qquad \dot{x}(t) = L(x_t),$$

under the condition

$$(2.8) \qquad \det \Delta(\lambda) \ne 0 \quad \text{for} \quad \operatorname{Re} \lambda \ge 0,$$

where L is a continuous linear operator on $\mathcal{B}$ into $\mathbb{E}$, and $\Delta(\lambda)$ is the characteristic matrix defined in Section 5.2. At first, to apply the results in Chapter 5, we assume that the space $\mathcal{B}$ satisfies Axiom (C1). Let β be the number defined by Relation (4.3.3). Suppose that $\beta < 0$, which holds if $|S_0(t)| \to 0$ as $t \to \infty$. In this case, if Condition (2.8) holds, from Theorem 3.2 and Proposition 5.3.2 the solution semigroup $T(t)$ of Equation (2.7) has a negative type number. Hence the zero solution is exponentially asymptotically stable in $\mathcal{B}$.

Consider the case $\beta \ge 0$. For example, let $g(\theta) = 1 + |\theta|$, and take the space C^0_g. Since $\omega(\lambda)b$ lies in $\mathcal{B}$ for $\operatorname{Re} \lambda \ge 0$ and for

$b \in \mathbb{E}$, from Theorem 5.2.1 we have that $\beta \geq 0$. On the other hand, since $|S_0(t)| \equiv 1$ from Theorem 1.4.2, we have that $\beta \leq 0$. Thus $\beta = 0$ for this space. Then, under Condition (2.8), Theorems 5.2.1 and 5.2.2 show that the type number of $T(t)$ is equal to zero, which does not imply the asymptotic stability of the zero solution. However, we will show this stability by using the property that C_g^0 is a fading memory space. First of all, we observe the following representation of L.

Proposition 2.7. Suppose that $\mathscr{B}$ is a fading memory space, and that L is a continuous linear operator on $\mathscr{B}$ into $\mathbb{E}$. Then there exists a matrix function $\eta(\theta)$, $-\infty < \theta \leq 0$, such that $\eta(\theta)$ is of bounded variation on $(-\infty, 0]$ and that

$$L(\varphi) = \int_{-\infty}^{0} d\eta(\theta)\varphi(\theta) := \lim_{R\to\infty} \int_{-R}^{0} d\eta(\theta)\varphi(\theta) \quad \text{for} \quad \varphi \quad \text{in} \quad BC.$$

Proof. From Theorem 3.4.2, the above representation holds good for φ in C_{00} and for some $\eta(\theta)$ which is locally of bounded variation. Since $K(r) \equiv K$ and $M(r) \equiv M$, constants, for the fading memory space, from estimate (3.4.3) we have

$$\operatorname{Var}(\eta, (-\infty, 0]) \leq c|L|KM.$$

Let $\varphi \in BC$. Define $\varphi^m \in C_{00}$ as in (1.1). Then $L(\varphi^m) \to L(\varphi)$ as $m \to \infty$, and

$$L(\varphi^m) = \int_{-m}^{0} d\eta(\theta)\varphi(\theta) + \int_{-m-1}^{-m} d\eta(\theta)\varphi^m(\theta).$$

The second integral converges to zero as $m \to \infty$ since $\lim_{m\to\infty} \operatorname{Var}(\eta, [-m-1, -m]) = \lim_{m\to\infty} \operatorname{Var}(\eta, [-m-1, 0]) - \lim_{m\to\infty} \operatorname{Var}(\eta, [-m, 0]) = 0$, and since $\varphi^m(\theta)$ are uniformly bounded for $\theta \leq 0$ and for $m = 1, 2, \ldots$. Thus we have the representation in the theorem.

From this proposition, the characteristic matrix $\Delta(\lambda)$ is well defined for λ with $\operatorname{Re}\lambda \geq 0$. If we set $\zeta(s) = -\eta(-s)$ for $s \geq 0$, then

$$\Delta(\lambda) = \lambda I - \int_{0}^{\infty} e^{-\lambda s} d\zeta(s) \quad \text{when} \quad \operatorname{Re}\lambda \geq 0.$$

Theorem 2.8 (Naito [8]). If $\mathscr{B}$ is a fading memory space,

Condition (2.8) is a necessary and sufficient condition that the zero solution of (2.7) is asymptotically stable in $\mathcal{B}$. If $\mathcal{B}$ is a uniform fading memory space, Condition (2.8) is a necessary and sufficient condition that the zero solution of (2.7) is uniformly asymptotically stable in $\mathcal{B}$.

<u>Proof.</u> If $\det \Delta(\alpha) = 0$ for some α with $\operatorname{Re}\alpha \geq 0$, then for any nontrivial vector a such that $\Delta(\lambda)a = 0$ the function $x(t) = \exp(\alpha t)a$ is a solution, of Equation (2.7), which does not tend to zero vector as $t \to \infty$. Hence the zero solution is not asymptotically stable in $\mathbb{E}$. Assume that Condition (2.8) holds. It suffices to show the stability in $\mathbb{E}$. Let $X(t)$ be the fundamental matrix of Equation (2.7): that is, $X(t)$ is locally absolutely continuous on $[0,\infty)$, and is a matrix solution of

$$(2.9)\qquad \dot{X}(t) = \int_{-t}^{0} d\eta(\theta)X(t+\theta)\ \left(= \int_{0}^{t} d\xi(s)X(t-s)ds\right) \quad \text{a.e. in } [0,\infty)$$

with the initial value $X(0) = I$. Let $x(t,\varphi)$ be the solution of Equation (2.7) such that $x_0 = \varphi \in \mathcal{B}$. Since the equation is autonomous, the variation of constants formula in Theorem 4.1.8 becomes

$$x(t,\varphi) = \varphi(0) + \int_{0}^{t} X(t-s)L(S(s)\varphi)ds \quad \text{for } t \geq 0.$$

Set $\psi = \varphi - \varphi(0)\chi$. Then $\psi \in \mathcal{B}_0$ and $\varphi = \psi + \varphi(0)\chi$ as in (1.4). Hence $S(s)\varphi = S_0(s)\psi + S(s)(\varphi(0)\chi) = S_0(s)\psi + \varphi(0)\chi$. Since ξ is of bounded variation on $[0,\infty)$, the limit $\xi(\infty) = \lim_{s\to\infty}\xi(s)$ exists, and $L(\varphi(0)\chi) = [\xi(\infty)-\xi(0)]\varphi(0)$. Hence, we have that

$$\int_{0}^{t} X(t-s)L(S(s)\varphi)ds$$

$$= \int_{0}^{t} X(t-s)L(S_0(s)\psi)ds + \int_{0}^{t} X(t-s)ds[\xi(\infty)-\xi(0)]\varphi(0).$$

In the case that ξ is of bounded variation on $[0,\infty)$, it is well known, cf. Gripenberg, Londen and Staffans [1], that, if Condition (2.8) holds, $X(t)$ is integrable over $[0,\infty)$. Since $X(t)$ satisfies (2.9), $\dot{X}(t)$ is also integrable over $[0,\infty)$; as a result, $X(t) \to 0$ as $t \to \infty$, cf. Miller [1]. Hence the integrations of both sides of (2.9) yields the relation

$$-I = [\zeta(\infty)-\zeta(0)]\int_0^\infty X(s)ds.$$

It then follows that the matrices in the right hand side are commutative; thus

$$w(t,\varphi) := \int_0^t X(t-s)ds[\zeta(\infty)-\zeta(0)]\varphi(0) \to -\varphi(0) \quad \text{as} \quad t \to \infty$$

uniformly for $|\varphi| \le 1$.

If $\mathscr{B}$ is a fading memory space, $|S_0(s)|$ is bounded for $s \ge 0$. This implies that the integral

$$v(t,\varphi) := \int_0^t X(t-s)L(S_0(s)\psi)ds$$

is bounded uniformly for $t \in [0,\infty)$ and for $|\varphi| \le 1$. Thus the zero solution of Equation (2.7) is uniformly stable in $\mathbb{E}$. Observe that a convolution of functions f and g converges to zero as $t \to \infty$ provided that f is integrable over $[0,\infty)$ and $g(t) \to 0$ as $t \to \infty$. Hence we have that $v(t,\varphi) \to 0$ as $t \to \infty$ for each $\varphi \in \mathscr{B}$: that is, the zero solution of (2.7) is asymptotically stable in $\mathbb{E}$.

If $\mathscr{B}$ is a uniform fading memory space, $|S_0(s)| \to 0$ as $s \to \infty$. This implies that $v(t,\varphi) \to 0$ as $t \to \infty$ uniformly for $|\varphi| \le 1$: that is, the zero solution of (2.7) is uniformly asymptotically stable in $\mathbb{E}$.

Corollary 2.9. If Condition (2.8) holds, and if $\mathscr{B}$ is a uniform fading memory space, the zero solution of Equation (2.7) is exponentially asymptotically stable in $\mathscr{B}$: that is, there exist positive constants C and α such that, for every $\varphi \in \mathscr{B}$,

$$|x_t(\varphi)| \le Ce^{-\alpha t}|\varphi| \quad \text{for} \quad t \ge 0.$$

7.3. Fading memory spaces and relative compactness of positive orbits.

In this section, we shall discuss the relative compactness in $\mathscr{B}$ for positive orbits of bounded solutions of FDEs.

Let $\mathcal{F}$ be a subset of $C([0,\infty),E)$. $\mathcal{F}$ is said to be uniformly bounded if $\sup\{|x(t)| : t \ge 0,\ x \in \mathcal{F}\} < \infty$. Also, $\mathcal{F}$ is said to be uniformly equicontinuous if for any $\varepsilon > 0$ there exists a $\delta(\varepsilon) > 0$ such that $|x(t_1) - x(t_2)| < \varepsilon$ for all $x \in \mathcal{F}$ whenever $t_1, t_2 \ge 0$ and $|t_1 - t_2| < \delta(\varepsilon)$. For any set $\mathscr{W}$ in $\mathscr{B}$ and any set $\mathcal{F}$ in

$C([0,\infty),E)$, we set

$$\tilde{X}(\mathcal{W},\mathcal{F}) = \{x(\cdot) : (-\infty,\infty) \to E \mid x_0 \in \mathcal{W} \text{ and } x|_{[0,\infty)} \in \mathcal{F}\}$$

and

$$X(\mathcal{W},\mathcal{F}) = \{x_t : t \geq 0,\ x \in \tilde{X}(\mathcal{W},\mathcal{F})\}.$$

The following result will often be used in the following chapters.

Theorem 3.1 (Murakami & Naito [1]). Suppose $\mathcal{B}$ is a fading memory space. If $\mathcal{W}$ is a compact set in $\mathcal{B}$ and if $\mathcal{F}$ is a uniformly bounded and uniformly equicontinuous set in $C([0,\infty),E)$, then the set $X(\mathcal{W},\mathcal{F})$ is relatively compact in $\mathcal{B}$.

Proof. We must prove that any given sequence $\{(x^k)_{t_k}\}$, $t_k \geq 0$, and $x^k(\cdot) \in \tilde{X}(\mathcal{W},\mathcal{F})$, contains a convergent subsequence. Taking a subsequence if necessary, we may assume that $t_k \to t_0 \leq \infty$ and $(x^k)_0 := \varphi^k \to \varphi$ in $\mathcal{B}$ as $k \to \infty$. Let

$$(x^k)_{t_k} = (y^k)_{t_k} + S_0(t_k)\psi^k$$

be the decomposition of $(x^k)_{t_k}$ corresponding to (1.4), where $\psi^k = \varphi^k - \varphi^k(0)\chi$. Then $\psi^k \to \psi := \varphi - \varphi(0)\chi$ in $\mathcal{B}$ as $k \to \infty$. Clearly, $\xi^k := (y^k)_{t_k}$ lies in BC, and the sequence $\{\xi^k\}$ is uniformly bounded and equicontinuous on $(-\infty,0]$. From the Ascoli-Arzéla theorem and (ii) of Proposition 1.1, we can assume that $\{\xi^k\}$ is a convergent sequence in $\mathcal{B}$. On the other hand, since $\sup_{t\geq 0} |S_0(t)| < \infty$, we have $|S_0(t_k)\psi^k - S_0(t_k)\psi|_{\mathcal{B}} \to 0$ as $k \to \infty$. If $t_0 < \infty$, then $S_0(t_k)\psi \to S_0(t_0)\psi$ as $k \to \infty$; if $t_0 = \infty$, then $S_0(t_k)\psi \to 0$ as $k \to \infty$ by the assumption that $\mathcal{B}$ is a fading memory space. As a result, $\{S_0(t_k)\psi^k\}$ is a convergent sequence in $\mathcal{B}$. Therefore, the sequence $\{(x^k)_{t_k}\}$ has the desired property.

Corollary 3.2. Suppose $\mathcal{B}$ is a fading memory space. If u is a bounded solution on $[0,\infty)$ of System (2.1) such that $\sup_{t\geq 0} |\dot{u}(t)| < \infty$, then the positive orbit $\{u_t : t \geq 0\}$ of u is relatively compact in $\mathcal{B}$.

For a sequence of bounded orbits, one can get a more applicable result than Theorem 3.1. To formulate the result, we need to extend the notion of uniform equicontinuity of a family of functions so that it is valid for a family of functions which are not necessary defined on a common interval. Let $\mathbb{N}$ be the set of all positive integers. A family $\{x^k(\cdot)\}_{k\in\mathbb{N}}$ of $\mathbb{E}$-valued functions defined on $I_k = [0,s_k)$, $0 < s_k \le \infty$, is said to be uniformly equicontinuous on I_k, if for any $\varepsilon > 0$ there exist a $\delta > 0$ and $N_0 > 0$ such that $|x^k(t) - x^k(s)|_{\mathbb{E}} < \varepsilon$ for all $t, s \in I_k$ with $|t - s| < \delta$ whenever $k \ge N_0$. If the family $\{x^k(\cdot)\}_{k\in\mathbb{N}}$ is uniformly equicontinuous on I_k with $\sup_k|x^k(0)| < \infty$, then it is uniformly bounded on $J \cap I_k$ for each compact interval J in $I = [0,s)$, where $s := \liminf_{k\to\infty} s_k$. Hence, the Ascoli-Arzéla theorem implies that a uniformly equicontinuous family $\{x^k(\cdot)\}_{k\in\mathbb{N}}$ on I_k with $\sup_k|x^k(0)| < \infty$ has a subfamily which converges compactly on $I = [0,s)$.

Recall that $C_I(\varphi)$ denotes the family of functions mapping $(-\infty,0]$ into $\mathbb{E}$ with $x_0 = \varphi$, and continuous on I.

Corollary 3.3. Let $\mathscr{B}$ be a fading memory space, and assume that $\mathscr{W} = \{\varphi^k : k \in \mathbb{N}\}$ is a relatively compact subset of $\mathscr{B}$, x^k is a function in $C_{I_k}(\varphi^k)$ for some interval $I_k = [0,s_k)$, $0 < s_k \le \infty$, and that the family $\{x^k|_{I_k}\}_{k\in\mathbb{N}}$ is uniformly bounded and uniformly equicontinuous on I_k. Then the set $\{(x^k)_t : t \in I_k,\ k = 1,2,\cdots\}$ is relatively compact in $\mathscr{B}$. More precisely; for any sequence $\{\tau_k\} \subset \mathbb{R}^+$, $\tau_k \in I_k$, there exists a subsequence $\{\tau_{k(\ell)}\} \subset \{\tau_k\}$ such that the family of $\mathscr{B}$-valued functions $\{(x^{k(\ell)})_{\tau_{k(\ell)}+t}\}$ is convergent uniformly on $[0,\tau]$ for any τ such that $0 \le \tau < \liminf_{k\to\infty}(s_k - \tau_k)$.

Proof. By the preceding remark and the proof of Theorem 3.1, there exists a subsequence $\{\tau_{k(\ell)}\} \subset \{\tau_k\}$ such that the sequence $\{(x^{k(\ell)})_{\tau_{k(\ell)}}\}$ is convergent in $\mathscr{B}$. Then the desired result follows from Axiom (A) and the uniform equicontinuity.

Next, we shall try to characterize fading memory spaces by the relative compactness of positive orbits.

Proposition 3.4. Suppose $\mathscr{B}$ satisfies Axioms (B) and (C2). If the set $X(\mathscr{W},\mathscr{F})$ is relatively compact in $\mathscr{B}$ for any compact set $\mathscr{W}$ in $\mathscr{B}$ and for any uniformly bounded and uniformly equicontinuous set $\mathscr{F}$ in $C([0,\infty),\mathbb{E})$, then $\mathscr{B}$ is a fading memory space whenever it satisfies Axiom (C1).

Proof. We shall prove the theorem by a contradiction. Suppose $\mathcal{B}$ is not a fading memory space. Then there exist an $\varepsilon > 0$, a $\varphi \in \mathcal{B}_0$, $|\varphi|_{\mathcal{B}} = 1$, and a sequence $\{t_k\}$, $t_k \to \infty$ as $k \to \infty$, such that

$$|S_0(t_k)\varphi|_{\mathcal{B}} \geq \varepsilon, \quad k = 1,2,\cdots.$$

Since $S_0(t_k)\varphi \in X(\{\varphi\},\mathcal{F})$, where $\mathcal{F} = \{x : [0,\infty) \to \mathbb{E} \mid |x(t_1)| \leq 1$ and $|x(t_1) - x(t_2)| \leq |t_1 - t_2|$ for $t_1, t_2 \geq 0\}$, we may assume that $\{S_0(t_k)\varphi\}$ is a convergent sequence by our assumption. Note that $[S_0(t_k)\varphi](\theta) = 0$ for all $\theta \in [-t_k,0]$; hence, $[S_0(t_k)\varphi](\theta)$ converges to the zero function compactly on $(-\infty,0]$. Thus, Axiom (C1) implies $|S_0(t_k)\varphi|_{\mathcal{B}} \to 0$ as $k \to \infty$, which is a contradiction.

As the following example shows, one cannot necessarily remove the additional axiom (C1) in Proposition 3.4.

Example 3.5. Consider the space C_g^∞, where $g(s) = 1 + |s|$, $s \leq 0$. The space C_g^∞ satisfies Axioms (B) and (C2) but not Axiom (C1); see Theorem 1.3.3. Now, let $\mathcal{W}$ and $\mathcal{F}$ be the ones as in Proposition 3.4, and let $\{x^n\} \subset \tilde{X}(\mathcal{W},\mathcal{F})$ and $\{t_n\} \subset [0,\infty)$ be given. Taking a subsequence if necessary, we may assume that the sequences $\{(x^n)_0\}$ and $\{x^n(t_n)\}$ are convergent in C_g^∞ and $\mathbb{E}$, respectively. Then one can easily see that the sequence $\{(x^n)_{t_n}\}$ is a Cauchy sequence in C_g^∞. Thus, the assumptions of Proposition 3.4 hold, because C_g^∞ is complete. However, as noted in Example 1.6, C_g^∞ is not a fading memory space.

Next, we shall make the similar consideration for uniform fading memory spaces. The following result is a characterization for a uniform fading memory space by means of the relative compactness of $X(\mathcal{W},\mathcal{F})$.

Theorem 3.6 (Murakami & Naito [1]). Let $\mathcal{B}$ be a fading memory space. Then $\mathcal{B}$ is a uniform fading memory space if and only if the following condition is satisfied:

For any bounded set $\mathcal{W}$ in $\mathcal{B}$ and for any uniformly bounded and uniformly equicontinuous set $\mathcal{F}$ in $C([0,\infty),\mathbb{E})$, each sequence $\{(x^k)_{t_k}\}$ in $X(\mathcal{W},\mathcal{F})$ such that $t_k \to \infty$ as $k \to \infty$ has a convergent subsequence in $\mathcal{B}$.

Proof. We shall prove the "if" part by a contradiction.

Suppose that the condition is satisfied, but $\mathcal{B}$ is not a uniform fading memory space. Then there exist an $\varepsilon > 0$ and sequences $\{t_k\}$, $t_k \to \infty$ as $k \to \infty$, and $\{\varphi^k\} \subset \mathcal{B}_0$, $|\varphi^k|_{\mathcal{B}} = 1$, such that

$$|S_0(2t_k)\varphi^k|_{\mathcal{B}} \geq \varepsilon, \quad k = 1,2,\cdots.$$

By the assumption, we may suppose that $S_0(t_k)\varphi^k \to \psi$ in $\mathcal{B}$ as $k \to \infty$ for some ψ. Since $\sup_{t\geq 0} |S_0(t)| < \infty$, it follows that $S_0(2t_k)\varphi^k - S_0(t_k)\psi \to 0$ as $k \to \infty$; hence $S_0(2t_k)\varphi^k \to 0$ as $k \to \infty$ because $\mathcal{B}$ is a fading memory space. This is a contradiction. This completes the proof of the "if" part.

Next, we shall prove the "only if" part. Set $y^k(t) = x^k(t)$ for $t \geq 0$ and $y^k(t) = x^k(0)$ for $t < 0$. Since $|S_0(t_k)| \to 0$ as $k \to \infty$, the relation

$$(x^k)_{t_k} = (y^k)_{t_k} + S_0(t_k)[(x^k)_0 - x^k(0)\chi]$$

shows that $\{(x^k)_{t_k}\}$ and $\{(y^k)_{t_k}\}$ converges simultaneously. Since $\{y^k(t_k+s)\}$ is uniformly bounded and uniformly equicontinuous for s in $(-\infty,0]$, it has a subsequence convergent in $\mathcal{B}$ from Proposition 1.1. Hence, $\{(x^k)_{t_k}\}$ has also a subsequence convergent in $\mathcal{B}$.

By the same arguments as in the proof of Corollary 3.3 and Theorem 3.6, we obtain the following result.

<u>Corollary 3.7.</u> Let $\mathcal{B}$ be a fading memory space. Then $\mathcal{B}$ is a uniform fading memory space if and only if the following condition is satisfied:

If $\mathcal{W} = \{\varphi^k : k \in \mathbb{N}\}$ is a bounded subset of $\mathcal{B}$, x^k is a function in $C_{I_k}(\varphi^k)$ for some interval $I_k = [0,s_k)$, $0 < s_k \leq \infty$, and the family $\{x^k|_{I_k}\}_{k\in\mathbb{N}}$ is uniformly bounded and uniformly equicontinuous on I_k, then for any sequence $\{\tau_k\} \subset \mathbb{R}^+$, $\tau_k \in I_k$, such that $\tau_k \to \infty$ as $k \to \infty$, there exists a subsequence $\{\tau_{k(\ell)}\} \subset \{\tau_k\}$ such that the family of $\mathcal{B}$-valued functions $\{(x^{k(\ell)})_{\tau_{k(\ell)}+t}\}$ is convergent uniformly on $[0,\tau]$ for any τ such that $0 \leq \tau < \liminf_{k\to\infty} (s_k-\tau_k)$.

7.4. Bounded solutions of linear periodic systems. A function $x : \mathbb{R} \to E$ is said to be $\mathbb{R}$-bounded with respect to the seminorm $|\cdot|_{\mathscr{B}}$ (resp. $\mathbb{R}$-bounded with respect to the norm $|\cdot|_E$) if $\sup_{t\in\mathbb{R}} |x_t|_{\mathscr{B}} < \infty$ (resp. $\sup_{t\in\mathbb{R}} |x(t)|_E < \infty$). If a continuous function is $\mathbb{R}$-bounded with respect to the seminorm $|\cdot|_{\mathscr{B}}$, then it is $\mathbb{R}$-bounded with respect to the norm $|\cdot|_E$ by Axiom (A-ii). On the other hand, if the space $\mathscr{B}$ satisfies Axiom (C2), then the contrary also holds by Proposition 1.1. Therefore, in this case, there is no need to distinguish the two notions of boundedness; hereafter, we will say it to be $\mathbb{R}$-bounded, in short.

Consider a linear periodic functional differential equation

$$\dot{x}(t) = L(t, x_t) + h(t), \tag{4.1}$$

where $h : \mathbb{R} \to E$ is a locally integrable function and $L : \mathbb{R} \times \mathscr{B} \to E$ is a continuous function such that $L(t,\varphi)$ is linear in φ and that $L(t+\omega,\varphi) = L(t,\varphi)$ for all $(t,\varphi) \in \mathbb{R} \times \mathscr{B}$ and for some constant $\omega > 0$. A solution of Equation (4.1) defined on $\mathbb{R}$ which is $\mathbb{R}$-bounded is called an $\mathbb{R}$-bounded solution of Equation (4.1). Throughout this section, we assume that $E = \mathbb{C}^n$ and that $\mathscr{B}$ is a uniform fading memory space satisfying Axiom (C1). Notice that $\beta < 0$ by Theorem 4.3.5. The purpose of this section is to give a condition under which Equation (4.1) has at least one $\mathbb{R}$-bounded solution when h is $\mathbb{R}$-bounded and continuous.

Recall that $\hat{P}(0)$ is the operator induced by the operator $P(0) = T(\omega,0)$, where $T(t,s)$ is the solution operator defined by the homogeneous equation

$$\dot{x}(t) = L(t, x_t) \tag{4.2}$$

as in Section 6.2.

Theorem 4.1 (Murakami [4]). Suppose $\mathscr{B}$ is a uniform fading memory space satisfying Axiom (C1). Then Equation (4.1) has a unique $\mathbb{R}$-bounded solution x for each $\mathbb{R}$-bounded continuous function h if and only if the set $\Sigma_0 = \{\mu \in \sigma(\hat{P}(0)) : |\mu| = 1\}$ is empty. In this case, we have the estimate

$$\sup_{t\in\mathbb{R}} |x_t|_{\mathscr{B}} \le \bar{K}|h|_\infty,$$

where $\bar{K}$ is a constant independent of h, and $|h|_\infty = \sup_{t\in\mathbb{R}} |h(t)|$.

In order to prove the theorem, we need the following lemmas.

Lemma 4.2. Let X be a normed space and U be a closed subspace of X. Then, the mapping $f : U^{**} \to U^{\perp\perp}$ defined by

$$<x^*, f(u^{**})> = <x^*|_U, u^{**}>, \quad x^* \in X^*, \ u^{**} \in U^{**},$$

is an isometric isomorphism from U^{**} onto $U^{\perp\perp}$, where $U^{\perp\perp} = (U^\perp)^\perp$ and $x^*|_U$ denotes the restriction of x^* to U.

Proof. This lemma can be verified by employing almost the same argument as in Rudin [2, pp. 91-92]. First, observe that f defines a mapping from U^{**} into $U^{\perp\perp}$. Indeed, if $x^* \in U^\perp$, then $x^*|_U = 0$; hence $<x^*, f(u^{**})> = <x^*|_U, u^{**}> = 0$ for all $u^{**} \in U^{**}$ and $x^* \in U^\perp$, which shows $f(U^{**}) \subset U^{\perp\perp}$. Clearly, the mapping f is linear and $|f(u^{**})| \le |u^{**}|$ for all $u^{**} \in U^{**}$. Now, we claim that the mapping f is injective. Suppose that $f(u^{**}) = 0$ for some $u^{**} \in U^{**}$. For each $u^* \in U^*$, select an extension $x^* \in X^*$ of u^* such that $|x^*| = |u^*|$, which is possible by the Hahn-Banach theorem. Then $<u^*, u^{**}> = <x^*, f(u^{**})> = 0$; hence $u^{**} = 0$. This proves the claim. Next, we prove that the mapping f is surjective. Let any $x^{**} \in U^{\perp\perp}$ be given. Set

$$u^{**}(u^*) = <x^*, x^{**}>$$

for $u^* \in U^*$, where x^* is the extension of u^* given in the preceding argument. Clearly, u^{**} defines a linear functional on U^*. Moreover, we have that $|u^{**}(u^*)| \le |x^{**}||x^*| = |x^{**}||u^*|$ for $u^* \in U^*$; hence $u^{**} \in U^{**}$, $|u^{**}| \le |x^{**}|$ and $f(u^{**}) = x^{**}$. Thus, f is surjective. At the same time, this proves that $|f(u^{**})| = |u^{**}|$ for all $u^{**} \in U^{**}$, because $|x^{**}| = |f(u^{**})| \le |u^{**}| \le |x^{**}|$.

Lemma 4.3. Let X and Y be normed spaces, and $T : X \to Y$ be a bounded linear operator satisfying $T(U) \subset V$, where U and V are closed subspaces of X and Y, respectively. Then $T^*(V^\perp) \subset U^\perp$ and $T^{**}(U^{\perp\perp}) \subset V^{\perp\perp}$. Furthermore, if $f : U^{**} \to U^{\perp\perp}$ and $g : V^{**} \to V^{\perp\perp}$ are the isometric isomorphisms given in Lemma 4.2, then the following relation holds:

$$(T^{**}|_{U^{\perp\perp}}) \circ f = g \circ (T|_U)^{**}.$$

In particular, we have $|T^{**}|_{U^{\perp\perp}}| = |T|_U|$.

Proof. The first part of the lemma is clear. We shall prove the second part of the lemma. Let $u^{**} \in U^{**}$ and $y^* \in Y^*$. Then

$$\langle y^*,((T^{**}|_{U^{\perp\perp}})\circ f)(u^{**})\rangle = \langle T^*y^*,f(u^{**})\rangle = \langle (T^*y^*)|_U,u^{**}\rangle$$

and

$$\begin{aligned}\langle y^*,(g\circ(T|_U)^{**}(u^{**})\rangle &= \langle y^*|_V,(T|_U)^{**}(u^{**})\rangle \\ &= \langle (T|_U)^*(y^*|_V),u^{**}\rangle.\end{aligned}$$

On the other hand, for any $u \in U$ we have

$$\langle (T|_U)^*(y^*|_V),u\rangle = \langle y^*|_V,(T|_U)u\rangle = \langle y^*,Tu\rangle = \langle T^*y^*,u\rangle,$$

which shows that $(T|_U)^*(y^*|_V) = (T^*y^*)|_U$. Consequently,

$$\langle y^*,((T^{**}|_{U^{\perp\perp}})\circ f)(u^{**})\rangle = \langle y^*,(g\circ(T|_U)^{**})(u^{**})\rangle$$

for all $y^* \in Y^*$ and $u^{**} \in U^{**}$; hence

$$(T^{**}|_{U^{\perp\perp}})\circ f = g\circ(T|_U)^{**}.$$

Finally, we obtain that $|T^{**}|_{U^{\perp\perp}}| = |T|_U|$, because

$$\begin{aligned}|T^{**}|_{U^{\perp\perp}}| &= |(T^{**}|_{U^{\perp\perp}})\circ f| = |g\circ(T|_U)^{**}| = |(T|_U)^{**}| = |(T|_U)^*| \\ &= |T|_U|.\end{aligned}$$

Lemma 4.4. Let $x(t)$ be an $\mathbb{R}$-bounded solution of Equation (4.2). Then $x(t)$ can be written as

$$x(t) = \sum_{j=1}^{q} e^{\nu_j t}[\hat{\Phi}_j(t)](0)b_j \tag{4.3}$$

for some constants $b_1,\cdots,b_q$, where $\Sigma_0 = \{e^{\omega\nu_1},\cdots,e^{\omega\nu_q}\}$, and $\hat{\Phi}_j(t)$ denotes the basis vector associated with the generalized eigenspace $\mathcal{N}((\hat{P}(0) - e^{\omega\nu_j}\hat{I})^{k_j})$. In particular, if the set Σ_0 is empty, then $x(t) \equiv 0$ on $\mathbb{R}$.

Proof. Since $\beta < 0$, we can obtain the decomposition of the space $\hat{\mathcal{B}}$ corresponding to the set $\Gamma(0) = \{\lambda \in \sigma(\hat{P}(0)) : |\lambda| \geq 1\}$;

$$\hat{\mathscr{B}} = \hat{E}(s) \oplus \hat{F}(s), \qquad s \in \mathbb{R},$$

by Theorem 6.2.1. Employing the same notations as in Section 6.2, first of all, we shall show that $|\pi_F(s)\hat{x}_s|_{\hat{\mathscr{B}}} = 0$ for all $s \in \mathbb{R}$. Suppose not. Then there exists an $s \in \mathbb{R}$ such that $|\pi_F(s)\hat{x}_s|_{\hat{\mathscr{B}}} \neq 0$. Since $\hat{T}(s,\theta)\pi_F(0) = \pi_F(s)\hat{T}(s,\theta)$ for $s \geq 0$ by (2.13) in Section 4.2, it follows that

$$\hat{T}(s,\theta)\pi_F(0)\hat{x}_\theta = \pi_F(s)\hat{x}_s \quad \text{for all} \quad \theta \leq s. \tag{4.4}$$

On the other hand,

$$|\hat{T}(s,\theta)|_{\hat{F}(\theta)}|_{\hat{\mathscr{B}}} \leq \ell e^{-\nu(s-\theta)}, \qquad s \geq \theta, \tag{4.5}$$

by (v) in Theorem 6.2.1, where ℓ and ν are some positive constants. Hence

$$|\pi_F(s)\hat{x}_{s-n\omega}|_{\hat{\mathscr{B}}} = |\pi_F(s-n\omega)\hat{x}_{s-n\omega}|_{\hat{\mathscr{B}}} \geq e^{n\omega\nu}|\pi_F(s)\hat{x}_s|_{\hat{\mathscr{B}}}/\ell$$

by (4.4) and (4.5), because $\pi_F(t) = \pi_F(t+\omega)$, $t \in \mathbb{R}$. Letting $n \to \infty$ in the above, there arises a contradiction to the fact that $x(t)$ is $\mathbb{R}$-bounded. Therefore, we have $|\pi_F(s)\hat{x}_s|_{\hat{\mathscr{B}}} = 0$; hence $\hat{x}_s \in \hat{E}(s)$ for all $s \in \mathbb{R}$.

Now, select a vector $b \in \mathbb{C}^d$ so that $\hat{x}_0 = \hat{\Phi}(0)b$. We claim that

$$\hat{x}_t = \hat{\Phi}(t)e^{Gt}b \quad \text{for all} \quad t \in \mathbb{R}. \tag{4.6}$$

Indeed, this claim holds for all $t \geq 0$, because of the uniqueness of solutions of Equation (4.2) for the initial conditions. Since $\hat{x}_t \in \hat{E}(t)$, there is a vector $c \in \mathbb{C}^d$ such that $\hat{x}_t = \hat{\Phi}(t)c$. Then $\hat{x}_0 = \hat{T}(0,t)\hat{x}_t = \hat{T}(0,t)\hat{\Phi}(t)c = \hat{\Phi}(0)e^{-Gt}c$ by (iii) in Theorem 6.2.1. Thus, $b = e^{-Gt}c$, which implies Relation (4.6).

Now, Relation (4.6) means that $\hat{x}_t$ is of the form $\sum_{j=1}^{m} e^{\nu_j t}\hat{\Phi}_j(t)b_j(t)$, where $b_j(t)$ is some vectors whose components are polynomials of t, and $\operatorname{Re}\nu_j > 0$ for $j = q+1,\cdots,m$. Since $x(t)$ is $\mathbb{R}$-bounded, we must have that $b_j(t)$ is constant for $j = 1,\cdots,q$, and $b_j(t) \equiv 0$ for $j = q+1,\cdots,m$. This proves the first part of the lemma. The remainder of the lemma follows from (4.3).

Now, we shall prove Theorem 4.1. Suppose $\Sigma_0 = 0$, and let

$$\hat{\mathscr{B}} = \hat{E}(s) \oplus \hat{F}(s), \qquad s \in \mathbb{R},$$

be the decomposition of $\hat{\mathscr{B}}$ associated with the set $\Lambda(0) = \{\mu \in \sigma(\hat{P}(0)) : |\mu| \geq 1\}$ ensured in Theorem 6.2.1. Lemma 5.3.6 implies that

$$\hat{\mathscr{B}} = \hat{E}(s)^{\perp\perp} \oplus \hat{F}(s)^{\perp\perp}, \qquad s \in \mathbb{R},$$

and $\pi_E^{**}(s)\mathscr{B}^{**} = E(s)^{\perp\perp}$ and $\pi_F^{**}(s)\mathscr{B}^{**} = F(s)^{\perp\perp}$. Since $\dim E(s) = d < \infty$, it follows that $E(s)^{\perp\perp} = E(s)$ by Lemma 5.3.6. Hence, the second dual space $\mathscr{B}^{**}$ of $\mathscr{B}$ is decomposed as

$$\mathscr{B}^{**} = E(s) \oplus F(s)^{\perp\perp}, \qquad s \in \mathbb{R},$$

and the projections $\bar{\pi}_E(t): \mathscr{B}^{**} \to E(t)$ and $\bar{\pi}_F(t): \mathscr{B}^{**} \to F(t)^{\perp\perp}$ are given by $\bar{\pi}_E(t) = \pi_E^{**}(t)$ and $\bar{\pi}_F(t) = \pi_F^{**}(t)$. Lemma 4.3 and (4.5) imply that

$$(4.7) \qquad \left|T^{**}(t,s)|_{\hat{F}(s)^{\perp\perp}}\right| \leq \ell e^{-\nu(t-s)}, \qquad t \geq s.$$

Thus,

$$\begin{aligned} &\left|\int_a^b T^{**}(t,s)\pi_F^{**}(s)\Gamma h(s)ds\right| \\ &= \sup_{\psi\in\mathscr{B}^*,\ |\psi|\leq 1} \left|\left\langle\psi, \int_a^b T^{**}(t,s)\pi_F^{**}(s)\Gamma h(s)ds\right\rangle\right| \\ &\leq \sup_{\psi\in\mathscr{B}^*,\ |\psi|\leq 1} \left|\int_a^b |\langle\psi, T^{**}(t,s)\pi_F^{**}(s)\Gamma h(s)\rangle| ds\right| \\ &\leq \bar{M}\left|\int_a^b \ell e^{-\nu(t-s)}ds\right| \cdot |h|_\infty \to 0 \quad \text{as} \quad (a,b) \to (-\infty,-\infty) \end{aligned}$$

by (4.7), where $\bar{M} = \sup_{t\in\mathbb{R}} \sum_{j=1}^{n} |\pi_F^{**}(t)\gamma_j|$ is finite by Proposition 6.2.2. Hence, the integral $\int_a^t T^{**}(t,s)\pi_F^{**}(s)\Gamma h(s)ds$ converges in $\mathscr{B}^{**}$ as $a \to -\infty$. Set

$$(4.8) \qquad \hat{\eta}_F(t) = \lim_{a\to-\infty} \int_a^t T^{**}(t,s)\pi_F^{**}(s)\Gamma h(s)ds.$$

Clearly, $\hat{\eta}_F(t) \in \hat{F}(t)$ for all $t \in \mathbb{R}$. Moreover, $\hat{\eta}_F(t)$ satisfies Equation (2.6) in Section 6.2. Indeed,

$$\hat{T}(t,\sigma)\hat{\eta}_F(\sigma) + \int_{*\sigma}^{t} T^{**}(t,s)\pi_F^{**}(s)\Gamma h(s)ds$$

$$= \lim_{a\to-\infty} \hat{T}(t,\sigma)\int_{*a}^{\sigma} T^{**}(\sigma,s)\pi_F^{**}(s)\Gamma h(s)ds + \int_{*\sigma}^{t} T^{**}(t,s)\pi_F^{**}(s)\Gamma h(s)ds$$

$$= \lim_{a\to-\infty} \int_{*a}^{\sigma} T^{**}(t,s)\pi_F^{**}(s)\Gamma h(s)ds + \int_{*\sigma}^{t} T^{**}(t,s)\pi_F^{**}(s)\Gamma h(s)ds$$

$$= \lim_{a\to-\infty} \int_{*a}^{t} T^{**}(t,s)\pi_F^{**}(s)\Gamma h(s)ds$$

$$= \hat{\eta}_F(t)$$

for all $t \geq \sigma$.

Notice that $\sigma(e^{G\omega}) = \{\mu_1,\cdots,\mu_m\}$ with $|\mu_i| > 1$ for $i = 1,\cdots,m$ by the assumption $\Sigma_0 = \phi$. Therefore, there exist positive constants $\bar{\ell}$ and $\bar{\nu}$ such that

$$(4.9) \qquad |\hat{\Phi}(t)e^{G(t-s)}[\tilde{\Psi}(s)](0^-)h(s)| \leq \bar{\ell}e^{\bar{\nu}(t-s)}|h|_\infty$$

for all $t \leq s$ and all $\mathbb{R}$-bounded continuous function h, since

$$\sup_{s\in\mathbb{R}}|[\tilde{\Psi}(s)](0^-)| = \sup_{-\omega\leq s<0} |[\tilde{\Psi}(s)](0^-)|$$

$$= \sup_{-\omega\leq s<0} |e^{Gs}y(s,0,\tilde{\Psi}(0))| < \infty$$

by (2.4) in Section 6.2. Consequently, if h is $\mathbb{R}$-bounded and continuous, then the integral $\int_t^\infty \hat{\Phi}(t)e^{G(t-s)}[\tilde{\Psi}(s)](0^-)h(s)ds$ converges. Set

$$\hat{\eta}_E(t) = \int_t^\infty \hat{\Phi}(t)e^{G(t-s)}[\tilde{\Psi}(s)](0^-)h(s)ds.$$

Clearly, $\hat{\eta}_E(t) \in \hat{E}(t)$ for all $t \in \mathbb{R}$. Furthermore, $\hat{\eta}_E(t)$ satisfies Equation (2.5) in Section 6.2. Indeed,

$$\hat{\Phi}(t)e^{G(t-\sigma)}\langle\Psi(\sigma),\hat{\eta}_E(\sigma)\rangle - \int_\sigma^t \hat{\Phi}(t)e^{G(t-s)}[\tilde{\Psi}(s)](0^-)h(s)ds$$

$$= \hat{\Phi}(t)e^{G(t-\sigma)}\langle\Psi(\sigma),\hat{\Phi}(\sigma)\rangle\int_\sigma^\infty e^{G(\sigma-s)}[\tilde{\Psi}(s)](0^-)h(s)ds$$

$$- \int_\sigma^t \hat{\Phi}(t)e^{G(t-s)}[\tilde{\Psi}(s)](0^-)h(s)ds$$

$$= \hat{\Phi}(t)e^{G(t-\sigma)}\int_\sigma^\infty e^{G(\sigma-s)}[\tilde{\Psi}(s)](0^-)h(s)ds$$

$$- \int_\sigma^t \hat{\Phi}(t)e^{G(t-s)}[\tilde{\Psi}(s)](0^-)h(s)ds$$

$$= \int_t^\infty \hat{\Phi}(t)e^{G(t-s)}[\tilde{\Psi}(s)](0^-)h(s)ds$$

$$= \hat{\eta}_E(t)$$

for all $t \geq \sigma$. Therefore, if we set $\hat{\eta}(t) = \hat{\eta}_E(t) + \hat{\eta}_F(t)$, then $\Pi(t,\sigma)\hat{\eta}(\sigma) = \hat{\eta}(t)$ for all $t \geq \sigma > -\infty$ by (2.3), (2.5) and (2.6) in Section 6.2. Then, Theorem 4.2.9 yields that $x(t) := [\hat{\eta}(t)](0)$ is a solution of Equation (4.1) defined on $\mathbb{R}$. Moreover, (4.7) and (4.9) imply that

$$|x_t|_{\mathcal{B}} = |\hat{\eta}(t)|_{\hat{\mathcal{B}}} \leq |\hat{\eta}_E(t)|_{\hat{\mathcal{B}}} + |\hat{\eta}_F(t)|_{\hat{\mathcal{B}}} \leq \bar{K}|h|_\infty, \quad t \in \mathbb{R},$$

where $\bar{K} = (\ell\bar{M}/\nu + \bar{\ell}/\bar{\nu})$. The uniqueness of $\mathbb{R}$-bounded solutions is a direct consequence of Lemma 4.3. This completes the proof of the "if" part of the theorem.

Next, we shall prove the "only if" part of the theorem. To do this, at first, we consider the case where $\mathcal{B}$ satisfies the additional condition;

(*) if $\psi_1, \psi_2 \in \mathcal{B}^*$ and $\tilde{\psi}_1 = \tilde{\psi}_2$, then $\psi_1 = \psi_2$.

If $x(t)$ is a solution of Equation (4.1), then the function $u(t)$ defined by $\hat{\Phi}(t)u(t) = \Pi_E(t)\hat{x}_t$ satisfies Equation (2.7) in Section 6.2 by Corollary 6.2.7. Suppose that $\Sigma_0 \neq \phi$. Then there exist a $\delta \in \sigma(G)$, Re $\delta = 0$, and a nonzero row vector b such that $bG = \delta b$. We claim that $b[\tilde{\Psi}(s)](0^-) \neq 0$ on $\mathbb{R}$. For otherwise,

$$b\cdot y(s,0,\tilde{\Psi}(0)) = be^{-Gs}[\tilde{\Psi}(s)](0^-) = be^{-\delta s}[\tilde{\Psi}(s)](0^-) = 0$$

for $s < 0$ by (2.4) in Section 6.2; hence $[b\Psi(0)]^{\sim} = 0$ by (2.5) in Section 4.2. Thus, $b\Psi(0) = 0$ by Condition (*), which contradicts $b \neq 0$. Now, since $y(s,0,\tilde{\Psi}(0))$ is left continuous for $s < 0$ by Theorem 4.1.4, the function $b[\tilde{\Psi}(s)](0^-) = b\cdot y(s,0,\tilde{\Psi}(0))e^{\delta s}$ is ω-periodic and left continuous in $s \in \mathbb{R}$. Hence there exists a continuous ω-periodic function $k(\cdot) : \mathbb{R} \to \mathbb{C}^n$ such that $\int^{\infty} b[\tilde{\Psi}(t)](0^-)k(t)dt = \infty$. Set $h(t) = k(t)e^{\delta t}$. Then $h(t)$ is a bounded continuous function, and $|bu(t)| = |bu(t)e^{-\delta t}| \to \infty$ as $t \to \infty$ for any solution $u(t)$ of Equation (2.7) in Section 6.2, since $(d/dt)(bu(t)e^{-\delta t}) = -b[\tilde{\Psi}(t)](0^-)k(t)$ a.e.. Consequently, Equation (4.1) has no $\mathbb{R}$-bounded solutions for this particular function $h(t)$. This proves the "only if" part of the theorem in the case where $\mathcal{B}$ satisfies Condition (*).

Next, we shall prove the assertion without assuming Condition (*). Let C_{00} be the set of all continuous functions on $(-\infty,0]$ with compact support, and denote by $\mathcal{B}^{\circ}$ the closure of C_{00} in $\mathcal{B}$. Since $\mathcal{B}$ is a uniform fading memory space, so is the space $\mathcal{B}^{\circ}$. Consider the equation

$$\dot{x}(t) = L^{\circ}(t,x_t) \tag{4.10}$$

on $\mathbb{R} \times \mathcal{B}^{\circ}$ associated with Equation (4.2), where $L^{\circ}(t,\varphi) := L(t,\varphi)$ for $(t,\varphi) \in \mathbb{R} \times \mathcal{B}^{\circ}$. Since the space BC is contained in $\mathcal{B}^{\circ}$ by Proposition 1.1, we have that x is an $\mathbb{R}$-bounded solution of Equation (4.1) if and only if it is an $\mathbb{R}$-bounded solution of Equation

$$\dot{x}(t) = L^{\circ}(t,x_t) + h(t) \tag{4.11}$$

on $\mathbb{R} \times \mathcal{B}^{\circ}$. Denote by $T^{\circ}(t,s)$ the solution operator of Equation (4.10), and denote by $\hat{T}^{\circ}(t,s)$ the operator induced by $T^{\circ}(t,s)$. Clearly, $T^{\circ}(t,s) = T(t,s)|_{\mathcal{B}^{\circ}}$ and $\hat{T}^{\circ}(t,s) = \hat{T}(t,s)|_{\hat{\mathcal{B}}^{\circ}}$. Now, suppose that $\Sigma_0 \neq \emptyset$. Then, from Corollary 6.2.3 there exists a nontrivial solution of Equation (4.2) defined on $\mathbb{R}$ of the form $x(t) = p(t)e^{\lambda t}$, where $p(t+\omega) = p(t)$ on $\mathbb{R}$ and $e^{\lambda\omega} \in \sigma(\hat{T}(\omega,0))$ with $\mathrm{Re}\,\lambda = 0$. Clearly, $x(t)$ is an $\mathbb{R}$-bounded solution of Equation (4.2); hence, it is an $\mathbb{R}$-bounded solution of Equation (4.10), and $\hat{T}^{\circ}(\omega,0)\hat{x}_t = e^{\lambda\omega}\hat{x}_t$ for all $t \in \mathbb{R}$. Thus, $e^{\lambda\omega} \in \{\mu \in \sigma(\hat{T}^{\circ}(\omega,0)) : |\mu| = 1\}$, from which it follows that there exists an $\mathbb{R}$-bounded continuous function $h(t)$ for which Equation (4.11) has no $\mathbb{R}$-bounded solutions, because $\mathcal{B}^{\circ}$ clearly satisfies Condition (*). Therefore, Equation (4.1) has no $\mathbb{R}$-bounded solutions for this function $h(t)$. This completes the proof of the "only if" part of the theorem.

7.5. Periodic solutions of linear periodic systems. Let $P_\omega = \{h : \mathbb{R} \to \mathbb{C}^n$ is continuous and ω-periodic$\}$. In this section, we assume that the space $\mathcal{B}$ is a uniform fading memory space satisfying Axiom (C1), again, and discuss the existence of solutions in P_ω of Equation (4.1) when h is in P_ω. As in the previous section, let

$$\hat{\mathcal{B}} = \hat{E}(s) \oplus \hat{F}(s), \qquad s \in \mathbb{R},$$

be the decomposition of the space $\hat{\mathcal{B}}$ corresponding to the set $\Gamma(0) = \{\lambda \in \sigma(\hat{P}(0)) : |\lambda| \geq 1\}$. By Corollary 6.2.7, we know that the coordinate $u(t) = \langle\Psi(t), \hat{x}_t(\sigma,\varphi,h)\rangle$ of $\hat{x}_t(\sigma,\varphi,h)$ with respect to the basis vector $\hat{\Phi}(t)$ satisfies the equation

$$\text{(5.1)} \qquad \dot{u}(t) = Gu(t) - [\tilde{\Psi}(t)](0^-)h(t) \quad \text{a.e.} \quad t \in \mathbb{R}.$$

Proposition 5.1. Let h be in P_ω. Then Equation (4.1) has a solution in P_ω if and only if Equation (5.1) has an ω-periodic solution.

Proof. As seen in the proof of Theorem 4.1, the function $\hat{\eta}_F : \mathbb{R} \to \hat{\mathcal{B}}$ defined by Relation (4.8) is a solution of Equation (2.6) in Section 6.2. Since h is in P_ω, we easily have that $\hat{\eta}_F(t+\omega) = \hat{\eta}_F(t)$, $t \in \mathbb{R}$. If $x(t)$ is a solution of Equation (4.1) such that $x(t)$ is in P_ω, then $\hat{T}(t,\sigma)(\Pi_F(\sigma)\hat{x}_\sigma - \hat{\eta}_F(\sigma)) = \Pi_F(t)\hat{x}_t - \hat{\eta}_F(t)$, $t \in \mathbb{R}$, by (2.6) in Section 6.2. Therefore, the same argument as in the proof of Lemma 4.4 shows that $\Pi_F(t)\hat{x}_t = \hat{\eta}_F(t)$, $t \in \mathbb{R}$. So, in order to investigate the existence of ω-periodic solutions of Equation (4.1), it is necessary only to consider Equation (2.5) in Section 6.2, and equivalently Equation (5.1). Indeed, if $u(t)$ is an ω-periodic solution of Equation (5.1), Theorem 4.2.9 implies that the function $x(t)$ defined by the relation

$$x(t) = [\hat{\eta}_E(t) + \hat{\eta}_F(t)](0), \qquad \hat{\eta}_E(t) = \hat{\Phi}(t)u(t),$$

is an ω-periodic solution of Equation (4.1), and vice versa.

In the case where $\mathcal{B} = C([-r,0]) \times L^p(g)$, Chow and Hale [1] has pointed out that the following theorem holds good by applying some fixed point theorem. Also, Shin [1] has extended their method to the general case of $\mathcal{B}$. We will give another proof by applying Proposition 5.1.

Theorem 5.2. Let $h \in P_\omega$, and suppose that $\mathcal{B}$ is a uniform fading memory space satisfying Axiom (C1). Then Equation (4.1) has a solution in P_ω whenever it has an $\mathbb{R}$-bounded solution.

Proof. By Proposition 5.1, it suffices to prove that Equation (5.1) has an ω-periodic solution. Note that Equation (5.1) has an $\mathbb{R}$-bounded solution if Equation (4.1) has an $\mathbb{R}$-bounded solution. Hence, the theorem follows from the well known theorem for linear systems of ordinary differential equations; see, e.g., Yoshizawa [2, Theorem 15.4].

The following theorem is related to (Halanay [1, Theorem 4.23]) for linear periodic functional differential equations with a finite delay r, $r < \omega$ (also, refer to Hale [1, Lemma 4.1.1], [5, Theorem 9.1.2]).

Theorem 5.3 (Murakami [4]). Suppose that all the conditions in Theorem 5.2 are satisfied. Then, the number of linearly independent ω-periodic solutions of Equation (4.2) is equal to the dimension of the space $\mathcal{A}_\omega = \{y(\cdot,\omega,\tilde{\psi}) : \psi \in \mathcal{B}^* \text{ and } y(\cdot,\omega,\tilde{\psi}) \text{ is } \omega\text{-periodic}\}$, where $y(\cdot,\omega,\tilde{\psi})$ is the solution of the adjoint equation

$$y(s) + \int_s^\omega y(\alpha)\eta(\alpha,s-\alpha)d\alpha = \tilde{\psi}(s), \qquad s \le \omega,$$

of Equation (4.2). Furthermore, Equation (4.1) has a solution in P_ω if and only if

$$\int_0^\omega y(t)h(t)dt = 0$$

for all $y(\cdot) \in \mathcal{A}_\omega$.

Proof. First of all, we shall prove the theorem under the additional condition (*) introduced in the proof of Theorem 4.1. From (4.6) it follows that a solution $x(t)$ of Equation (4.2) is ω-periodic if and only if $\hat{x}_t = \hat{\Phi}(t)e^{Gt}a$, $t \in \mathbb{R}$, for a vector a satisfying $Ma = a$, where $M = e^{G\omega}$. On the other hand, $y(\cdot,\omega,\tilde{\psi})$, $\psi \in \mathcal{B}^*$, is in $\mathcal{A}_\omega$ if and only if $\tilde{S}(-\omega)y_\omega(\omega,\tilde{\psi}) = y_\omega(\omega,\tilde{\psi})$, which is equivalent to that

$$[T^*(\omega,0)\psi]^\sim = (I + \Omega(0))\tilde{S}(-\omega)(I + \Omega(\omega))^{-1}\tilde{\psi}$$

$$= (I + \Omega(0))\tilde{S}(-\omega)y_\omega(\omega,\tilde{\psi})$$

$$= (I + \Omega(0))y_{\omega}(\omega, \tilde{\psi})$$

$$= \tilde{\psi}$$

by Theorem 4.2.2 and Relation (2.8) in Section 4.2, where $\tilde{S}(\cdot)$ and $\Omega(\cdot)$ are the ones defined by (2.4) and (2.7), respectively, in Section 4.2. Also, the last equation in the above means that $T^*(\omega,0)\psi = \psi$ by Condition (*). Thus, $y(\cdot,\omega,\tilde{\psi})$, $\psi \in \mathscr{B}^*$, is in $\mathscr{A}_{\omega}$ if and only if ψ can be written as the form $\psi = b\Psi(\omega)$ $(= b\Psi(0))$ for a row vector b satisfying $bM = b$. Therefore, the first part of the theorem follows from the well known result of linear algebra.

Next, we shall prove the second part of the theorem. It is well known (e.g., Halanay [1], Hale [1]) that the linear system of ordinary differential equations (5.1) has an ω-periodic solution if and only if

$$\int_0^{\omega} be^{-Gt}\Lambda(t)h(t)dt = 0$$

for all row vectors b for which be^{-Gt} is ω-periodic, that is, $bM = b$. On the other hand, since $y(t+\omega,\omega,\tilde{\psi}) = y(t,0,\tilde{\psi})$ for all $t < 0$ and $\tilde{\psi} \in BV$ by Relation (2.6) in Section 4.2, $be^{-Gt}\Lambda(t)$ for all b for which be^{-Gt} is ω-periodic coincides with the functions in $\mathscr{A}_{\omega}$, because

$$-be^{-Gt}\Lambda(t) = y(t,0,[b\tilde{\Psi}(0)]) = y(t+\omega,\omega,[b\tilde{\Psi}(0)]), \quad t < 0,$$

by (2.4) in Section 6.2. This proves the second part of the theorem under Condition (*).

Finally, we shall prove the theorem without assuming Condition (*). As in the proof of Theorem 4.1, let $\mathscr{B}^{\circ}$ be the closure of C_{00} in $\mathscr{B}$. For each $\psi^{\circ} \in (\mathscr{B}^{\circ})^*$, there exists a $\psi \in \mathscr{B}^*$ such that $\psi|_{\mathscr{B}^{\circ}} = \psi^{\circ}$ by the Hahn-Banach theorem. Let $\tilde{\psi}$ and $\tilde{\psi}^{\circ}$ be the elements in BV corresponding to ψ and ψ°, respectively. Then $\int_{-\infty}^0 [d\tilde{\psi}(\theta)]\varphi(\theta) = \int_{-\infty}^0 [d\tilde{\psi}^{\circ}(\theta)]\varphi(\theta)$ for all $\varphi \in C_{00}$; hence $\tilde{\psi} = \tilde{\psi}^{\circ}$ by Theorem 3.4.2. Thus, the adjoint equation of Equation (4.2) is equal to the one of Equation (4.10), and the set $\mathscr{A}_{\omega}$ is identical with the set $\mathscr{A}_{\omega}^{\circ} = \{y(\cdot,\omega,\tilde{\psi}^{\circ}) : \psi^{\circ} \in (\mathscr{B}^{\circ})^*$ and $y(\cdot,\omega,\tilde{\psi}^{\circ})$ is ω-periodic$\}$. On the other hand, by the same reason as in the proof of Theorem 4.1, we see that x is an ω-periodic solution of Equation (4.2) (resp. Equation (4.1)) if and only if it is an ω-periodic solution of Equation (4.10) (resp. Equation (4.11)). These observations complete the proof of the theorem, because the counterparts of the theorem hold for the space $\mathfrak{B}^{\circ}$.

The following theorem is a direct consequence of Theorem 5.3; cf. Hale [1, Theorem 4.1.1], [5, Theorem 9.1.1]).

<u>Theorem 5.4.</u> Suppose that all the assumptions in Theorem 5.2 are satisfied. Then, a necessary and sufficient condition that Equation (4.1) has an ω-periodic solution whenever h is in P_ω is that Equation (4.2) has no nontrivial ω-periodic solutions. In this case, this ω-periodic solution of Equation (4.1) is unique for $h \in P_\omega$.

CHAPTER 8

STABILITIES IN PERTURBED SYSTEMS AND LIMITING EQUATIONS

8.1. The hull of functions, its compactness and limiting equations.

Let X and Y be pseudo-metric spaces with pseudo-metric ρ_X and ρ_Y, respectively. Let $C(X,Y)$ be the space of all continuous functions mapping X into Y. We denote by $C^c(X,Y)$ the topological space $C(X,Y)$ equipped with the compact open topology. If X is separable, the space $C(X,Y)$ can be equipped with a metric. In fact, for a countable dense set $K_\infty = \{a_1, a_2, \cdots, a_m, \cdots\}$ in X, set

$$\rho(f,g) = \sum_{m=1}^{\infty} \frac{1}{2^m} \frac{\rho_Y(f(a_m),g(a_m))}{1+\rho_Y(f(a_m),g(a_m))}, \qquad f,\ g \in C(X,Y).$$

Clearly, ρ defines a metric on $C(X,Y)$. Henceforth, we denote by $C^\rho(X,Y)$ the space $C(X,Y)$ equipped with the metric ρ.

A family $\mathcal{F}$ of $C(X,Y)$ is said to be equicontinuous at $x_0 \in X$, if for any $\varepsilon > 0$ there exists a $\delta = \delta(x_0,\varepsilon) > 0$ such that

$$\rho_X(x,x_0) < \delta \quad \text{and} \quad f \in \mathcal{F} \quad \text{imply} \quad \rho_Y(f(x),f(x_0)) < \varepsilon.$$

If the family $\mathcal{F}$ is equicontinuous at each $x_0 \in X$, then it is said to be equicontinuous on X. In particular, if the family $\mathcal{F}$ is equicontinuous on X, and if one can choose a $\delta > 0$ independent of x_0 in the above, then the family $\mathcal{F}$ is said to be uniformly equicontinuous on X.

In what follows, we shall present some conditions for a family $\mathcal{F}$ in $C(X,Y)$ to be relatively (sequentially) compact. The following result is easy; so, we omit the proof.

<u>Proposition 1.1.</u> A family $\mathcal{F}$ in $C(X,Y)$ is equicontinuous on X if and only if it is uniformly equicontinuous on each compact set in X.

<u>Proposition 1.2.</u> If a family $\mathcal{F}$ in $C(X,Y)$ is equicontinuous on X, then the closure $\bar{\mathcal{F}}^c$ of $\mathcal{F}$ in $C^c(X,Y)$ is also equicontinuous on X.

Proof. Let $x_0 \in X$. For any $\varepsilon > 0$. we let $\delta = \delta(x_0, \varepsilon) > 0$ be the number in the definition of equicontinuity of $\mathcal{F}$. Let $g \in \bar{\mathcal{F}}^c$. Fix an $x_1 \in X$ such that $\rho_X(x_1, x_0) < \delta$. Then there exists an $f \in \mathcal{F}$ such that $\rho_Y(f(x_0), g(x_0)) < \varepsilon$ and $\rho_Y(f(x_1), g(x_1)) < \varepsilon$. Hence $\rho_Y(g(x_1), g(x_0)) < 3\varepsilon$. This proves the equicontinuity of $\bar{\mathcal{F}}^c$ on X.

Proposition 1.3. Suppose X is separable. If a family $\mathcal{F}$ in $C(X, Y)$ is equicontinuous, then the topologies of $\mathcal{F}$ induced from $C^c(X, Y)$ and $C^\rho(X, Y)$ are equivalent to each other.

Proof. Denote by $\mathcal{F}^c$ and $\mathcal{F}^\rho$ the space $\mathcal{F}$ equipped with the topologies induced from $C^c(X, Y)$ and $C^\rho(X, Y)$, respectively. We must prove that the identity map i from $\mathcal{F}^c$ into $\mathcal{F}^\rho$ is homeomorphic. First, we shall establish the continuity of i. To do this, let $g \in \mathcal{F}$ and let $V_\varepsilon(g) = \{h \in \mathcal{F} : \rho(g, h) < \varepsilon\}$ be the ε-neighborhood of g in $\mathcal{F}^\rho$. Select a positive integer N so that $\sum_{m=N}^{\infty} 1/2^m < \varepsilon/2$. Then, the set $U_N(g) := \{h \in \mathcal{F}: \rho_N(h, g) < \varepsilon/2\}$, where $\rho_N(h, g) = \sup\{\rho_Y(h(a_i), g(a_i)): i = 1, \cdots, N\}$, is a neighborhood of g in $\mathcal{F}^c$, and it satisfies $U_N(g) \subset V_\varepsilon(g)$, which proves the continuity of i.

Next, we shall establish the continuity of the inverse map of i. Let $U_{W,\varepsilon}(g) = \{h \in \mathcal{F}: \sup_{x \in W} \rho_Y(g(x), h(x)) < \varepsilon\}$ be any neighborhood of g in $\mathcal{F}^c$, where W is a compact set in X. We must show that there exists a $\gamma > 0$ such that the set $V_\gamma(g) = \{h \in \mathcal{F}: \rho(g, h) < \gamma\}$ is contained in $U_{W,\varepsilon}(g)$. To do this, we first claim that for any $\eta > 0$, there exists a $\delta(\eta) > 0$ such that if $\rho_X(x, y) < \delta(\eta)$ and $x \in W$, then $\rho_Y(f(x), f(y)) < \eta/4$ for all $f \in \mathcal{F}$. Indeed, if not, then there exist an $\eta > 0$ and sequences $\{f^k\} \subset \mathcal{F}$, $\{x_k\} \subset W$, $\{y_k\} \subset X$ such that $\rho_X(x_k, y_k) < 1/k$ and $\rho_Y(f^k(x_k), f^k(y_k)) \geq \eta/4$. $k = 1, 2, \cdots$. Then $W^* := W \cup \{y_1, y_2, \cdots\}$ is a compact set in X. Since $\mathcal{F}$ is uniformly equicontinuous on W^* by Proposition 1.1, we have that $\rho_Y(f^k(x_k), f^k(y_k)) \to 0$ as $k \to \infty$, which is a contradiction. Thus, the claim holds. Let $\{O_1, \cdots, O_{p_\varepsilon}\}$ be an open covering of W with diameter $\delta(\varepsilon)$. Since $K_\infty = \{a_1, a_2, \cdots\}$ is dense in X, there is an integer $k(\ell)$ such that $a_{k(\ell)} \in O_\ell$ for any ℓ with $1 \leq \ell \leq p_\varepsilon$. Set $N = \max\{k(\ell) : \ell = 1, \cdots, p_\varepsilon\}$. Select a constant $\gamma > 0$ so that $\gamma < \min(1/2^{N+1}, \varepsilon/2^{N+2})$. Let $h \in V_\gamma(g)$. Then $\rho_Y(g(a_i), h(a_i))/[1 + \rho_Y(g(a_i), h(a_i))] < 2^N \gamma$ for $i = 1, \cdots, N$; hence $\rho_N(g, h) \leq 2^{N+1} \gamma < \varepsilon/2$. Now, let $x \in W$. Then there exists an $a_k \in K_\infty$ with $1 \leq k \leq N$ such that $\rho_X(x, a_k) < \delta(\varepsilon)$; hence

$\rho_Y(f(x),f(a_k)) < \varepsilon/4$ for all $f \in \mathcal{F}$ by the preceding assertion. Thus, $\rho_Y(g(x),h(x)) \le \rho_Y(g(x),g(a_k)) + \rho_Y(g(a_k),h(a_k)) + \rho_Y(h(a_k),h(x)) \le \varepsilon/4 + \rho_N(g,h) + \varepsilon/4 < \varepsilon$, which shows that $h \in U_{W,\varepsilon}(g)$. Consequently, $V_\gamma(g) \subset U_{W,\varepsilon}(g)$. This is the desired result.

Theorem 1.4. Suppose Y is complete. If a family $\mathcal{F}$ is a relatively compact set in $C^c(X,Y)$, then it is equicontinuous, and moreover $\mathcal{F}(x) := \{f(x) : f \in \mathcal{F}\}$ is a relatively compact set in Y for each $x \in X$. If X is separable, then the contrary also holds.

Proof. To prove the first assertion, we may assume that $\mathcal{F}$ is a compact set in $C^c(X,Y)$. First of all, we shall prove that $\mathcal{F}(x)$ is a totally bounded set for each $x \in X$; then it is a relatively compact set in Y. Now, for any $f \in \mathcal{F}$ and any $\varepsilon > 0$, we set $V_\varepsilon(f,\{x\}) = \{g \in \mathcal{F} : \rho_Y(g(x),f(x)) < \varepsilon\}$. Then $V_\varepsilon(f,\{x\})$ is a neighborhood of f in $\mathcal{F}^c$, and $\mathcal{F} = \bigcup_{f\in\mathcal{F}} V_\varepsilon(f,\{x\})$. Since $\mathcal{F}$ is compact, we can choose a finite covering, say, $V_\varepsilon(f^1,\{x\}),\cdots, V_\varepsilon(f^n,\{x\})$, of $\mathcal{F}$. Thus, $\mathcal{F}(x) = \bigcup_{i=1}^{n} \{g(x) : g \in V_\varepsilon(f^i,\{x\})\}$, which shows that $\mathcal{F}(x)$ is totally bounded. Next, we shall establish the equicontinuity of $\mathcal{F}$ on X. If this is not the case, then there exist an $x_0 \in X$, an $\varepsilon_0 > 0$ and sequences $\{x_n\} \subset X$ and $\{f^n\} \subset \mathcal{F}$ such that $\rho_X(x_n,x_0) < 1/n$ and $\rho_Y(f^n(x_n),f^n(x_0)) \ge \varepsilon_0$, $n = 1,2,\cdots$. Since $S = \{x_0,x_1,\cdots\}$ is a compact set in X, and since $\mathcal{F}$ is compact in $C^c(X,Y)$, we may assume that $\{f^n\}$ converges to an $f^0 \in \mathcal{F}$ uniformly on S. Hence there exists an integer $n_0 > 0$ such that $\rho_Y(f^n(x_k),f^0(x_k)) < \varepsilon_0/3$ if $n > n_0$ and $k = 0,1,2,\cdots$. On the one hand, since f^0 is continuous, we can choose an integer $k_0 > 0$ so that $\rho_Y(f^0(x_k),f^0(x_0)) < \varepsilon_0/3$ if $k > k_0$. Hence, if $n > n_0$ and $k > k_0$, then

$$\rho_Y(f^n(x_k),f^n(x_0)) \le \rho_Y(f^n(x_k),f^0(x_k)) + \rho_Y(f^0(x_k),f^0(x_0)) + \rho_Y(f^0(x_0),f^n(x_0)) < \varepsilon_0,$$

a contradiction. Hence, the family $\mathcal{F}$ must be equicontinuous on X.

Next, we shall prove the second assertion. Since $\mathcal{F}$ is equicontinuous, it suffices to show that $\mathcal{F}$ is relatively compact in $C^p(X,Y)$ by Proposition 1.3. Let $\{f^k\}_{k=1}^{\infty}$ be any sequence in $\mathcal{F}$.

Since $\{f^k(a_m) : k = 1,2,\cdots\}$ is a relatively compact set in Y for each $a_m \in K_\infty$, by the diagonalization procedure we can select a subsequence $\{f^{k(i)}\}$ of $\{f^k\}$ so that the sequence $\{f^{k(i)}(a_m)\}_{i=1}^{\infty}$ converges for all $m = 1,2,\cdots$. Let $\varepsilon > 0$ and $x \in X$. Since $\mathcal{F}$ is equicontinuous at x and the set K_∞ is dense in X, there exists an $a_m \in K_\infty$ such that

$$\rho_Y(f^k(x), f^k(a_m)) < \varepsilon/3 \quad \text{for all } k = 1, 2, \cdots.$$

Fix such an a_m in K_∞. Since $\{f^{k(i)}(a_m)\}_{i=1}^{\infty}$ is a convergent sequence, there is an integer $i_0 > 0$ such that

$$\rho_Y(f^{k(i)}(a_m), f^{k(j)}(a_m)) < \varepsilon/3 \quad \text{if } i, j > i_0;$$

consequently,

$$\rho_Y(f^{k(i)}(x), f^{k(j)}(x)) < \varepsilon \quad \text{if } i, j > i_0.$$

Hence the sequence $\{f^{k(i)}(x)\}$ is convergent, because Y is complete. Denote by $f(x)$ its limit function. Clearly, f is in $C(X,Y)$ and $\rho_Y(f^{k(i)}(a_m), f(a_m)) \to 0$ as $i \to \infty$ for each $m = 1,2,\cdots$. Thus, $\rho(f^{k(i)}, f) \to 0$ as $i \to \infty$, which is the desired result.

Henceforth, we denote by J either $\mathbb{R}^+ = [0,\infty)$ or $\mathbb{R} = (-\infty,\infty)$. Obviously, the product space $J \times X$ is a pseudo-metric space. Recall that $C(J\times X,Y)$ be the space of all continuous functions mapping $J \times X$ into Y. In what follows, the space $C^c(J\times X,Y)$ will also be written as $C(J\times X,Y)$, for simplicity. Now, for $\tau \in J$ and $f \in C(J\times X,Y)$, we define the τ-translation f^τ of f by

$$f^\tau(t,x) = f(t+\tau,x), \qquad (t,x) \in J \times X.$$

Clearly, f^τ is in $C(J\times X,Y)$, too. Set

$$T(f) = \{f^\tau : \tau \in J\}$$

and

$$H(f) = \text{the closure of } T(f) \text{ in } C(J\times X,Y).$$

The set $H(f)$ is called the hull of f. Clearly, the hull $H(f)$ is invariant with respect to the τ-translation; that is, $g^{\tau} \in H(f)$ whenever $g \in H(f)$ and $\tau \in J$. Consider the space $UC(J\times X, Y)$ whose elements are all of functions in $C(J\times X, Y)$ continuous uniformly on $J \times W$ for any compact set W in X. If f is in $UC(J\times X, Y)$, then $T(f)$ is a family equicontinuous on $J \times X$ by Proposition 1.1; hence, so is the hull $H(f)$ by Proposition 1.2. If, in addition, X is separable, then the hull $H(f)$ is metrizable by Proposition 1.3. In this case, for any $g \in H(f)$ one can choose a sequence $\{\tau_m\} \subset J$ so that $f(t+\tau_m, x) \to g(t,x)$ as $m \to \infty$ uniformly on any compact subset of $J \times X$. Henceforth, if this situation occurs, we write as

$$f^{\tau_m} \to g \quad \text{compactly on} \quad J \times X,$$

for simplicity. In particular, we denote by $\Omega(f)$ the set of all elements g in $H(f)$ for which one can choose a sequence $\{\tau_m\} \subset J$ so that $\tau_m \to \infty$ as $m \to \infty$ and $f^{\tau_m} \to g$ compactly on $J \times X$. Clearly, $\Omega(f)$ is a closed subset of $H(f)$ and is invariant with respect to the τ-translation.

Now, let $\mathscr{B}$ be a fading memory space which is separable, and consider a system of FDEs

$$\text{(E)} \qquad \dot{x}(t) = F(t, x_t),$$

where $F \in UC(J\times\mathscr{B}, \mathbb{E})$. Throughout the remainder of this chapter, we will assume the following boundedness condition on F:

(E1) $\sup\{|F(t,\varphi)| : (t,\varphi) \in J \times \mathscr{B},\ |\varphi|_{\mathscr{B}} \le N\}\ (:= \ell(N)) < \infty$ for each $N > 0$.

From Theorem 1.4, we know that the hull $H(F)$ is compact in $C(J\times\mathscr{B}, \mathbb{E})$. If $G \in H(F)$, the system

$$\text{(G)} \qquad \dot{x}(t) = G(t, x_t), \qquad t \in J,$$

is called an equation in the hull of System (E). In particular, if $G \in \Omega(F)$, then it is called a limiting equation of System (E). System (E) is said to be regular, if any solution of each limiting equation of System (E) is unique for the initial conditions. Clearly, if $F(t,\varphi)$ is locally Lipschitzian in φ with Lipschitzian constant independent of t, then System (E) is regular.

In the following, we will always assume the following condition, too;

(E2) System (E) has a solution u on J such that

$$\sup_{t\in J} |u(t)|_{\mathbb{E}} < \infty.$$

In veirtue of Proposition 7.1.1. and 7.1.5, Condition (E2) implies that $\sup_{t\in J}|u_t|_{\mathcal{B}} < \infty$; hence, $\sup_{t\in J}|\dot{u}(t)| < \infty$ by Condition (E1). Thus, the orbit $\{u_t : t \in J\}$ of u is relatively compact in $\mathcal{B}$ by Proposition 7.1.1 and Corollary 7.3.2. Consider the function $U : J \to \mathcal{B}$ defined by

$$U(t) = u_t, \qquad t \in J.$$

Since $U \in C(J,\mathcal{B})$ by Axiom (A1), one can consider the hull $H(U)$ of U in $C(J,\mathcal{B})$. From the fact that $u \in UC(J,\mathbb{E})$, it follows that $U \in UC(J,\mathcal{B})$; hence, the hull $H(U)$ is metrizable by Proposition 1.3. Furthermore, since the set $\{U(t) : t \in J\}$ is relatively compact in $\mathcal{B}$, the hull $H(U)$ is compact in $C(J,\mathcal{B})$ by Theorem 1.4. Therefore, for any sequence $\{\tau_m'\} \subset J$ there exist a sequence $\{\tau_m\} \subset \{\tau_m'\}$ and a function $V \in C(J,\mathcal{B})$ such that

(1.1) $\quad U(t+\tau_m) \to V(t)$ as $m \to \infty$ in $C(J,\mathcal{B})$.

Set

$$v(t) = [V(t)](0) \quad \text{for} \quad t \in J$$

and, in the case of $J = \mathbb{R}^+$, extend the function v on $\mathbb{R}$ as

$$v_0 = V(0).$$

Relation (1.1) is equivalent that $u(t+\tau_m) \to v(t)$ as $m \to \infty$ in $C(J,\mathbb{E})$ and $u_{\tau_m} \to v_0$ in $\mathcal{B}$ as $m \to \infty$. So, for the sake of simplicity, we will dare to use the notation $H(u)$ in the place of $H(U)$, and write Relation (1.1) as

$$u^{\tau_m} \to v \quad \text{compactly on } J.$$

Next, we denote by $H(u,F)$ the closure of the set $\{(u^\tau,F^\tau) : \tau\in J\}$ in the product space $C(J,\mathcal{B}) \times C(J\times\mathcal{B},\mathbb{E})$. The space $H(u,F)$ is metrizable and compact. Moreover, it is invariant with respect to the τ-translation. Let $(v,G) \in H(u,F)$. There exists a sequence

$\{\tau_m\} \subset J$ such that $(u_{t+\tau_m}, F^{\tau_m}(t,\varphi)) \to (v_t, G(t,\varphi))$ as $m \to \infty$ in $C(J,\mathcal{B}) \times C(J\times\mathcal{B},E)$. In this case, we abbreviate it as

$$(u^{\tau_m}, F^{\tau_m}) \to (v,G) \quad \text{compactly on } J \times \mathcal{B}.$$

Recall that $\Omega(u,F)$ is the set of all $(v,G) \in H(u,F)$ for which there exists a sequence $\{\tau_m\} \subset J$ with $\tau_m \to \infty$ as $m \to \infty$ such that $(u^{\tau_m}, F^{\tau_m}) \to (v,G)$ compactly on $J \times \mathcal{B}$. The set $\Omega(u,F)$ is closed in $H(u,F)$ and invariant with respect to the τ-translation. In particular, it is metrizable and compact, too. Now, if $(v,G) \in H(u,F)$, then v is a solution on J of System (G). Indeed, let J_0 be a compact interval in J, and fix a sequence $\{\tau_m\} \subset J$ such that $(u^{\tau_m}, F^{\tau_m}) \to (v,G)$ compactly on $J \times \mathcal{B}$. Since the set $W := \{u_{\tau_m+t}, v_t : t \in J_0, m = 1,2,\cdots\}$ is compact in $\mathcal{B}$, the sequence $\{F(t+\tau_m, u_{t+\tau_m})\}$ converges to $G(t,v_t)$ as $m \to \infty$ uniformly for $t \in J_0$. Hence, letting $m \to \infty$ in the relation

$$u(t+\tau_m) = u(\tau_m) + \int_0^t F(s+\tau_m, u_{s+\tau_m})ds, \qquad t \in J_0,$$

we have

$$v(t) = v(0) + \int_0^t G(s,v_s)ds, \qquad t \in J_0.$$

This proves the above claim. Summarizing the above results, we have the following theorem, which will be used later on.

<u>Theorem 1.5.</u> Let $\mathcal{B}$ be a separable fading memory space, and suppose Conditions (E1) and (E2) are satisfied. Given any sequences $\{(x^k,G^k)\} \subset H(u,F)$ (resp. $(x^k,G^k) \subset \Omega(u,F)$) and $\{s_k\} \subset J$, there exist subsequences $\{(x^{k_j},G^{k_j})\} \subset \{(x^k,G^k)\}$ and $\{s_{k_j}\} \subset \{s_k\}$ such that $((x^{k_j})^{s_{k_j}}, (G^{k_j})^{s_{k_j}}) \to (y,G)$ compactly on $J \times \mathcal{B}$ for some $(y,G) \in H(u,F)$ (resp. $(y,G) \in \Omega(u,F)$). In this case, y is a solution on J of System (G).

<u>8.2. Stabilities for equations in the hull.</u>

Throughout the remainder of this chapter, we assume that $\mathcal{B}$ is a separable fading memory space. In this section, we consider System (E) as $J = \mathbb{R}^+ = [0,\infty)$ under Conditions (E1) and (E2), and discuss some stability properties for System (G), an equation in the hull H(F).

Definition 2.1. The solution u is said to be uniformly stable (US) in H(F) (resp. $\Omega(F)$), if for any $(v,G) \in H(u,F)$ (resp. $\Omega(u,F)$), the solution v of System (G) is US in $\mathscr{B}$ with a common number $\delta(\cdot)$. The solution u is said to be weakly uniformly asymptotically stable (WUAS) in H(F) (resp. $\Omega(F)$), if for any $(v,G) \in H(u,F)$ (resp. $\Omega(u,F)$), the solution v of System (G) is WUAS in $\mathscr{B}$ with a common pair $(\delta_0,\delta(\cdot))$. Furthermore, the solution u is said to be uniformly asymptotically stable (UAS) in H(F) (resp. $\Omega(F)$), if for any $(v,G) \in H(u,F)$ (resp. $\Omega(u,F)$), the solution v of System (G) is UAS in $\mathscr{B}$ with a common triple $(\delta_0,\delta(\cdot),T(\cdot))$.

In the same manner as in Section 7.2, one can define the stabilities "in $\mathbb{F}$". However, from Proposition 7.2.5 and Theorem 7.2.6, the concepts of stabilities in $\mathscr{B}$ and stabilities in $\mathbb{F}$ are equivalent to each other, whenever the stabilities in $\mathscr{B}$ happen to occur. So, we shall often omit the suffix "in $\mathscr{B}$" in the above stabilities.

The following results are due to Hino [9], which show that under some assumptions on System (E), the stability properties for System (E) are inherited to the equations in the hull (for the ODE case, see [Yoshizawa [2]).

Proposition 2.2. Suppose System (E) is regular. If the solution u of System (E) is US (resp. UAS), then it is US (resp. UAS) in H(F).

Proposition 2.3. Suppose System (E) is periodic. If the solution u of System (E) is US (resp. WUAS, resp. UAS), then it is US (resp. WUAS, resp. UAS) in H(F).

Proof of Proposition 2.2. Suppose the solution u is US with a pair $(\varepsilon,\delta(\varepsilon))$. Let $(v,G) \in H(u,F)$. Then there exists a sequence $\{\tau_m\} \subset \mathbb{R}^+$ such that $(u^{\tau_m},F^{\tau_m}) \to (v,G)$ compactly on $\mathbb{R}^+ \times \mathscr{B}$. If the sequence $\{\tau_m\}$ is bounded, then $(v,G) = (u^{\tau},F^{\tau})$ for some $\tau \in \mathbb{R}^+$; hence the solution v of System (G) is US with the pair $(\varepsilon,\delta(\varepsilon))$. Thus, without loss of generality, we may assume that $\tau_m \to \infty$ as $m \to \infty$; so, $(v,G) \in \Omega(u,F)$.

Now, let $s \in \mathbb{R}^+$, $\varepsilon > 0$ and $|\varphi - v_s|_{\mathscr{B}} < \delta(\varepsilon/2)/2$. Put $x^m(t) =$

$x(t+\tau_m, s+\tau_m, \varphi, F)$. Then $x^m(t)$ is the solution of the system

$$(2.1) \qquad \dot{x}(t) = F(t+\tau_m, x_t)$$

through (s,φ). Since $u^m(t) := u(t+\tau_m)$ is a solution of System (2.1) which is US with the pair $(\varepsilon, \delta(\varepsilon))$, and since $|(u^m)_s - (x^m)_s|_{\mathscr{B}} \le |(u^m)_s - v_s|_{\mathscr{B}} + |v_s - \varphi|_{\mathscr{B}} < \delta(\varepsilon/2)$ for sufficiently large m, we have

$$(2.2) \qquad |(u^m)_t - (x^m)_t|_{\mathscr{B}} < \varepsilon/2, \qquad t \ge s,$$

for sufficiently large m. In particular, we see that the sequence of functions $\{x^m(t)\}$ is uniformly bounded and equicontinuous on each compact subset of $[s,\infty)$ with $(x^m)_s = \varphi$. Hence, by Axiom (A) and the Ascoli-Arzéla theorem, we may assume that the sequence $\{(x^m)_t\}$ converges to y_t in $\mathscr{B}$ compactly on $[s,\infty)$, where y is a solution of System (G) through (s,φ) and it is uniquely determined on $[s,\infty)$ by the assumption. Consequently, if m is sufficiently large, then

$$(2.3) \qquad |(x^m)_t - y_t|_{\mathscr{B}} < \varepsilon/4 \quad \text{and} \quad |(u^m)_t - v_t|_{\mathscr{B}} < \varepsilon/4$$

for each $t \ge s$. Hence, $|v_s - \varphi|_{\mathscr{B}} < \delta(\varepsilon/2)/2$ implies $|v_t - y_t|_{\mathscr{B}} < \varepsilon$ for all $t \ge s$ by (2.2) and (2.3). Thus, v is US with the pair $(\varepsilon, \delta(\varepsilon/2)/2)$; hence, u is US in $H(F)$.

Next, suppose u is UAS with a triple $(\delta_0, \delta(\cdot), T(\cdot))$. Let $(v,G) \in H(u,F)$. By employing the same argument as in the above, we easily see that, if $|v_s - \varphi|_{\mathscr{B}} < \delta_0/2$, then $|v_t - y_t|_{\mathscr{B}} < \varepsilon$ for $t \ge s+T(\varepsilon/2)$. Hence, v is UAS with a common triple $(\delta_0/2, \delta(\cdot/2)/2, T(\cdot/2))$ for each $(v,G) \in H(u,F)$; thus, u is UAS in $H(F)$.

<u>Proof of Proposition 2.3.</u> Suppose $F(t+\omega,\varphi) = F(t,\varphi)$ for an $\omega > 0$. Let (v,G), $\{\tau_m\}$ and $u^m(t)$ be the ones in the proof of Proposition 2.2. Set $\tau_m = N_m\omega + \sigma_m$, where N_m is a nonnegative integer and $0 \le \sigma_m < \omega$. We may assume that $\sigma_m \to \sigma$ as $m \to \infty$. Then $0 \le \sigma \le \omega$ and $G(t,\varphi) = F(t+\sigma,\varphi)$.

Now, suppose the solution u of System (E) is US with a pair $(\varepsilon, \delta(\varepsilon))$. Let any $\varepsilon > 0$ be given, and assume that $|v_s - \varphi|_{\mathscr{B}}$ $(=: k)$ $< \delta(\varepsilon)$ for some $s \in \mathbb{R}^+$. Since u is uniformly continuous on $\mathbb{R}^+$, so is the $\mathscr{B}$-valued function u_t. Therefore, if m is sufficiently large, we have $|(u^m)_s - v_s|_{\mathscr{B}} < (\delta(\varepsilon)-k)/2$ and $|u_{s+\sigma+N_m\omega} -$

$u_{s+\sigma_m+N_m\omega}|_{\mathscr{B}} < (\delta(\varepsilon)-k)/2$, consequently $|u_{s+\sigma+N_m\omega} - \varphi|_{\mathscr{B}} \le$ $|u_{s+\sigma+N_m\omega} - u_{s+\sigma_m+N_m\omega}|_{\mathscr{B}} + |(u^m)_s - v_s|_{\mathscr{B}} + |v_s - \varphi|_{\mathscr{B}} < \delta(\varepsilon)$. Notice that $u(t+\sigma+N_m\omega)$ is a solution of System (G) which is US with the pair $(\varepsilon, \delta(\varepsilon))$. Thus,

$$|u_{t+\sigma+N_m\omega} - x_t(s,\varphi,G)|_{\mathscr{B}} < \varepsilon, \qquad t \ge s,$$

for all large m. On the one hand, for an arbitrary $\gamma > 0$, we have

$$|v_s - u_{s+\sigma+N_m\omega}|_{\mathscr{B}} \le |v_s - (u^m)_s|_{\mathscr{B}} + |u_{s+\sigma_m+N_m\omega} - u_{s+\sigma+N_m\omega}|_{\mathscr{B}} < \delta(\gamma)$$

for all large m; hence

$$|v_t - u_{t+\sigma+N_m\omega}|_{\mathscr{B}} < \gamma, \qquad t \ge s,$$

for all large m. Thus, if $|v_s - \varphi|_{\mathscr{B}} < \delta(\varepsilon)$, then $|v_t - x_t(s,\varphi,G)|_{\mathscr{B}} < \varepsilon+\gamma$ for all $t \ge s$. Since γ is arbitrary, we have $|v_t - x_t(s,\varphi,G)|_{\mathscr{B}} \le \varepsilon$ for all $t \ge s$ whenever $|v_s - \varphi|_{\mathscr{B}} < \delta(\varepsilon)$. This proves that u is US in $H(F)$.

Next, suppose the solution u is WUAS (resp. UAS) with $(\delta_0, \delta(\cdot))$ (resp. $(\delta_0, \delta(\cdot), T(\cdot))$. Assume that $|v_s - \varphi|_{\mathscr{B}} < \delta_0$ for some $s \in \mathbb{R}^+$. Employing the same argument as in the proof of the first part, we have that $|u_{s+\sigma+N_m\omega} - \varphi|_{\mathscr{B}} < \delta_0$ and $|u_{s+\sigma+N_m\omega} - v_s|_{\mathscr{B}} < \delta_0$ for sufficiently large m. Hence, if m is sufficiently large, then $|u_{t+\sigma+N_m\omega} - x_t(s,\varphi,G)|_{\mathscr{B}} \to 0$ as $t \to \infty$ (resp. $< \varepsilon$ for all $t \ge s+T(\varepsilon)$) and $|u_{t+\sigma+N_m\omega} - v_t|_{\mathscr{B}} \to 0$ as $t \to \infty$ (resp. $< \varepsilon$ for $t \ge s+T(\varepsilon)$). Thus, if $|v_s - \varphi|_{\mathscr{B}} < \delta_0$ for some $s \in \mathbb{R}^+$, then $|x_t(s,\varphi,G) - v_t|_{\mathscr{B}} \to 0$ as $t \to \infty$ (resp. $< 2\varepsilon$ for $t \ge s+T(\varepsilon)$). This shows that u is WUAS (resp. UAS) in $H(F)$.

In Proposition 2.2, the similar implication is not true for the "WUAS", as the simple example $\dot{x} = -x/(t+1)$ shows. Furthermore, in Propositions 2.2 and 2.3, one cannot remove the regularity or periodicity assumption on System (E). In fact, even for an almost periodic equation, the uniform stability and the uniform asymptotic stability are not necessarily inherited without these assumptions. The following example due to Kato [1] shows this fact; also, refer to Yoshizawa [2, pp. 143-145].

<u>Kato's example.</u> Let $f(x)$ be defined by

$$f(x) = \begin{cases} 2\sqrt{|nx - 1|} & \frac{2}{2n+1} \le x \le \frac{2}{2n-1} \quad (n = 1, 2, \cdots) \\ 0 & x = 0 \\ -f(-x) & x < 0. \end{cases}$$

Then, we can see that the zero solution of the system

$$\dot{x} = f(x)$$

is not unique for initial conditions. Let $a(t) = \sum_{k=0}^{\infty} a_k(t)$ be the almost periodic function, where $a_0(t) = 1$ and $a_k(t)$ is a periodic function of period 2^k such that

$$a_k(t) = \begin{cases} 0 & 0 < t < 2^{k-1} \\ -\frac{1}{2^k} & 2^{k-1} \le t \le 2^k \end{cases} \qquad (k = 1, 2, \cdots)$$

(the continuity can be given by a slight modification). Since $\sum_{k=0}^{\infty} a_k(t) = 1-(1/2+1/2^2+ \cdots + 1/2^m)$ for $2^{m-1} \le t \le 2^m$, we have $a(t+2^m-1) \to 0$ as $m \to \infty$ uniformly for $0 \le t \le 1$. Now, consider the ordinary differential equation

$$\dot{x} = F(t,x), \tag{2.4}$$

where

$$F(t, x) = \begin{cases} f(x) - ca(t)\sqrt{x} & x \ge 0, \\ -F(t,-x) & x < 0 \end{cases}$$

for a constant $c > 2\sqrt{2}$.

Since $a(t) > 0$ for all t, we have $F(t,1/m) < 0$ for all t and $m = 1,2,\cdots$; hence, any solution of (2.4) cannot cross the line $x = 1/m$ upwards for $m = 1,2,\cdots$. Consequently, the zero solution of (2.4) is US. Notice that $a(t) \ge 1/2$ for $2m < t < 2m+1$ and $f(x) \le \sqrt{2x}$. Then

$$F(t,x) \le (\sqrt{2} - \frac{c}{2})\sqrt{x} < 0 \quad \text{for} \quad x > 0 \quad \text{and} \quad 2m < t < 2m+1;$$

so, comparing with the solution of $\dot{x} = (2\sqrt{2}-c)\sqrt{x}\,/2$, we can see that every solution of (2.4) starting from a neighborhood of $x = 0$ tends to zero as $t \to \infty$ and, more precisely, that the zero solution of (2.4) is UAS. On the one hand, since $a(t+2^m-1) \to 0$ as $m \to \infty$ for $0 \le t \le 1$, there exists a $G(t,x) \in \Omega(F) \subset H(F)$ such that $G(t,x) = f(x)$ for all $t \in [0,1]$. Therefore, the zero solution of $\dot{x} = G(t,x)$ is not unique for initial conditions; consequently, it is not uniformly stable.

<u>8.3. Stabilities in perturbed systems via limiting equations.</u> We consider a system of FDEs

$$\text{(E)} \qquad \dot{x}(t) = F(t,x_t)$$

and its perturbed system

$$\text{(P)} \qquad \dot{x}(t) = F(t,x_t) + g(t,x_t),$$

where F and g are in $C(\mathbb{R}^+ \times \mathscr{B}, E)$. The purpose of this section is to investigate some stability properties for System (P), assuming some stability properties for the limiting equations of System (E). To do this, we need some definitions and preparations.

Let $C(\mathbb{R}^+, E)$ denote the space of all continuous functions mapping $\mathbb{R}^+ = [0,\infty)$ into E, and let T, L and M be the normed spaces in $C(\mathbb{R}^+, E)$ with the norms $|\cdot|_T$, $|\cdot|_L$ and $|\cdot|_M$, respectively, where

$$|p|_T = \sup_{t \ge 0} |p(t)|, \quad |p|_L = \int_0^\infty |p(t)|dt, \quad |p|_M = \sup_{t \ge 0} \int_t^{t+1} |p(s)|ds.$$

Clearly, $\{0\}$, T and L are subspaces of M; more precisely, $|p|_M \le |p|_X$ for each $p \in X$. Here and hereafter, X denotes the one of $\{0\}$, T, L or M.

Let $u : \mathbb{R} \to E$ be a function which is continuous on $\mathbb{R}^+$ with $u_0 \in \mathscr{B}$. In the below, we define stabilities of the function u for System (E). It should be noticed that u is not necessarily a solution of System (E).

<u>Definition 3.1.</u> The function u is said to be eventually stable under X-perturbations for System (E) (Ev XS for (E), in

short), if for any $\varepsilon > 0$ there exist $\alpha = \alpha(\varepsilon) \geq 0$ and $\delta = \delta(\varepsilon) > 0$ such that $\sigma \geq \alpha(\varepsilon)$, $|u_\sigma - \varphi|_{\mathcal{B}} < \delta(\varepsilon)$ and $|p|_X < \delta(\varepsilon)$ imply $|u_t - x_t(\sigma,\varphi,F+p)|_{\mathcal{B}} < \varepsilon$ for all $t \geq \sigma$, where $x(t,\sigma,\varphi,F+p)$ denotes any solution of the system $\dot{x}(t) = F(t,x_t) + p(t)$ through (σ,φ). In particular, if one can choose $\alpha(\varepsilon) \equiv 0$, then the function u is said to be stable under X-perturbations for System (E) (XS for (E), in short).

Definition 3.2. The function u is said to be eventually asymptotically stable under X-perturbations for System (E) (Ev XAS for (E), in short), if it is Ev XS for (E), and moreover, there exist $\delta_0 > 0$ and $\alpha_0 \geq 0$ with the property that for any $\varepsilon > 0$ there exist $T(\varepsilon) > 0$ and $\gamma(\varepsilon) > 0$ such that $\sigma \geq \alpha_0$, $|u_\sigma - \varphi|_{\mathcal{B}} < \delta_0$ and $|p|_X < \gamma(\varepsilon)$ imply $|u_t - x_t(\sigma,\varphi,F+p)|_{\mathcal{B}} < \varepsilon$ for all $t \geq \sigma + T(\varepsilon)$. In particular, the function u is said to be asymptotically stable under X-perturbations for System (E) (XAS for (E), in short), if it is XS for (E) and if one can choose $\alpha_0 = 0$.

If $X = \{0\}$, the stability under X-perturbations and the asymptotic stability under X-perturbations correspond to the uniform stability (in $\mathcal{B}$) and the uniform asymptotic stability (in $\mathcal{B}$), respectively (see Section 7.2). On the one hand, if $X = T$, L or M, then the stability under X-perturbations (resp. the asymptotic stability under X-perturbations) is called the total stability (TS) (resp. the total asymptotic stability (TAS)), the integral stability (IS) (resp. the integral asymptotic stability (IAS)), or the M-stability (MS) (resp. the M-asymptotic stability (MAS)), respectively. The implications (Ev)MS $\Rightarrow$ (Ev)TS $\Rightarrow$ (Ev)US and (Ev)MS $\Rightarrow$ (Ev)IS $\Rightarrow$ (Ev)US follow from the fact that $M \supset T \cup L$ and $T \cap L \supset \{0\}$. Moreover, the similar relationships hold among UAS, TAS, IAS and MAS.

If the function u is XS for (E), then it is the unique solution of System (E) for the initial conditions; that is, $u(t) \equiv x(t,s,u_s,F)$ for all $t \geq s \geq 0$. The following result shows that the prefix "Ev" can be deleted in Definitions 3.1 and 3.2, whenever u is the unique solution of System (E) for the initial conditions.

Proposition 3.3. Suppose u is the unique solution of System (E) for the initial conditions. If u is Ev XS (resp. Ev XAS) for (E), then it is XS (resp. XAS) for (E).

Proof. Let an $a > 0$ be given. Since the set $\{(t,u_t) : 0 \le t \le a\}$ is compact in $\mathbb{R}^+ \times \mathscr{B}$, there exist constants $\gamma > 0$ and $\ell > 0$ such that $(s,\psi) \in W$ implies $|F(s,\psi)| \le \ell$, where $W = \{(s,\psi) \in \mathbb{R}^+ \times \mathscr{B} : |t - s| \le \gamma$ and $|\psi - u_t|_{\mathscr{B}} \le \gamma$ for some $t \in [0,a]\}$. Since W is a closed set in $\mathbb{R}^+ \times \mathscr{B}$, by the Tietze extension theorem there exists a continuous function $G(s,\psi)$ on $\mathbb{R}^+ \times \mathscr{B}$ such that $|G(s,\psi)| \le \ell$ on $\mathbb{R}^+ \times \mathscr{B}$ and $G = F$ on W.

Now, we shall prove the following claim:

(*) for any $\varepsilon > 0$, there exists a $\delta_1(\varepsilon,a) > 0$, $0 < \delta_1(\varepsilon,a) < \varepsilon$, such that if $s \in [0,a]$, $|u_s - \varphi|_{\mathscr{B}} < \delta_1(\varepsilon,a)$ and $|p|_X < \delta_1(\varepsilon,a)$, then $\sup_{t\in[s,a]} |u_t - x_t(s,\varphi,F+p)|_{\mathscr{B}} < \varepsilon$.

If the claim (*) is false, then there exist an $\varepsilon > 0$ and sequences $\{s_k\} \subset [0,a]$, $\{\varphi^k\} \subset \mathscr{B}$, $\{p^k\} \subset X$ and $\{t_k\} \subset [s_k,a]$ such that $|u_{s_k} - \varphi^k|_{\mathscr{B}} < 1/k$, $|p_k|_X < 1/k$ and $|u_{t_k} - x_{t_k}(s_k,\varphi^k,F+p^k)|_{\mathscr{B}} = \varepsilon$, where we may assume that $\varepsilon \le \gamma$ and $|u_t - x_t(s_k,\varphi^k,F+p^k)|_{\mathscr{B}} < \varepsilon$ on $[s_k,t_k)$. Since $|G(s,\psi)| \le \ell$ on $\mathbb{R}^+ \times \mathscr{B}$, $x(t,s_k,\varphi^k,F+p^k)$ has a continuous extension beyond t_k up to t_k+1 as a solution of $\dot{x}(t) = G(t,x_t) + p^k(t)$, which we shall denote by $x^k(t)$, again. We may assume that $s_k \to s_0 \in [0,a]$ and $t_k \to t_0 \in [s_0,a]$ as $k \to \infty$. Hence, $u_{s_k} \to u_{s_0}$ and $\varphi^k \to u_{s_0}$ as $k \to \infty$. Since $x^k(t+s_k)$ is a solution of $\dot{x}(t) = G(t+s_k,x_t) + p^k(t+s_k)$ on $[0, t_k-s_k+1]$ through $(0,\varphi^k)$, the sequence of functions $\{x^k(t+s_k)\}$ is uniformly bounded and equicontinuous on any compact subset of $[0,t_0-s_0+1)$. By the Ascoli-Arzéla theorem, the sequence $\{x^k(t+s_k)\}$ has a subsequence, which will be denoted by $\{x^k(t+s_k)\}$ again, converging to a function y, $y_0 = u_{s_0}$, compactly on $[0,t_0-s_0+1)$. Since $(x^k)_{t+s_k} \to y_t$ as $k \to \infty$ compactly on $[0,t_0-s_0+1)$ by Axiom (A) and $|p^k|_X \to 0$ as $k \to \infty$, we can easily see that $y(t)$ is a solution on $[0,t_0-s_0+1)$ of $\dot{x}(t) = G(t+s_0,x_t)$ through $(0,u_{s_0})$. Moreover, we have

$$(3.1) \qquad |u_{s_0} - y_{t_0-s_0}|_{\mathscr{B}} = \varepsilon, \text{ and } |u_{t+s_0} - y_t|_{\mathscr{B}} \le \varepsilon \text{ on } [0,t_0-s_0].$$

Hence, $(t+s_0,y_t) \in W$ for all $t \in [0,t_0-s_0]$; so, $y(t)$ is a solution on $[0,t_0-s_0]$ of $\dot{x}(t) = F(t+s_0,x_t)$ through $(0,u_{s_0})$. By the assumption of this proposition, we have $y(t-s_0) = x(t,s_0,u_{s_0},F) = u(t)$ for all $t \ge s_0$; in particular, $|u_{t_0} - y_{t_0-s_0}|_{\mathscr{B}} = 0$ by Axiom (A), which contradicts (3.1). Thus, the claim (*) must be true.

Now, suppose that $u(t)$ is Ev XS for (E), and let $\delta(\varepsilon)$ and $\alpha(\varepsilon)$ be the numbers given in Definition 3.1. Then $\bar{\delta}(\varepsilon) = \delta_1(\delta(\varepsilon),\alpha(\varepsilon))$ satisfies the condition for $u(t)$ to be XS for (E). In the similar way, we can verify the assertion on the asymptotic stability.

Now, as in the preceding section, we assume that $\mathscr{B}$ is a separable fading memory space and that Conditions (E1) and (E2) are satisfied. Moreover, we assume the following condition:

$$\text{(E3)} \qquad \sup_{0\le\alpha\le 1} \left|\int_t^{t+\alpha} g(s,\varphi)ds\right| \to 0 \quad \text{as} \quad t \to \infty \quad \text{for each} \quad \varphi \quad \text{in} \quad \mathscr{B},$$

and moreover, for each $r > 0$ there exist functions $b^r(s)$, $b^r(0) = 0$, $b^r(s)$ nondecreasing in s, $b^r(s) \to 0$ as $s \to 0$, and $h^r(t)$, $\int_t^{t+1} |h^r(\tau)|d\tau \to 0$ as $t \to \infty$, such that

$$|g(t,\varphi) - g(t,\psi)| \le b^r(|\varphi - \psi|_{\mathscr{B}}) + h^r(t)$$

for all (t,φ) and (t,ψ) in $\mathbb{R}^+ \times \mathscr{B}$ with $|\varphi|_{\mathscr{B}} \le r$ and $|\psi|_{\mathscr{B}} \le r$.

The function g satisfying Condition (E3) is said to be diminishing; cf. Strauss & Yorke [1,2]. For instance, $g(t,\varphi) = \varphi(0)\cdot\sin t^2$ satisfies Condition (E3). Note that u is a solution of System (E) by Condition (E2), while it is not necessarily a solution of System (P). In the following theorem, we will derive the stability under X-perturbations for System (P) from some stability properties for the limiting equations of System (E); cf. Hino [9].

Theorem 3.4 (Murakami [1]). If u is WUAS in $\Omega(F)$, then it is Ev MS for (P); consequently, it is Ev XS for (P). In addition, if u is the unique solution of System (P) for the initial conditions, then it is MS for (P); consequently, it is XS for (P).

Proof. Suppose u is WUAS in $\Omega(F)$ with a pair $(\delta_0,\delta(\cdot))$, and that it is not Ev MS for (P). Then, there exist a constant ε, $0 < \varepsilon < \delta_0$, and sequences $\{\varepsilon_k\}$, $\varepsilon_k < \varepsilon$ and $\varepsilon_k \to 0$ as $k \to \infty$, $\{s_k\}$, $s_k \to \infty$ as $k \to \infty$, $\{t_k\}$, $t_k > s_k$, $\{p^k\} \subset M$, $|p^k|_M < \varepsilon_k$, $\{\varphi^k\} \subset \mathscr{B}$ and $\{x(\cdot,s_k,\varphi^k,F+g+p^k)\}$ such that

$$(3.2) \qquad |\varphi^k - u_{s_k}|_{\mathscr{B}} < \varepsilon_k$$

and

(3.3) $\qquad |x_{t_k}(s_k,\varphi^k,F+g+p^k) - u_{t_k}|_{\mathscr{B}} = \varepsilon$ and

$$|x_t(s_k,\varphi^k,F+g+p^k) - u_t|_{\mathscr{B}} < \varepsilon \quad \text{on } [s_k,t_k).$$

Moreover, there exists a sequence $\{\tau_k\}$, $s_k < \tau_k < t_k$, such that

(3.4) $\qquad |x_{\tau_k}(s_k,\varphi^k,F+g+p^k) - u_{\tau_k}|_{\mathscr{B}} = \delta(\varepsilon/2)/2$

and

(3.5) $\qquad \delta(\varepsilon/2)/2 \le |x_t(s_k,\varphi^k,F+g+p^k) - u_t|_{\mathscr{B}} \le \varepsilon$ on $[\tau_k,t_k]$

by (3.2) and (3.3). Since the set $\{u_t : t \ge 0\}$ is relatively compact in $\mathscr{B}$, we may assume that $u_{s_k} \to \varphi^0$ as $k \to \infty$ for a φ^0 in $\mathscr{B}$; hence, $\varphi^k \to \varphi^0$ as $k \to \infty$ by (3.2). Moreover, since $\sup_{t\ge 0} |u_t|_{\mathscr{B}} < \infty$, (3.5) implies that

(3.6) $\qquad |x_t(s_k,\varphi^k,F+g+p^k)|_{\mathscr{B}} \le r/2$ on $[\tau_k,t_k]$

for some constant r and for all $k = 1,2,\cdots$. Now, we shall establish the following claim:

(3.7) $\qquad \sup\{|x_t(s_k,\varphi^k,F+g+p^k)|_{\mathscr{B}} : t \in [\tau_k, t_k+a],\ k = 1,2,\cdots\} \le r$ for some $a > 0$.

Indeed, if we set $x^k(t) = x(t,s_k,\varphi^k,F+g+p^k)$ and $a_k = \sup\{t > 0 : |(x^k)_{\tau+t_k}|_{\mathscr{B}} \le r$ for all $\tau \in [0,t]\}$, then $a_k > 0$ for each k. If $\sigma, \tau \in [s_k,t_k+a_k]$ with $0 \le \sigma - \tau < 1$, then

$$|x^k(\sigma) - x^k(\tau)|$$

$$\le \int_\tau^\sigma |F(s,(x^k)_s)|ds + \int_\tau^\sigma |g(s,(x^k)_s)|ds + \int_\tau^\sigma |p^k(s)|ds$$

$$\le L(r)(\sigma-\tau) + |p^k|_M + \int_\tau^\sigma |g(s,(x^k)_s) - g(s,0)|ds + |\int_\tau^\sigma g(s,0)ds|$$

$$\le [L(r)+b^r(r)](\sigma-\tau) + \varepsilon_k + \sup_{t\ge s_k} \int_t^{t+1} |h^r(s)|ds$$

$$+ \sup_{0 \le \alpha \le 1,\ t \ge s_k} \left| \int_t g(s,0)ds \right|$$

by Conditions (E1) and (E3). Since $\sup_{t \ge s_k} \int_t^{t+1} |h^r(s)|ds \to 0$ and $\sup_{0 \le \alpha \le 1,\ t \ge s_k} \left| \int_t^{t+\alpha} g(s,0)ds \right| \to 0$ as $k \to \infty$ by Condition (E3), from (3.6) and Axiom (A) we see that $\inf_k a_k > 0$; this proves (3.7). Moreover, by the above argument, we have that

(3.8) the sequence $\{x^k(t)\}$ is uniformly equicontinuous on $[s_k, t_k+a]$.

Now, set $2r_k = t_k - \tau_k$, and suppose $r_k \to \infty$ as $k \to \infty$. From Theorem 1.5, we may assume that $(u^{\tau_k+r_k}, F^{\tau_k+r_k}) \to (v,G)$ compactly on $\mathbb{R}^+ \times \mathscr{B}$ for a $(v,G) \in \Omega(u,F)$. Moreover, since $(x^k)_{s_k} = \varphi^k \to \varphi$ as $k \to \infty$, in virtue of Corollary 7.3.3 and Relations (3.7) and (3.8), we can assume that $(x^k)_{t+\tau_k+r_k} \to y_t$ compactly on $\mathbb{R}^+$ for some function $y : \mathbb{R} \to E$. From a rather lengthy calculation (see Appendix), it follows that

$$\int_{\tau_k+r_k}^{t+\tau_k+r_k} g(s,(x^k)_s)ds \to 0 \quad \text{as} \quad k \to \infty \tag{3.9}$$

for each $t \in \mathbb{R}^+$. Hence, by the standard argument, we see that $y(t)$ is a solution on $\mathbb{R}^+$ of System (G). Letting $k \to \infty$ in (3.5), we have $\delta(\varepsilon/2)/2 \le |y_t - v_t|_{\mathscr{B}} \le \varepsilon$ on $\mathbb{R}^+$. Since $\varepsilon < \delta_0$ and u is WUAS in $\Omega(F)$, we must have $\delta(\varepsilon/2)/2 \le |y_t - v_t|_{\mathscr{B}} \to 0$ as $t \to \infty$, a contradiction. Thus, $r_k \nrightarrow \infty$ as $k \to \infty$. Taking a subsequence if necessary, we may assume that $r_k \to r_0 < \infty$ as $k \to \infty$ for a constant r_0. Moreover, by the same reason as in the preceding paragraph, we may assume that $(u^{\tau_k}, F^{\tau_k}) \to (\bar{v}, \bar{G})$ compactly on $\mathbb{R}^+ \times \mathscr{B}$ and that a subsequence of $\{(x^k)_{\tau_k+t}\}$ converges to z_t in $\mathscr{B}$ compactly on $[0, 2r_0+a)$ for some solution z of System $(\bar{G})$. Then, by (3.3) and (3.4), we have

$$|z_{2r_0} - \bar{v}_{2r_0}|_{\mathscr{B}} = \varepsilon \quad \text{and} \quad |z_0 - \bar{v}_0|_{\mathscr{B}} = \delta(\varepsilon/2)/2 < \delta(\varepsilon/2).$$

Since u is WUAS in $\Omega(F)$, $|z_0 - \bar{v}_0|_{\mathscr{B}} < \delta(\varepsilon/2)$ yields that $|z_{2r_0} - \bar{v}_{2r_0}|_{\mathscr{B}} < \varepsilon/2$, a contradiction. This contradiction shows that u is

Ev MS for (P).

The second part of the theorem is a direct consequence of Proposition 3.3.

Next, we shall establish the asymptotic stability result under X-perturbations. To do so, we must restrict our consideration to a uniform fading memory space $\mathscr{B}$; cf. Theorem 7.2.6.

<u>Theorem 3.5 (Murakami [1])</u>. In addition to the assumptions in Theorem 3.4, suppose $\mathscr{B}$ is a uniform fading memory space. Then u is Ev XAS for (P). Furthermore, if u is the unique solution of System (P) for the initial conditions, then it is XAS for (P).

<u>Proof.</u> Suppose that u is WUAS in $\Omega(F)$ with the pair $(\delta_1, \delta_1(\cdot))$. By Theorem 3.4, u is Ev XS for (P). Let $\delta(\cdot)$ and $\alpha(\cdot)$ be the ones in Definition 3.1, and set $\delta_0 = \delta(\delta_1/2)$ and $\alpha_0 = \alpha(\delta_1/2)$. For these δ_0, α_0 and any $\varepsilon > 0$, we shall show that there exist the numbers $T(\varepsilon)$ and $\gamma(\varepsilon)$ which satisfy the condition in Definition 3.2. Indeed, if this is not the case, then there exist an $\varepsilon > 0$, sequences $\{\tau_k\}$, $\tau_k \ge \alpha_0$, $\{t_k\}$, $t_k \ge \tau_k + 3k$, $\{\varphi^k\} \subset \mathscr{B}$, $|\varphi^k - u_{\tau_k}|_{\mathscr{B}} < \delta_0$, $\{p^k\} \subset X$, $|p^k|_X < \min(1/k, \delta_0)$, and $\{x(\cdot, \tau_k, \varphi^k, F+g+p^k)\}$ such that

$$(3.10) \qquad |x_{t_k}(\tau_k, \varphi^k, F+g+p^k) - u_{t_k}|_{\mathscr{B}} \ge \varepsilon, \qquad k = 1, 2, \cdots.$$

In virtue of Theorem 1.5, we may assume that $(u^{\tau_k+2k}, F^{\tau_k+2k}) \to (v, G)$ compactly on $\mathbb{R}^+$ for some $(v, G) \in H(u, F)$. Since $\tau_k + 2k \to \infty$ as $k \to \infty$, we have $(v, G) \in \Omega(u, F)$. Since u is Ev XAS for (P), $\tau_k \ge \alpha(\delta_1/2)$, $|\varphi^k - u_{\tau_k}|_{\mathscr{B}} < \delta(\delta_1/2)$ and $|p^k|_X < \delta(\delta_1/2)$ yield that

$$(3.11) \qquad |(x^k)_t - u_t|_{\mathscr{B}} < \delta_1/2$$

for all $t \ge \tau_k$ and $k = 1, 2, \cdots$, where $x^k(t) = x(t, \tau_k, \varphi^k, F+g+p^k)$. Select an integer $k_0 = k_0(\varepsilon)$ so that $\tau_k + k \ge \alpha(\varepsilon)$ and $|p^k|_X < 1/k < \delta(\varepsilon)$ if $k \ge k_0$. Then, by (3.10) and the fact that u is Ev XS for (P), we get

$$(3.12) \qquad |(x^k)_t - u_t|_{\mathscr{B}} \ge \delta(\varepsilon) \quad \text{for} \quad t \in [\tau_k + k, \tau_k + 3k]$$

if $k \ge k_0$. Note that $\sup\{|(x^k)_{t+\tau_k+k}|_{\mathscr{B}} : t \ge 0,\ k = 1, 2, \cdots\} < \infty$ by (3.11). By repeating the same argument as in the proof of Theorem

3.4, one sees that the family of functions $\{x^k(t+\tau_k+k)\}$ is uniformly equicontinuous on $\mathbb{R}^+$. Hence, in virtue of Corollary 7.3.7, we may assume that $(x^k)_{\tau_k+2k+t} \to y_t$ compactly on $\mathbb{R}^+$, where $y : \mathbb{R} \to E$ is a solution of System (G). Observing that

$$\delta(\varepsilon) \le |(x^k)_{\tau_k+2k+t} \quad u_{\tau_k+2k+t}|_{\mathscr{B}} < \delta_1/2, \qquad t \in [0,k],\ k \ge k_0,$$

by (3.11) and (3.12), and letting $k \to \infty$ in the above, we have

$$(3.13) \qquad \delta(\varepsilon) \le |y_t - v_t|_{\mathscr{B}} \le \delta_1/2 \quad \text{for all} \quad t \in \mathbb{R}^+.$$

Hence, we must have $|y_t - v_t|_{\mathscr{B}} \to 0$ as $t \to \infty$, because $|y_0 - v_0|_{\mathscr{B}} < \delta_1$ and u is WUAS in $\Omega(F)$. This is a contradiction to (3.13); thus, u is Ev XAS for (P).

The second part of the theorem follows from Proposition 3.3.

Next, we consider the case where $F(t,0) = 0$ on $\mathbb{R}^+$. $u(t) \equiv 0$ is clearly a solution of System (E). Denote it by 0. 0 is said to be uniformly attracting in $\Omega(F)$, if there exists a $\delta_0 > 0$ and for any $\varepsilon > 0$ there exists a $T(\varepsilon) > 0$ such that $t_0 \in \mathbb{R}^+$, $|\varphi|_{\mathscr{B}} < \delta_0$ and $G \in \Omega(F)$ imply $|x_t(t_0,\varphi,G)|_{\mathscr{B}} < \varepsilon$ for all $t \ge t_0+T(\varepsilon)$. 0 is said to be unique in $\Omega(F)$ for the initial conditions, if $t_0 \in \mathbb{R}^+$ and $G \in \Omega(F)$ imply $x(t,t_0,0,G) \equiv 0$ for all $t \ge t_0$. Clearly, 0 is unique in $\Omega(F)$ for the initial conditions if System (E) is regular.

The following result extends the result of Bondi et al. [1] to an FDE case.

<u>Theorem 3.6</u>. Let $F(t,0) = 0$ on $\mathbb{R}^+$. If 0 is uniformly attracting in $\Omega(F)$ and unique in $\Omega(F)$ for the initial conditions, then it is UAS in $\Omega(F)$; consequently, it is Ev XS for (P). In addition, if $\mathscr{B}$ is a uniform fading memory space and if 0 is the unique solution of System (P) for the initial conditions, then 0 is XAS for (P).

<u>Proof.</u> Suppose 0 is uniformly attracting in $\Omega(F)$ with a pair $(\delta_0, T(\cdot))$. We shall verify that 0 is US in $\Omega(F)$. Then the assertions of this theorem follow from Theorems 3.4 and 3.5. To show that 0 is US in $\Omega(F)$, it suffices to prove that for any $\varepsilon > 0$ there exists a $\delta(\varepsilon) > 0$, $\delta(\varepsilon) < \min(\delta_0, \varepsilon)$, such that $|x_t(t_0,\varphi,G)|_{\mathscr{B}} < \varepsilon$ on $[t_0, t_0+T(\varepsilon)]$ if $t_0 \in \mathbb{R}^+$, $|\varphi|_{\mathscr{B}} < \delta(\varepsilon)$ and $G \in \Omega(F)$. Assume the contrary holds. Then there exists an $\varepsilon > 0$ and

sequences $\{\tau_k\}$, $\tau_k \in \mathbb{R}^+$, $\{r_k\}$, $0 \le r_k \le T(\varepsilon)$, $\{G^k\} \subset \Omega(F)$ and $\{\varphi^k\} \subset \mathscr{B}$, $|\varphi^k|_{\mathscr{B}} < \min(\delta_0, \varepsilon)$ and $|\varphi^k|_{\mathscr{B}} \to 0$ as $k \to \infty$, such that

$$(3.14) \qquad |x_{\tau_k+r_k}(\tau_k, \varphi^k, G^k)|_{\mathscr{B}} = \varepsilon \quad \text{and} \quad |x_t(\tau_k, \varphi^k, G^k)|_{\mathscr{B}} < \varepsilon$$

$$\text{on } [\tau_k, \tau_k+r_k).$$

Taking a subsequence if necessary, we may assume that $r_k \to r_0 \in [0,T(\varepsilon)]$ as $k \to \infty$ and $(G^k)^{\tau_k} \to \bar{G}$ compactly on $\mathbb{R}^+ \times \mathscr{B}$ for a $\bar{G} \in \Omega(F)$. By the same argument as in the proof of Theorem 3.4, we see that there exists an $a > 0$ such that $|x_t(\tau_k, \varphi^k, G^k)|_{\mathscr{B}} < 2\varepsilon$ on $[\tau_k, \tau_k+r_k+a]$, $k = 1,2,\cdots$; consequently, $|\dot{x}(t+\tau_k, \tau_k, \varphi^k, G^k)| \le \ell(2\varepsilon)$ on $[0,r_k+a]$, $k = 1,2,\cdots$, by Condition (E1). Therefore, taking a subsequence, we may assume that $x_{t+\tau_k}(\tau_k, \varphi^k, G^k) \to y_t$ in $\mathscr{B}$ as $k \to \infty$ compactly on $[0,r_0+a)$, where y is a solution on $[0,r_0+a)$ of System ($\bar{G}$) through $(0,0)$. Letting $k \to \infty$ in (3.14), we have $|y_{r_0}|_{\mathscr{B}} = \varepsilon$, which contradicts the assumption that 0 is unique in $\Omega(F)$ for the initial conditions. This contradiction shows that 0 is US in in $\Omega(F)$.

In the above theorems, we assumed the stability properties for "all" limiting equations to deduce some stability properties for System (P). On the one hand, the following example suggests that without any additional condition, we cannot always deduce some nice stability properties for (P) from stability properties for "one" limiting equation.

<u>Example 3.7.</u> Let

$$f^1(t) = \begin{cases} 4t & \text{if } 0 \le t \le 1/4 \\ 1 & \text{if } 1/4 < t \le 3/4 \\ 4 - 4t & \text{if } 3/4 < t \le 1, \end{cases}$$

and let

$$f^{k+1}(t) = \begin{cases} f^k(t) & \text{if } 0 \le t \le 2^k - 1 \\ f^k(t-2^k+1) & \text{if } 2^k - 1 < t \le 2^{k+1} - 2 \\ 0 & \text{if } 2^{k+1} - 2 < t \le 2^{k+1} - 1, \end{cases}$$

for $k = 1,2,\cdots$, and define a function $f(t)$ by $f(t) = f^k(t)$ if

$t \in [0, 2^k-1]$, $k = 1, 2, \cdots$. Consider a scalar ODE

$$\dot{x}(t) = -f(t)x. \tag{3.15}$$

Clearly, the function $F(t,x) = -f(t)x$ satisfies the condition (E1) with $\mathscr{B} = \mathbb{R}$. On the other hand, for all $t \in [0, 2^k-1]$ and $x \in \mathbb{R}$, we have

$$F(t+2^k-1, x) = -f^{k+1}(t+2^k-1-2^k+1)x = -f^k(t)x = -f(t)x = F(t,x),$$

and hence $F \in \Omega(F)$. Notice that the solution $x(t, t_0, x_0, F+p)$ of $\dot{x} = F(t,x) + p(t)$ through (t_0, x_0) is given by

$$x(t, t_0, x_0, F+p) = x_0 \exp(-\int_{t_0}^{t} f(s)ds) + \int_{t_0}^{t} \exp(-\int_{s}^{t} f(\tau)d\tau)p(s)ds.$$

So, one can easily check that 0 is WUAS for (3.15). However, 0 is not UAS for (3.15), because $x(t+2^{n+1}-1-n, 2^{n+1}-1-n, x_0, F) = x_0$ on $[0, n]$. Furthermore, 0 is not TS for (3.15), because $x(2^{n+1}-1, 2^{n+1}-1-n, 0, F+p) = np$ for each constant function p.

Next, we shall derive some stability properties for (E) from stability properties for "one" limiting equation. To do this, we consider the case where the hull $H(F)$ is minimal; that is,

(E4) $\quad H(F) = \Omega(G)$ for all $G \in H(F)$.

Clearly, Condition (E4) implies that $F \in \Omega(G)$ for every $G \in \Omega(F)$. If $F(t, \varphi)$ is the restriction to $\mathbb{R}^+ \times \mathscr{B}$ of an almost periodic function in t uniformly for $\varphi \in \mathscr{B}$, then it satisfies Condition (E4); see Kato [1]. On the other hand, the function F in Example 3.7 has not this property. To see this, observe that $f(t) = 0$ for $2^k-k \le t \le 2^k-1$, $k = 1, 2, \cdots$. Hence, if we set $\tau_k = 2^k-k$, then $\{F^{\tau_k}\}$ converges to 0 compactly; that is, 0 is in $\Omega(F)$. If $u(t) \equiv 0$ and $X = \{0\}$, we can obtain the conclusion similar to the one in Theorem 3.5 under a weaker assumption. The following result is due to Murakami [1]. For the ODE case, see Yoshizawa [2] and Seifert [1].

<u>Theorem 3.8.</u> Let $\mathscr{B}$ be a uniform fading memory space, and suppose Condition (E4) holds and $F(t, 0) \equiv 0$ on $\mathbb{R}^+$. If the zero solution of System (E) is WUAS, then it is UAS.

Proof. Suppose the zero solution of System (E) is WUAS with a pair $(\delta_1, \delta(\cdot))$, and set $\delta_0 = \delta(\delta_1/2)$. For this δ_0, we shall show that there exists the number $T(\cdot)$ which satisfies Definition 3.2 (with $u(t) \equiv 0$ and $X = \{0\}$). Suppose this is not the case. By repeating the same argument as in the proof of Theorem 3.5 (with $u(t) \equiv 0$ and $X = \{0\}$), for some $\varepsilon > 0$ we have

$$(3.16) \qquad \delta(\varepsilon) \le |y_t|_{\mathscr{B}} \le \delta_1/2 \quad \text{for all } t \in \mathbb{R}^+,$$

where y is a solution on $\mathbb{R}^+$ of System (G) with $G \in \Omega(F)$. Since $F \in \Omega(G)$ by Condition (E4), there exists a sequence $\{s_k\}$, $s_k \to \infty$ as $k \to \infty$, and a $(z,F) \in \Omega(y,G)$ such that $(y^{s_k}, G^{s_k}) \to (z,F)$ compactly on $\mathbb{R} \times \mathscr{B}$. Set $x^k(t) = y(t+s_k)$ for $t \in \mathbb{R}$. Then

$$\delta(\varepsilon) \le |(x^k)_t|_{\mathscr{B}} \le \delta_1/2 \quad \text{for all } t \ge -s_k$$

by (3.16). Letting $k \to \infty$ in the above, we get

$$\delta(\varepsilon) \le |z_t|_{\mathscr{B}} \le \delta_1/2 \quad \text{for all } t \in \mathbb{R}.$$

This relation yields a contradiction by the same reason as in the proof of Theorem 3.5, which proves the theorem.

Remark 3.9. In Theorem 3.8, we cannot necessarily ensure the stability under X-perturbations with $X \ne \{0\}$. In fact, Kato's example in the previous section shows this fact.

8.4. Stabilities in the original system and perturbed systems. The purpose of this section is to discuss some stability properties for the perturbed system (P), assuming some stability properties for the original system (E). It is quite natural to ask whether "UAS for (E)" implies "Ev UAS for (P)". The following example due to Strauss and Yorke [2] shows that this is not always the case even for ODEs.

Example 4.1. Let $f\colon (0,1] \to \mathbb{R}$ be a function defined by

$$f(x) = -2\pi(2^{-n} - x)^{1/2}(x - 2^{-n-1})^{1/2} \quad \text{if } 2^{-n-1} \le x \le 2^{-n},$$
$$n = 0,1,2,\cdots,$$

and define

$$F(t,x) = \begin{cases} \min\{f(x),\ -xe^{-t^2}\} & \text{if } 0 < x < 3/4, \\ -2\pi x/3 & \text{if } x \le 0 \text{ or if } x \ge 3/4. \end{cases}$$

Since the function $f(x)$ and xe^{-t^2} is uniformly continuous in $x \in (0,1]$ and in $(t,x) \in \mathbb{R}^+\times[0,1]$, respectively, the function $F(t,x)$ is uniformly continuous on $\mathbb{R}^+ \times \mathbb{R}$. Moreover, it satisfies $|F(t,x)| \le 2\pi|x|$ for all $(t,x) \in \mathbb{R}^+\times \mathbb{R}$. Hence, the scalar equation

$$(4.1) \qquad \dot{x} = F(t,x)$$

satisfies Condition (E1). Observe that for each $t_0 \ge 0$, there are many solutions of

$$(4.2) \qquad \dot{z} = f(z)$$

through $(t_0, 2^{-n})$, one of which is $z(t) = 2^{-n}$ and another of which is

$$z(t) = 2^{-n} - 2^{-n-1}\sin^2\pi(t-t_0)$$

for $t_0 \le t \le t_0+1/2$. For $0 < x_0 \le 1$, let $z^*(t,t_0,x_0)$ denote the solution of (4.2) through (t_0,x_0) which is strictly decreasing for all $t \ge t_0$. For example, for any nonnegative integer n_0, we have

$$z^*(t,t_0,2^{-n_0}) = 2^{-n_0-n} - 2^{-n_0-n-1}\sin^2\pi(t-t_0)$$

for $t_0+n/2 \le t \le t_0+(n+1)/2$. In particular, if for each $\xi \in (0,1]$, we denote by $n_0(\xi)$ the smallest nonnegative integer such that $\xi \le 2^{-n_0(\xi)}$, then there exists at least one solution $z(t,t_0,x_0)$ of (4.2) through (t_0,x_0) such that

$$z(t,t_0,x_0) \le z^*(t,t_0,2^{-n_0(x_0)}) \le 2^{-n_0(x_0)}\cdot 2^{1-2(t-t_0)}$$

$$\le 4x_0\cdot 4^{-(t-t_0)}$$

for all $t \ge t_0$. By the comparison theorem (see, e.g., Hale [1]), the solution of (4.1) satisfies, for $0 < x_0 < 3/4$, $x(t,t_0,x_0,F) \le$

$4x_0 \cdot 4^{-(t-t_0)}$. More precisely, we have $|x(t,t_0,x_0,F)| \le 4|x_0| \cdot 2^{-(t-t_0)}$ for $t \ge t_0 \ge 0$ and all $x \in \mathbb{R}$, which implies that 0 is UAS for (4.1). Nevertheless, for every $\beta > 0$ and every locally Lipschitz continuous $\gamma(x)$ for $x > 0$, 0 is not attracting for $\dot{x} = F(t,x) + \gamma(x)e^{-\beta t}$, because its right-hand side is positive for large t and $x = 2^{-n}$. Moreover, the function $g(t,x) = \gamma(x)e^{-\beta t}$ is clearly diminishing.

The above example shows that without any additional assumption on $F(t,\varphi)$, stability properties for the original system (E) are not necessarily inherited to the perturbed system (P). Furthermore, Kato's example given in Section 8.2 tells us that the uniform asymptotic stability for (E) is not necessarily inherited to each limiting equation, even if $F(t,\varphi)$ is almost periodic in t uniformly for φ. So, in this section, we shall restrict our consideration to the case where System (E) is regular or periodic, and study some stability properties for the perturbed system (P) under Conditions (E1) through (E3).

The following result is a direct consequence of Theorem 3.5 and Propositions 2.2 and 2.3.

Theorem 4.2. Let $\mathscr{B}$ be a uniform fading memory space. Then the following statements hold:

(i) If System (E) is regular and if u is UAS for (E), then u is Ev XAS for (P);

(ii) If System (E) is periodic and if u is WUAS for (E), then u is Ev XAS for (P);

(iii) In either case of (i) or (ii), if, in addition, u is the unique solution of System (P) for the initial conditions, then the prefix "Ev" can be deleted.

Recall that X stands for the one of {0}, T, L or M. From Theorem 4.2 (with $g \equiv 0$), one obtains the following result on the relationships among UAS, TAS, IAS and MAS, which has been proved by Kato & Yoshizawa [1] for finite delay systems.

Theorem 4.3. Let $\mathscr{B}$ be a uniform fading memory space. If System (E) is regular or periodic, then the relation

$$\text{MAS} \Longleftrightarrow \text{IAS} \Longleftrightarrow \text{TAS} \Longleftrightarrow \text{UAS}$$

holds for System (E).

As easily seen, u is US in $\Omega(F)$ whenever it is MS for (E). Hence, from Kato's example, we see that the relation in the above theorem does not necessarily hold without the regularity or periodicity condition on System (E).

Next, in the case where $u(t) \equiv 0$ is not necessarily a solution of System (E), we shall deduce some stability properties for (P) under some stability assumptions on System (E). The following results intimately relate to those in Strauss & Yorke [2] for ODEs.

Theorem 4.4. Let $\mathscr{B}$ be a uniform fading memory space, and let System (E) be regular or periodic. If 0 is Ev UAS for (E), then it is Ev XAS for (P).

Theorem 4.5. Let $\mathscr{B}$ be a uniform fading memory space, System (E) be regular or periodic, and suppose that 0 is Ev UAS for (E). Then, 0 is Ev XAS for the system

$$\text{(P1)} \qquad \dot{x}(t) = F(t, x_t) + p(t)$$

with a continuous function p, if and only if p is diminishing; that is, $\sup_{0\le\alpha\le 1} |\int_t^{t+\alpha} p(s)ds| \to 0$ as $t \to \infty$.

In order to prove Theorem 4.4, it is sufficient by Theorem 3.5 to certify:

Proposition 4.6. Let System (E) be regular or periodic, and suppose that 0 is Ev UAS for (E). Then 0 is UAS in $\Omega(F)$.

Proof. Let $\delta(\cdot)$, $\alpha(\cdot)$, δ_0, α_0 and $T(\cdot)$ be the ones given for the eventual asymptotic stability of 0 for (E).

First, we consider the case where System (E) is regular. Let $\varepsilon > 0$, $\sigma \in \mathbb{R}^+$, $G \in \Omega(F)$ and $\varphi \in \mathscr{B}$ with $|\varphi|_{\mathscr{B}} < \delta(\varepsilon/2)$. There exists a sequence $\{t_k\}$, $t_k \to \infty$ as $k \to \infty$, such that $F^{t_k} \to G$ compactly on $\mathbb{R}^+ \times \mathscr{B}$. Since $|\varphi|_{\mathscr{B}} < \delta(\varepsilon/2)$ and $t_k \ge \alpha(\varepsilon/2)$, $k \ge k_0$, for some k_0, and since 0 is Ev UAS for (E), we have

$$\text{(4.3)} \qquad |x_{t_k+\sigma+t}(t_k+\sigma, \varphi, F)|_{\mathscr{B}} < \varepsilon/2 \quad \text{for all} \quad t \ge 0$$

if $k \ge k_0$. By the same argument as in the proof of Proposition 2.2, we may assume that the sequence $\{x_{t_k+\sigma+t}(t_k+\sigma,\varphi,F)\}$ converges to y_t compactly on $\mathbb{R}^+$, where $y(t)$ is the unique solution on $\mathbb{R}^+$ of $\dot{x}(t) = G(t+\sigma,x_t)$ through $(0,\varphi)$, that is, $y(t) = x(t+\sigma,\sigma,\varphi,G)$. Therefore, letting $k \to \infty$ in (4.3), we have $|x_{t+\sigma}(\sigma,\varphi,G)|_{\mathscr{B}} \le \varepsilon/2 < \varepsilon$. Thus, $\sigma \in \mathbb{R}^+$, $|\varphi|_{\mathscr{B}} < \delta(\varepsilon/2)$ and $G \in \Omega(F)$ imply that $|x_t(\sigma,\varphi,F)|_{\mathscr{B}} < \varepsilon$ for all $t \ge \sigma$; hence, 0 is US in $\Omega(F)$. Next, let $\varphi \in \mathscr{B}$ with $|\varphi|_{\mathscr{B}} < \bar{\delta}_0 := \min(\delta(1),\delta_0)$. By the same argument as in the above, we get $|x_{t_k+\sigma+t}(t_k+\sigma,\varphi,F)|_{\mathscr{B}} < 1$ for all $t \ge 0$ and

$$(4.4) \qquad |x_{t_k+\sigma+t}(t_k+\sigma,\varphi,F)|_{\mathscr{B}} < \varepsilon/2 \quad \text{for all} \quad t \ge T(\varepsilon/2),$$

if k is sufficiently large. Letting $k \to \infty$ in (4.4), we can conclude that $|x_{t+\sigma}(\sigma,\varphi,G)|_{\mathscr{B}} \le \varepsilon/2$ for all $t \ge T(\varepsilon/2)$ by the same argument as in the above. Hence, 0 is UAS in $\Omega(F)$.

Next, we consider the case where System (E) is periodic with a period $\omega > 0$. To prove that 0 is UAS in $\Omega(F)$, it suffices to show that $x(t,0,0,F) = 0$ for all $t \ge 0$ by Propositions 2.3 and 3.3. Now, let $v(t)$ be a solution of System (E) satisfying $|v_t|_{\mathscr{B}} \to 0$ as $t \to \infty$. Since $v(t+k\omega)$ is a solution of System (E) through $(0,v_{k\omega})$ and $v(t+k\omega) \to 0$ as $k \to \infty$ compactly on $\mathbb{R}^+$, 0 is a solution on $\mathbb{R}^+$ of System (E) by Theorem 1.5; thus, $F(t,0) \equiv 0$ for $t \ge 0$. Therefore, if we set

$$y^k(t) = \begin{cases} 0 & \text{if} \quad t \le k\omega \\ x(t-k\omega,0,0,F) & \text{if} \quad k\omega < t \end{cases}$$

for $k = 1,2,\cdots$, then $y^k(t)$ is a solution on $\mathbb{R}^+$ of System (E) for each k. Let an $\varepsilon > 0$ be given, and choose an integer $N = N(\varepsilon)$ so that $N\omega \ge \alpha(\varepsilon)$. Since 0 is Ev US for (E), we have $|(y^N)_t|_{\mathscr{B}} < \varepsilon$ for all $t \ge N\omega$; consequently, $|x_t(0,0,F)|_{\mathscr{B}} < \varepsilon$ for all $t \ge 0$. Since $\varepsilon > 0$ is arbitrary, we have $|x_t(0,0,F)|_{\mathscr{B}} = 0$; thus, $x(t,0,0,F) = 0$ for all $t \ge 0$, which is the desired result.

Finally, we shall prove Theorem 4.5. If p is diminishing, then 0 is Ev XAS for (P1) by Theorem 4.4. Next, suppose that 0 is Ev XAS for (P1). Since 0 is Ev XAS for (E), Proposition 4.6 implies that 0 is a solution of System (G) for all $G \in \Omega(F)$; hence, $G(t,0) \equiv 0$ for all $t \ge 0$. Let $\varepsilon > 0$. Then there exist a $T_1(\varepsilon) > 0$ and a $\delta_1(\varepsilon) > 0$, $\delta_1(\varepsilon) < \varepsilon$, such that $t \ge T_1(\varepsilon)$ and $|\varphi|_{\mathscr{B}} < \delta_1(\varepsilon)$ imply $|F(t,\varphi)| < \varepsilon$. Indeed, if this is not the case, then there exist sequences $\{\varphi^k\} \subset \mathscr{B}$, $|\varphi^k|_{\mathscr{B}} \to 0$ as $k \to \infty$, and $\{t_k\}$,

$t_k \to \infty$ as $k \to \infty$, such that

$$(4.5) \qquad |F(t_k, \varphi^k)| \geq \varepsilon.$$

Taking a subsequence if necessary, we may assume that $F^{t_k} \to G$ compactly on $\mathbb{R}^+ \times \mathscr{B}$ for some $G \in \Omega(F)$. Letting $k \to \infty$ in the above, we get $|G(t,0)| \geq \varepsilon$, a contradiction. Now, let $x(t)$ be a solution of System (P1) satisfying $|x_t|_{\mathscr{B}} \to 0$ as $t \to \infty$. Choose a $t_0(\varepsilon) > 0$ so that $|x_t|_{\mathscr{B}} < \delta_1(\varepsilon)$ for all $t \geq t_0(\varepsilon)$. Then, if $0 \leq \alpha \leq 1$ and $t \geq \max(t_0(\varepsilon), T_1(\varepsilon))$, we have $|\int_t^{t+\alpha} p(s)ds| \leq |x(t+\alpha)| + |x(t)| + \int_t^{t+\alpha} |F(s,x_s)|ds \leq H[|x_{t+\alpha}|_{\mathscr{B}} + |x_t|_{\mathscr{B}}] + \varepsilon \leq 2H\delta_1(\varepsilon) + \varepsilon$ by Axiom (A), which shows that p is diminishing. This completes the proof of the theorem.

<u>8.5. Application to some practical equations.</u> In this section, we shall present two examples which show how our approach can be employed for the analysis of practical equations.

First, we consider the second order scalar delay-differential equation

$$(5.1) \qquad \ddot{x}(t) + p(t,x(t),\dot{x}(t))\dot{x}(t) + q(x(t-r(t))) = G(t)x(t-r_1),$$

and its equivalent system

$$(5.2) \qquad \begin{cases} \dot{x}(t) = y(t), \\ \dot{y}(t) = -p(t,x(t),y(t))y(t) - q(x(t)) + G(t,x(t-r_1)) \\ \qquad\qquad + \displaystyle\int_{-r(t)}^{0} \dot{q}(x(t+s))y(t+s)ds, \end{cases}$$

where $0 \leq r(t)$, $r_1 \leq r < \infty$. Some authors have investigated such an equation. For instance, as an application of the invariance principle, the autonomous case ($r(t) = r$, $p(t,x,y) = a/r$, $q(x) = (b/r)\sin x$, $r < a/b$ and $G(t) \equiv 0$) was studied by Hale [5] (for ODE case, refer to Artstein [1]). We impose the following assumptions on System (5.2):

(H1) $p(t,x,y)$ and $r(t)$ are bounded and uniformly continuous on $\mathbb{R}^+ \times W$ for any bounded set W in $\mathbb{R}^2$, and $p(t,x,y) \geq$

$dr + e$, $|\dot{q}(x)| \le L < d$ and $xq(x) > 0$ $(0 < |x| \le c)$ for some positive constants c, d, e and L;

(H2) $G(t)$ is bounded and continuous on $\mathbb{R}^+$, and

$$\sup_{0\le\alpha\le 1} \left|\int_t^{t+\alpha} G(s)ds\right| \to 0 \text{ as } t \to \infty.$$

Let $C = C([-r,0]) = \{\varphi: [-r,0] \to \mathbb{R}^2$ is continuous on $[-r,0]\}$ be the Banach space with the norm $|\varphi|_C = \sup\{|\varphi(\theta)|: -r \le \theta \le 0\}$ for $\varphi \in C$. The space C is clearly a (separable) uniform fading memory space, and System (5.2) can be considered as a system of FDEs on the space C. Moreover, the limiting equations of System (5.2) (with $G \equiv 0$) are of the form

$$\text{(5.3)} \qquad \begin{cases} \dot{x}(t) = y(t) \\ \dot{y}(t) = -\tilde{p}(t,x(t),y(t))y(t) - q(x(t)) \\ \qquad\qquad + \displaystyle\int_{-\tilde{r}(t)}^{0} \dot{q}(x(t+s))y(t+s)ds, \end{cases}$$

where $\tilde{p}(t,x,y)$, $\tilde{r}(t)$ and $q(x)$ also satisfy (H1) with the same constants c, d, e and L.

<u>Theorem 5.1.</u> Suppose Conditions (H1) and (H2) hold. Then the origin $(0,0)$ is Ev XAS for System (5.2).

<u>Proof.</u> To prove the theorem, it suffices to show by Theorem 3.5 that $(0,0)$ is WUAS in $\Omega(F)$. Now, select a constant δ, $L/d < \delta < 1$, and consider a functional $V(\varphi)$ defined on the set $U_c :=$ $\{\varphi \in C : |\varphi|_C < c\}$ by

$$\text{(5.4)} \qquad V(\varphi) = 2\int_0^{\varphi^1(0)} q(s)ds + \{\varphi^2(0)\}^2 + d\delta\int_{-r}^0 \left[\int_s^0 \{\varphi^2(\tau)\}^2 d\tau\right]ds$$

for $\varphi = (\varphi^1,\varphi^2) \in U_c$. Clearly, we have

$$\text{(5.5)} \qquad a(\varphi^1(0),\varphi^2(0)) \le V(\varphi) \le b(|\varphi|), \qquad \varphi \in U_c,$$

where $a(x,y) := 2\int_0^x q(s)ds + y^2$ is a positive definite function for $(x,y) \in \mathbb{R}^2$, and $b(s) := (2L+1+d\delta r^2)s^2$ for $s \ge 0$. Let $\varphi \in U_c$. Then, we have

$$\dot{V}_{(5.3)}(t,\varphi) = 2q(\varphi^1(0))\varphi^2(0) + 2\varphi^2(0)[-\tilde{p}(t,\varphi^1(0),\varphi^2(0))\varphi^2(0)$$

$$-q(\varphi^1(0)) + \int_{-\tilde{r}(t)}^{0} \dot{q}(\varphi^1(s))\varphi^2(s)ds]$$

$$+ d\delta\int_{-r}^{0}\{(\varphi^2(0))^2 - (\varphi^2(s))^2\}ds$$

$$\le -e(\varphi^2(0))^2 - \int_{-r}^{0}[d(2-\delta)(\varphi^2(0))^2$$

$$-2L|\varphi^2(0)||\varphi^2(s)| + d\delta(\varphi^2(s))^2]ds$$

by Condition (H1). Here, the last integrand is a positive definite quadratic form in $|\varphi^2(0)|$ and $|\varphi^2(s)|$, since $L/\alpha < \delta < 1$. Thus, we have

(5.6) $$\dot{V}_{(5.3)}(t,\varphi) \le -e\{\varphi^2(0)\}^2 \le 0.$$

Therefore, from (5.5) and (5.6) it follows that (0,0) is US for (5.3) in $\mathbb{R}^2$; consequently, (0,0) is US for (5.3) (in C). Note that the construction of the functional $V(\varphi)$ is independent of the particular choice of a limiting equation. Hence, (0,0) is US in $\Omega(F)$. Now, select a constant $\delta_1 > 0$ so that if $\varphi = (\varphi^1,\varphi^2) \in U_{\delta_1}$ and $s \ge 0$, then the solution $x(t) = (x^1(t),x^2(t))$ of each limiting equation through (s,φ) remains in U_c for all $t \ge s$. This is possible, since (0,0) is US in $\Omega(F)$. To establish the theorem, it suffices to show that $(x^1(t),x^2(t)) \to (0,0)$ as $t \to \infty$. First, we claim that $x^2(t) \to 0$ as $t \to \infty$. Indeed, from (5.6) it follows that $V(t,x_t) - V(s,x_s) \le -e\int_s^t\{x^2(\tau)\}^2d\tau$, $t \ge s$. Hence, $\int^\infty\{x^2(\tau)\}^2d\tau < \infty$ because of $V(x_t) \ge 0$ for $t \ge 0$. On the one hand, since $|(d/dt)x^2(t)|$ is bounded on $[s,\infty)$, $x^2(t)$ is uniformly continuous on $[s,\infty)$. Combining these results, we have $x^2(t) \to 0$ as $t \to \infty$. Notice that $\lim_{t\to\infty} V(x_t)$ exists; for $V(x_t)$ is nonnegative and nonincreasing in t by (5.5) and (5.6). Then, from (5.4) it follows that $x^1(t) \to k$ as $t \to \infty$ for a constant k, $|k| \le c$. Integrating the second equation of (5.3) over $[m, m+t]$, $0 \le t \le 1$, and letting $m \to \infty$, we have

$$\int_0^t q(k)du = 0 \quad \text{for all } t \in [0,1].$$

Thus, we must obtain $k = 0$ by Condition (H1); hence, $(x^1(t), x^2(t)) \to (0,0)$ as $t \to \infty$. This completes the proof.

Next, we consider a system of integrodifferential equations

$$(5.7) \qquad (d/dt)x^i(t) = x^i(t)\{b_i(t) - a_i(t)x^i(t) - \sum_{j=1}^{n} a_{ij}(t)\int_{-\infty}^{t} K_{ij}(t,\tau)x^j(\tau)d\tau\},$$

$$i = 1,\cdots,n, \ t \geq 0, \ x = (x^1, \cdots, x^n) \in \mathbb{R}^n,$$

which denotes a model of the dynamics of an n-species system in mathematical ecology; cf. Gopalsamy [1], Hamaya [1] and Murakami [5]. We impose the following conditions on System (5.7);

(H3) for each $i,j = 1,\cdots,n$, the functions $a_{ij}(t)$, $a_i(t)$ and $b_i(t)$ are real-valued, uniformly continuous on $\mathbb{R}^+ = [0,\infty)$, and

$$a_{ij}^{\ell} \leq a_{ij}(t) \leq a_{ij}^{u}, \quad a_i^{\ell} \leq a_i(t) \leq a_i^{u}, \quad b_i^{\ell} \leq b_i(t) \leq b_i^{u}$$

on $\mathbb{R}^+$ for some positive constants a_{ij}^{ℓ}, a_{ij}^{u}, a_i^{ℓ}, a_i^{u}, b_i^{ℓ} and b_i^{u} satisfying

$$b_i^{\ell} > \sum_{j=1}^{n} a_{ij}^{u} b_j^{u} / a_j^{\ell}, \qquad i = 1,\cdots,n;$$

(H4) for each $i,j = 1,\cdots,n$, $K_{ij}(t,\tau)$ is a nonnegative continuous function on $\mathbb{R}^+ \times \mathbb{R}$ such that $K_{ij}(t,t+s)$ is bounded and uniformly continuous in $(t,s) \in \mathbb{R}^+ \times [-N,0]$ for each $N > 0$, and

$$K_{ij}(t,\tau) \leq K_{ij}^{*}(t-\tau), \qquad \tau \leq t,$$

for some continuous function $K_{ij}^{*}(s)$ on $\mathbb{R}^+$ satisfying $\int_0^{\infty} K_{ij}^{*}(s)ds = 1$ and $\int_0^{\infty} sK_{ij}^{*}(s)ds < \infty$.

Now, we shall set up System (5.7) as a system of FDEs on a fading memory space. By Condition (H4), we can select a sequence $\{T_m\}$, $T_m \geq T_m+1$, $T_1 \geq 1$, so that

$$\int_{T_m}^{\infty} sK_{ij}^{*}(s)ds < 1/[m^2(m+2)]$$

for all $i,j = 1,2,\cdots,n$ and $m = 1,2,\cdots$. Define a continuous function $g : (-\infty,0] \to [1,\infty)$ by

$$g(s) = \begin{cases} 1 & \text{if } s = 0, \\ m+1 & \text{if } s = -T_m, \\ \text{linear} & \text{if } -T_{m+1} \le s \le -T_m, \quad m = 1,2,\cdots \end{cases}$$

for $s \le 0$. For this function g, the space C_g^0 is a (separable) fading memory space; see Sections 1.3 and 7.1. For each $\varphi = (\varphi^1,\cdots,\varphi^n) \in C_g^0$, we set

$$f_{ij}(t,\varphi) = \int_{-\infty}^0 K_{ij}(t,t+s)\varphi^j(s)ds.$$

Then,

$$\begin{aligned} &|f_{ij}(t,\varphi)| \\ &\le \int_{-T_1}^0 K_{ij}(t,t+s)|\varphi^j(s)|ds + \sum_{m=1}^{\infty}\int_{-T_{m+1}}^{-T_m} K_{ij}(t,t+s)|\varphi^j(s)|ds \\ &\le \int_{-T_1}^0 K_{ij}^*(-s)ds\cdot g(-T_1)|\varphi|_g + \sum_{m=1}^{\infty}\int_{-\infty}^{-T_m} K_{ij}^*(-s)ds\cdot g(-T_{m+1})|\varphi|_g \\ &\le \int_0^{\infty} K_{ij}^*(s)ds\cdot 2|\varphi|_g + \sum_{m=1}^{\infty}\int_{T_m}^{\infty} sK_{ij}^*(s)ds\cdot(m+2)|\varphi|_g \\ &= \ell|\varphi|_g, \end{aligned}$$

where $\ell = 2+\sum_{m=1}^{\infty}(1/m^2)$. Thus, $f_{ij}(t,\varphi)$ is a real valued function on $R^+ \times C_g^0$ which is Lipschitz continuous in φ with a Lipschitz constant ℓ independent of $t \in R^+$. Let W be any bounded set in C_g^0, and fix any φ in W. Since

$$\begin{aligned} \int_{-\infty}^{-T_N} K_{ij}(t,t+\tau)|\varphi^j(\tau)|d\tau &\le \sum_{m=N}^{\infty}\int_{-T_{m+1}}^{-T_m} K_{ij}(t,t+\tau)d\tau\cdot g(-T_{m+1})|\varphi|_g \\ &\le \sum_{m=N}^{\infty}\int_{-\infty}^{-T_m} K_{ij}^*(-\tau)d\tau\cdot(m+2)|\varphi|_g \\ &\le \sum_{m=N}^{\infty}(1/m^2)|\varphi|_g \end{aligned}$$

for all $t \geq 0$ and $N = 1, 2, \cdots$, we have

$$|f_{ij}(t,\varphi) - f_{ij}(s,\psi)|$$

$$\leq \int_{-\infty}^{-T_N} K_{ij}(t,t+\tau)|\varphi^j(\tau)|d\tau + \int_{-\infty}^{-T_N} K_{ij}(s,s+\tau)|\varphi^j(\tau)|d\tau$$

$$+ \int_{-T_N}^{0} |K_{ij}(t,t+\tau) - K_{ij}(s,s+\tau)||\varphi^j(\tau)|d\tau$$

$$\leq 2|\varphi|_g \cdot \sum_{m=N}^{\infty} (1/m^2) + \int_{-T_N}^{0} |K_{ij}(t,t+\tau) - K_{ij}(s,s+\tau)|d\tau \cdot |\varphi|_g$$

for all $t, s \geq 0$ and $N = 1, 2, \cdots$. Hence, by Condition (H4), the function $f_{ij}(t,\varphi)$ is uniformly continuous in $t \in \mathbb{R}^+$ uniformly for $\varphi \in W$. Consequently, the function $f_{ij}(t,\varphi)$ is uniformly continuous on $\mathbb{R}^+ \times W$ for any bounded set W in C_g^0. So, if we set

$$F(t,\varphi) = (F_1(t,\varphi), \cdots, F_n(t,\varphi))$$

with

$$F_i(t,\varphi) = \varphi^i(0)\{b_i(t) - a_i(t)\varphi^i(0) - \sum_{j=1}^{n} f_{ij}(t,\varphi)\},$$

$$i = 1, \cdots, n,$$

then the function $F : \mathbb{R}^+ \times C_g^0 \to \mathbb{R}^n$ satisfies Condition (E1) by Condition (H3), and System (5.7) can be considered as a system of FDEs

$$\text{(E)} \qquad \dot{x}(t) = F(t, x_t)$$

on $\mathbb{R}^+ \times C_g^0$. As an immediate consequence of this fact, we conclude that for any $(\sigma,\varphi) \in \mathbb{R}^+ \times C_g^0$, there exists a (unique) local solution of System (5.7) through (σ,φ) and its solution is continuable to $t = \infty$ whenever it remains in a bounded set; see Theorem 2.2.7.

Now, for $i = 1, \cdots, n$, set

$$\beta_i = b_i^u / a_i^\ell$$

and

$$\alpha_i = [b_i^\ell - \sum_{j=1}^{n} (a_{ij}^u b_j^u / a_j^\ell)]/a_i^u.$$

By Condition (H3), we have $0 < \alpha_i < \beta_i$ for all $i = 1,\cdots,n$.

<u>Lemma 5.2.</u> Let $\varphi = (\varphi^1,\cdots,\varphi^n) \in C_g^0$ and let $x(t) = (x^1(t),\cdots,x^n(t))$ be the solution of System (5.7) through $(0,\varphi)$. Then the following statements hold:

(i) If $\varphi^i(s) \ge 0$ on $(-\infty,0)$ and $0 < \varphi^i(0) \le M_i$ for some constant M_i and all $i = 1,\cdots,n$, then

$$(5.8) \qquad 0 < x^i(t) \le \max(\beta_i, M_i)$$

for all $t \in \mathbb{R}^+$ and all $i = 1,\cdots,n$.

(ii) If $\alpha_i \le \varphi^i(s) \le \beta_i$ on $(-\infty,0]$ for all $i = 1,\cdots,n$, then

$$\alpha_i \le x^i(t) \le \beta_i$$

for all $t \in \mathbb{R}^+$ and all $i = 1,\cdots,n$.

<u>Proof.</u> First, we shall verify the claim (i). As long as $x(t)$ exists, we have

$$x^i(t) = \varphi^i(0)\cdot\exp[\int_0^t \{b_i(s) - a_i(s)x^i(s)$$

$$- \sum_{j=1}^{n} a_{ij}(s)\int_{-\infty}^{s} K_{ij}(s,\tau)x^j(\tau)d\tau\}ds] \quad i = 1,2,\cdots,n,$$

and consequently, $x^i(t) > 0$ for all $i = 1,\cdots,n$. Hence, if $x^i(t) > \max(\beta_i, M_i)$ at some value t, we obtain

$$(d/dt)x^i(t) \le x^i(t)\{b_i^u - a_i^\ell x^i(t)\} < 0.$$

Therefore, we easily see that $0 < x^i(t) \le \max(\beta_i, M_i)$ for all $i = 1,\cdots,n$ as long as $x(t)$ exists, because of $0 < \varphi^i(0) \le M_i$. Thus, $x(t)$ exists on $\mathbb{R}^+$, and relation (5.8) holds.

Next, we shall verify the claim (ii). By (i), we have $0 < x^i(t) \le \beta_i$ on $\mathbb{R}^+$; hence

$$(d/dt)x^i(t) \ge x^i(t)\{b_i^\ell - a_i^u x^i(t) - \sum_{j=1}^{n} a_{ij}^u \beta_j\}$$

for all $t \in \mathbb{R}^+$ and all $i = 1,\cdots,n$. Then, by Condition (H3) and the above inequality, we can see that $\alpha_i \le x^i(t)$ for all $t \in \mathbb{R}^+$ and all $i = 1,\cdots,n$, because of $\varphi^i(0) \ge \alpha_i$.

Next, we shall determine the form of the elements in $\Omega(F)$. From Conditions (H3) and (H4), we see that for any sequence $\{\tau_m'\}$, $\tau_m' \to \infty$ as $m \to \infty$, there exist a subsequence $\{\tau_m\}$ of $\{\tau_m'\}$ and functions $\tilde{a}_i(t)$, $\tilde{b}_i(t)$, $\tilde{a}_{ij}(t)$ and $\tilde{K}_{ij}(t,\tau)$ such that

$$(5.9)\quad \begin{cases} a_i(t+\tau_m) \to \tilde{a}_i(t),\ b_i(t+\tau_m) \to \tilde{b}_i(t),\ a_{ij}(t+\tau_m) \to \tilde{a}_{ij}(t) \\ \text{compactly on } \mathbb{R}^+, \text{ and } K_{ij}(t+\tau_m, t+\tau_m+\tau) \to \tilde{K}_{ij}(t,\tau) \\ \text{compactly on } \mathbb{R}^+ \times (-\infty, t] \text{ for all } i,j = 1,\cdots,n. \end{cases}$$

Clearly, the functions $\tilde{a}_i(t)$, $\tilde{b}_i(t)$, $\tilde{a}_{ij}(t)$ and $\tilde{K}_{ij}(t,\tau)$ satisfy Conditions (H3) and (H4) with the same a_i^ℓ, a_i^u, b_i^ℓ, b_i^u, a_{ij}^ℓ, a_{ij}^u and K_{ij}^* for all $i, j = 1,\cdots,n$. Set

$$(5.10)\qquad G(t,\varphi) = (G_1(t,\varphi),\cdots,G_n(t,\varphi)),$$

where

$$(5.11)\qquad G_i(t,\varphi) = \varphi^i(0)\{\tilde{b}_i(t) - \tilde{a}_i(t)\varphi^i(0) - \sum_{j=1}^n \tilde{a}_{ij}(t)\int_{-\infty}^0 \tilde{K}_{ij}(t,t+\tau)\varphi^j(\tau)d\tau\}$$

$$i = 1,\cdots,n,$$

for $\varphi = (\varphi^1,\cdots,\varphi^n) \in C_g^0$. By the preceding argument, we see that G is in $C(\mathbb{R}^+ \times C_g^0, \mathbb{R}^n)$ and that $F^{\tau_m} \to G$ compactly on $\mathbb{R}^+ \times C_g^0$. Thus, the set $\Omega(F)$ consists of all G in $C(\mathbb{R}^+ \times C_g^0, \mathbb{R}^n)$ such that Relations (5.9), (5.10) and (5.11) hold for some sequence $\{\tau_m\}$, $\tau_m \to \infty$ as $m \to \infty$.

Now, let $u(t) = (u^1(t),\cdots,u^n(t))$ be a solution on $\mathbb{R}^+$ of System (5.7) satisfying $u_0 \in BC$ and $\alpha_i \le u^i(s) \le \beta_i$ for all $s \le 0$ and all $i = 1,\cdots,n$. By Lemma 5.2, we have $\alpha_i \le u^i(t) \le \beta_i$ for all $t \in \mathbb{R}^+$ and all $i = 1,\cdots,n$. In the following, we shall investigate stability properties of the solution u of System (5.7) (with $\mathcal{B} = C_g^0$).

<u>Theorem 5.3.</u> Let Conditions (H3) and (H4) hold, and assume the following condition holds;

(H5) there exists a positive constant $\mu > 0$ such that

$$a_i^\ell > \sum_{j=1}^{n} a_{ji}^u + \mu \quad \text{for all} \quad i = 1,\cdots,n.$$

Then, the solution of System (5.7) is WUAS in $\Omega(F)$, and moreover, it is XS for System (5.7).

<u>Proof.</u> We shall show that the solution u is WUAS in $\Omega(F)$ in $\mathbb{R}^n$. Then the assertions of the theorem follow from Theorem 3.4, since u is the unique solution of System (E) for the initial conditions and the stabilities in $\mathbb{R}^n$ are equivalent to the stabilities in C_g^0.

Take any $(v,G) \in \Omega(u,F)$, $v = (v^1,\cdots,v^n)$. Clearly, $\alpha_i \le v^i(t) \le \beta_i$ on $\mathbb{R}$. For any $(\sigma,\varphi) \in \mathbb{R}^+ \times C_g^0$, let $x(t) = (x^1(t),\cdots,x^n(t))$ be the solution of System (G) through (σ,φ), and set

$$V(t) = \sum_{i=1}^{n} [\left|\log \frac{x^i(t)}{v^i(t)}\right| + \sum_{j=1}^{n}\int_0^\infty K_{ij}^*(s)\{\int_{t-s}^{t} \tilde{a}_{ij}(s+\tau)|x^j(\tau) - v^j(\tau)|d\tau\}ds].$$

As long as $|v^i(t) - x^i(t)| \le \alpha_i/2$ for all $i = 1,\cdots,n$, $V(t)$ is a real valued function and is continuous in t, because $\int_0^\infty sK_{ij}^*(s)g(-s)ds < \infty$, and moreover, if we set

$$r = \min\{2/(3\beta_i) : i = 1,\cdots,n\}$$

and

$$R = \sum_{i=1}^{n} [2/\alpha_i + \sum_{j=1}^{n} a_{ij}^u\cdot\int_0^\infty sK_{ij}^*(s)g(-s)ds],$$

then

$$\begin{aligned} r\cdot\sum_{i=1}^{n} |x^i(t) - v^i(t)| &\le V(t) \\ &\le \sum_{i=1}^{n} [2|x^i(t) - v^i(t)|/\alpha_i + \sum_{j=1}^{n}\int_0^\infty sK_{ij}^*(s)g(-s)ds\cdot a_{ij}^u|x_t - v_t|_g] \\ &\le R|x_t - v_t|_g. \end{aligned}$$

Furthermore, considering the each case where $x^i(t) > v^i(t)$, $x^i(t) < v^i(t)$ and $x^i(t) = v^i(t)$, we obtain

$$
\begin{aligned}
D^+V(t) \le \sum_{i=1}^{n} \{ & -\tilde{a}_i(t)|x^i(t) - v^i(t)| \\
& + \sum_{j=1}^{n} \tilde{a}_{ij}(t)\int_{-\infty}^{t} \tilde{K}_{ij}(t,\tau)|x^j(\tau) - v^j(\tau)|d\tau\} \\
& + \sum_{i,j=1}^{n} \int_0^{\infty} K^*_{ij}(s)\{\tilde{a}_{ij}(s+t)|x^j(t) - v^j(t)| \\
& \qquad - \tilde{a}_{ij}(t)|x^j(t-s) - v^j(t-s)|\}ds \\
\le - \sum_{i=1}^{n} [& a_i^{\ell} - \sum_{j=1}^{n} a_{ji}^{u} \cdot \int_0^{\infty} K^*_{ji}(s)ds]|x^i(t) - v^i(t)| \\
& + \sum_{i,j=1}^{n} \tilde{a}_{ij}(t)\int_{-\infty}^{t} |\tilde{K}_{ij}(t,\tau) - K^*_{ij}(t-\tau)||x^j(\tau) - v^j(\tau)|d\tau \\
\le -\mu & \sum_{i=1}^{n} |x^i(t) - v^i(t)| \le 0
\end{aligned}
$$

by Conditions (H3), (H4) and (H5). For any $\varepsilon > 0$, set

$$\delta(\varepsilon) = \min\{r\varepsilon/R,\ r\alpha_i/(2R):\ i = 1,\cdots,n\}.$$

If $|\varphi - v_\sigma|_g < \delta(\varepsilon)$, from the above relations it follows that

$$\sum_{i=1}^{n} |x^i(t) - v^i(t)| < R\delta(\varepsilon)/r \le \min\{\varepsilon,\ \alpha_i/2;\ i = 1,\cdots,n\}$$

as long as $|v^i(t) - x^i(t)| \le \alpha_i/2$ for all $i = 1,\cdots,n$. Consequently, we obtain

$$\sum_{i=1}^{n} |v^i(t) - x^i(t)| < \min\{\varepsilon,\ \alpha_i/2;\ i = 1,\cdots,n\} \le \varepsilon, \quad t \ge \sigma,$$

if $|\varphi - v_\sigma|_g < \delta(\varepsilon)$, which shows that v is US for (G) in R^n with $\delta(\cdot)$ which does not depend on the particular choice of $(v,G) \in \Omega(u,F)$.

Next, let $|\varphi - v_\sigma|_g < (r/R)\cdot\min\{\alpha_i/2;\ i = 1,\cdots,n\} =: \delta_0$ for some $\sigma \in R^+$. From the above calculations, it follows that

$$\sum_{i=1}^{n} |x^i(t) - v^i(t)| < \min\{\alpha_i/2;\ i = 1,\cdots,n\}$$

and

$$V(t) - V(\sigma) \le -\mu \sum_{i=1}^{n} \int_{\sigma}^{t} |x^i(s) - v^i(s)| ds$$

for all $t \ge \sigma$. Consequently, $\sum_{i=1}^{n} \int_{\sigma}^{\infty} |x^i(s) - v^i(s)| ds < \infty$, and hence $\sum_{i=1}^{n} |x^i(t) - v^i(t)| \to 0$ as $t \to \infty$ because the function $\sum_{i=1}^{n} |x^i(s) - v^i(s)|$ of s is uniformly continuous on $[\sigma,\infty)$. Therefore, the solution v of System (G) is WUAS in $\mathbb{R}^n$. Note that the above number δ_0 does not depend on the particular choice of $(v,G) \in \Omega(u,F)$. Thus, the solution u of System (E) is WUAS in $\Omega(F)$ in $\mathbb{R}^n$. This completes the proof.

From Lemma 5.2 and the proof of Theorem 5.3, it is not difficult to see that the solution u of System (5.7) is globally attracting; that is, if $x(t) = (x^1(t),\cdots,x^n(t))$ is a solution of System (5.7) such that $x_0 \in C_g^0$ and $x^i(s) \ge 0$ on $(-\infty,0)$ with $x^i(0) > 0$ for all $i = 1,\cdots,n$, then $|x_t - u_t|_g \to 0$ as $t \to \infty$. If one can choose an exponential function $e^{-\gamma s}$ with a constant $\gamma > 0$ as the function $g(s)$ (e.g., the case $\int_0^{\infty} sK_{ij}^*(s)e^{\gamma s} ds < \infty$ for a constant $\gamma > 0$), then Theorem 3.5 yields the asymptotic stability under X-perturbations in Theorem 5.3. Observe that the space C_g^0 constructed in this section is not necessarily a uniform fading memory space. Hence, without any additional condition, System (5.7) has no uniformly asymptotically stable solutions by Theorem 7.2.6.

In the following, we shall restrict our consideration to BC, the family of bounded continuous functions, as the initial functions, and obtain some kinds of stability properties for the solution u. Here, we recall that BC is the Banach space with the supremum norm $|\cdot|_{BC}$.

<u>Definition 5.4.</u> The solution u is said to be stable under X-perturbations for System (5.7) with respect to the initial functions in BC (XS for System (5.7) w.r.t. BC, in short), if for any $\varepsilon > 0$ there exists a $\delta(\varepsilon) > 0$ such that $\varphi \in BC$, $|u_\sigma - \varphi|_{BC} < \delta(\varepsilon)$ for some $\sigma \in \mathbb{R}^+$ and $|p|_X < \delta(\varepsilon)$ imply $|u_t - x_t(\sigma,\varphi,F+p)|_g < \varepsilon$ for all $t \ge \sigma$.

Definition 5.5. The solution u is said to be asymptotically stable under X-perturbations for System (5.7) with respect to the initial functions in BC (XAS for System (5.7) w.r.t. BC, in short), if it is XS for System (5.7) w.r.t. BC, and moreover, there exists a $\delta_0 > 0$ with the property that for any $\varepsilon > 0$ there exist $T(\varepsilon) > 0$ and $\gamma(\varepsilon) \ge 0$ such that $\varphi \in BC$, $|u_\sigma - \varphi|_{BC} < \delta_0$ for some $\sigma \in \mathbb{R}^+$ and $|p|_X < \gamma(\varepsilon)$ imply $|u_t - x_t(\sigma,\varphi,F+p)|_g < \varepsilon$ for all $t \ge \sigma + T(\varepsilon)$.

Theorem 5.6. Suppose Conditions (H3), (H4) and (H5) hold. Then the solution u is XAS for System (5.7) w.r.t. BC.

Proof. By Theorem 5.3, the solution u is WUAS in $\Omega(F)$ and XS for System (5.6). Let δ_0 and $\delta(\cdot)$ be the numbers ensured by the facts that u is WUAS in $\Omega(F)$ and XS for System (5.7), respectively. Set $\delta_1 = \delta(\delta_0/2)$. Obviously, u is XS for System (5.7) w.r.t. BC with the pair $(\varepsilon,\delta(\varepsilon))$. So, it suffices to prove that the number δ_1 satisfies the condition in Definition 5.5. Suppose this is not the case. Then there exist an $\varepsilon > 0$, sequences $\{\tau_m\}$, $\tau_m \ge 0$, $\{t_m\}$, $t_m \ge \tau_m + 2m$, $\{\varphi^m\} \subset BC$ and $\{p^m\} \subset X$ such that

$$(5.12)\qquad \begin{cases} |\varphi^m|_{BC} < \delta_1,\ |p^m|_X < \min(1/m,\ \delta_1) \text{ and} \\ |x_{t_m}(\tau_m,\varphi^m,F+p^m) - u_{t_m}|_g \ge \varepsilon \end{cases}$$

for all $m = 1,2,\cdots$. Without loss of generality, we may assume that $(u^{\tau_m+m},F^{\tau_m+m}) \to (v,G)$ compactly on $\mathbb{R}^+ \times \mathcal{B}$ for some $(v,G) \in \Omega(u,F)$. Since u is XS for System (5.7), we obtain, by (5.12),

$$(5.13)\qquad |x_t(\tau_m,\varphi^m,F+p^m) - u_t|_g \ge \delta(\varepsilon) \quad \text{on } [\tau_m,\tau_m+2m]$$

and

$$(5.14)\qquad |x_t(\tau_m,\varphi^m,F+p^m) - u_t|_g < \delta_0/2 \quad \text{on } [\tau_m,\infty)$$

for sufficiently large m. Set $z^m(t) = x(t+m+\tau_m,\tau_m,\varphi^m,F+p^m)$. By (5.14), it is easy to see that the sequence of functions $\{z^m(t)\}$ is uniformly bounded and equicontinuous on each compact interval in $\mathbb{R}$. Applying the Ascoli-Arzéla theorem and the diagonalization procedure, we may assume that $z^m(t) \to y(t)$ compactly on $\mathbb{R}$ for some continuous function y. Since $\{z^m(\cdot)\}$ and $y(\cdot)$ are uniformly bounded on $\mathbb{R}$, we have $(z^m)_t \to y_t$ in C_g^0 compactly on $\mathbb{R}^+$ because of the inequality

$$|(z^m)_t - y_t|_g$$

$$= \max(\sup_{\theta \le -N} |z^m(t+\theta) - y(t+\theta)|/g(\theta),\ \sup_{-N\le\theta\le 0} |z^m(t+\theta) - y(t+\theta)|/g(\theta))$$

$$\le [|z^m|_{BC} + |y|_{BC}]/g(-N) + \sup_{-N\le\theta\le 0} |z^m(t+\theta) - y(t+\theta)|$$

for all $N > 0$. Hence, we easily see that $y(t)$ is a solution on $\mathbb{R}^+$ of System (G), and moreover, letting $m \to \infty$ in (5.13) and (5.14), we have

(5.15) $$\delta(\varepsilon) \le |y_t - v_t|_g \le \delta_0/2 \quad \text{on } \mathbb{R}^+.$$

Then, we must have $|y_t - v_t|_g \to 0$ as $t \to \infty$ by (5.15), because u is WUAS in $\Omega(F)$. This is a contradiction to (5.15). This completes the proof.

CHAPTER 9.

EXISTENCE OF PERIODIC SOLUTIONS AND ALMOST PERIODIC SOLUTIONS

<u>9.1. Existence theorems for nonlinear systems</u>. As in the previous chapter, we assume that $\mathcal{B}$ is a separable fading memory space throughout this chapter. Recall that $C(\mathbb{R}\times\mathcal{B},E)$ is the set of all continuous functions mapping $\mathbb{R}\times\mathcal{B}$ into E. An $f(t,\varphi)\in C(\mathbb{R}\times\mathcal{B},E)$ is called an almost periodic function in t uniformly for $\varphi\in\mathcal{B}$, if for any $\varepsilon>0$ and any compact set W in $\mathcal{B}$, there exists a positive number $\ell(\varepsilon,W)$ such that any interval of length $\ell(\varepsilon,W)$ contains a τ for which

$$|f(t+\tau,\varphi)-f(t,\varphi)|_E\le\varepsilon \quad \text{for all } t\in\mathbb{R} \text{ and all } \varphi\in W.$$

Denote by AP the set of all functions $f(t,\varphi)$ which are almost periodic in t uniformly for $\varphi\in\mathcal{B}$. Moreover, for any $\omega>0$, denote $P_\omega=\{f\in C(\mathbb{R}\times\mathcal{B},E) : f(t+\omega,\varphi)=f(t,\varphi) \text{ for all } (t,\varphi)\in\mathbb{R}\times\mathcal{B}\}$. Clearly, $P_\omega\subset AP$.

Let $BU=BU(\mathbb{R}\times\mathcal{B},E)$ be the set of all bounded uniformly continuous functions on $\mathbb{R}\times W$ for any compact set W in $\mathcal{B}$. If $f(t,\varphi)\in AP$, then it belongs to the set BU; see, e.g. Yoshizawa [2, p.7, Theorem 1]; consequently, the hull $H(f)$ of $f(t,\varphi)$ is metrizable and compact in $C(\mathbb{R}\times\mathcal{B},E)$ by Proposition 8.1.3 and Theorem 8.1.4. Hence, for any sequence $\{\tau_m'\}\subset R$, there exist a sequence $\{\tau_m\}$ of $\{\tau_m'\}$ and a $g\in C(\mathbb{R}\times\mathcal{B},R^n)$ such that $f^{\tau_m}\to g$ compactly on $\mathbb{R}\times\mathcal{B}$; that is $f(t+\tau_m,\varphi)\to g(t,\varphi)$ as $m\to\infty$ uniformly on any compact subset of $\mathbb{R}\times\mathcal{B}$. In fact, we can assume that $g(t,\varphi)\in AP$ and $f(t+\tau_m,\varphi)\to g(t,\varphi)$ uniformly on $\mathbb{R}\times W$ for any compact set W in $\mathcal{B}$; see Yoshizawa [2, Theorem 2.2]. For the other properties of almost periodic functions, we refer the reader to Yoshizawa [2, pp. 5 - 20].

Now, consider a system of functional differential equations

$$\text{(E)} \qquad \dot{x}(t)=F(t,x_t), \quad t\in\mathbb{R}.$$

We assume the following conditions on System (E):

(H1) $F\in AP$, and for each $N>0$ there exists a number $L(N)>$

0 such that $|F(t,\varphi)| \le L(N)$ for all $(t,\varphi) \in \mathbb{R} \times \mathcal{B}$ with $|\varphi|_{\mathcal{B}} \le N$;

(H2) System (E) has a solution u on $\mathbb{R}^+ := [0,\infty)$ such that $\sup_{t\in\mathbb{R}^+} |u(t)| =: H < \infty$.

As was stated in Section 8.1, the closure of positive orbit of u,

$$\mathcal{O}^+(u) := \{u_t : t \in \mathbb{R}^+\}^-,$$

is compact in $\mathcal{B}$. Moreover, the hull $H(u,F)$ of (u,F) is compact in the product space $C(\mathbb{R}^+,\mathcal{B}) \times C(\mathbb{R}\times\mathcal{B},E)$, and if $(v,G) \in H(u,F)$, then v is a solution on $\mathbb{R}^+$ of

$$\text{(G)} \qquad \dot{x}(t) = G(t,x_t).$$

The set $\Omega(u,F)$ also has the same properties; see Theorem 8.1.5.

For any x in $BU(\mathbb{R},E)$, we set

$$|x|_\infty = \sup_{t\in\mathbb{R}} |x(t)|.$$

A function $x \in C(\mathbb{R},E)$ is called an $\mathbb{R}$-bounded solution of System (E), if $x \in BU(\mathbb{R},E)$ and it satisfies System (E) for all $t \in \mathbb{R}$.

<u>Proposition 1.1</u>. Assume Conditions (H1) and (H2). Then, for any given sequence $\{\tau_k'\} \subset \mathbb{R}^+$, $\tau_k' \to \infty$ as $k \to \infty$, there exist a sequence $\{\tau_k\}$ of $\{\tau_k'\}$ and a $(v,G) \in \Omega(u,F)$ such that

(1.1) $F(t+\tau_k,\varphi) \to G(t,\varphi)$ as $k \to \infty$ uniformly on $\mathbb{R} \times W$ for any compact set W in $\mathcal{B}$ and $u_{t+\tau_k} \to v_t$ as $k \to \infty$ on any compact subset of $\mathbb{R}$.

In this case, v is an $\mathbb{R}$-bounded solution of System (G) and it satisfies $|v|_\infty \le H$. In particular, System (E) has an $\mathbb{R}$-bounded solution x such that $|x|_\infty \le H$.

<u>Proof.</u> Since $F \in AP$ and $u \in BU(\mathbb{R}^+,E)$, there exist a sequence $\{\tau_k\}$ of $\{\tau_k'\}$, a $v \in BU(\mathbb{R},E)$, $|v|_\infty \le H$, and a $G \in \Omega(F)$ such that $F(t+\tau_k,\varphi) \to G(t,\varphi)$ as $k \to \infty$ uniformly on $\mathbb{R} \times W$ for any compact set W in $\mathcal{B}$ and $u(t+\tau_k) \to v(t)$ as $k \to \infty$ compactly on $\mathbb{R}$. Let J be any compact interval in $\mathbb{R}$. We must prove

(1.2) $\quad u_{t+\tau_k} \to v_t$ as $k \to \infty$ uniformly on J.

To do this, set

$$\tilde{u}(t) = \begin{cases} u(0) & \text{if } t < 0, \\ u(t) & \text{if } t \geq 0. \end{cases}$$

Then Axiom (C2) yields that

(1.3) $\quad |\tilde{u}_{t+\tau_k} - v_t|_{\mathscr{B}} \to 0$ as $k \to \infty$ uniformly on J.

Moreover, since $\mathscr{B}$ is a fading memory space, we have

(1.4) $\quad |S_0(t+\tau_k)(u_0-u(0)\chi)|_{\mathscr{B}} \to 0$ as $k \to \infty$ uniformly on J,

where $\chi(\theta) = 1$ for $\theta \leq 0$. Since $u_t = \tilde{u}_t + S_0(t)[u_0-u(0)\chi]$ for $t \geq 0$, (1.2) follows from (1.3) and (1.4). The assertion that v is a solution on $\mathbb{R}$ of System (G) can be proved by the standard argument; see, e.g., the paragraph foregoing Theorem 8.1.5.

The last part of the proposition follows from the first part of the proposition, because $F \in \Omega(F)$ by the fact $F \in AP$.

Henceforth, when (1.1) holds, we write it as

$$\text{"}(u^{\tau_m}, F^{\tau_m}) \to (v,G) \text{ compactly on } \mathbb{R}\text{"},$$

for simplicity; cf. Section 8.1. For any $G \in \Omega(F)$, we set

$$\Omega_G(u) = \{v \in C(\mathbb{R},E) : (u^{\tau_m}, F^{\tau_m}) \to (v,G) \text{ compactly on } \mathbb{R} \text{ for some sequence } \{\tau_m\} \text{ with } \tau_m \to \infty \text{ as } m \to \infty\}$$

and

$$\Omega(u) = \cup\{v \in \Omega_G(u) : G \in \Omega(F)\}.$$

In virtue of Proposition 1.1, the set $\Omega_G(u)$ is not empty, and if $v \in \Omega_G(u)$, then v is an $\mathbb{R}$-bounded solution of System (G).

Now, let $f(t)$ be a continuous function defined on $a \leq t < \infty$. $f(t)$ is said to be asymptotically almost periodic if it is a sum of a continuous almost periodic function $p(t)$ and a continuous function $q(t)$ defined on $a \leq t < \infty$ which tends to zero as $t \to \infty$; that is,

$$f(t) = p(t) + q(t).$$

It is well known (e.g., Yoshizawa [2, pp. 20-30]) that $f(t)$ is asymptotically almost periodic if and only if for any sequence $\{\tau_k'\}$ such that $\tau_k' \to \infty$ as $k \to \infty$, there exists a subsequence $\{\tau_k\}$ of $\{\tau_k'\}$ for which $f(t+\tau_k)$ converges uniformly on $a \le t < \infty$.

Theorem 1.2 (Hino [4]). Suppose the solution u is asymptotically almost periodic. If $G \in \Omega(F)$ and $v \in \Omega_G(u)$, then v is an almost periodic solution of System (G).

Proof. Since u is asymptotically almost periodic, it has the decomposition $u(t) = p(t) + q(t)$, where $p(t)$ is almost periodic and $q(t) \to 0$ as $t \to \infty$. Choose a sequence $\{\tau_k\}$, $\tau_k \to \infty$ as $k \to \infty$, so that $(u^{\tau_k}, F^{\tau_k}) \to (v, G)$ compactly on $\mathbb{R}$. By Proposition 1.1, v is an $\mathbb{R}$-bounded solution of System (G). Moreover, $p(t+\tau_k) \to v(t)$ as $k \to \infty$ compactly on $\mathbb{R}$. Hence $v \in AP$, because of $p \in AP$.

Now, we need some definitions of stabilities with respect to the hull.

Definition 1.3. The solution $u(t)$ is uniformly stable with respect to $\Omega_F(u)$ (in short, u.s.$\Omega_F(u)$), if for any $\varepsilon > 0$ there exists a $\delta(\varepsilon) > 0$ such that

$$|u_t - x_t|_{\mathscr{B}} < \varepsilon \quad \text{for all} \quad t \ge t_0,$$

whenever $t_0 \ge 0$, $x \in \Omega_F(u)$ and $|u_{t_0} - x_{t_0}|_{\mathscr{B}} < \delta(\varepsilon)$.

Definition 1.4. The solution $u(t)$ is uniformly asymptotically stable with respect to $\Omega_F(u)$ (in short, u.a.s.$\Omega_F(u)$), if it is u.s.$\Omega_F(u)$ and moreover, there exists a $\delta_0 > 0$ and for any $\varepsilon > 0$ there exists a $T(\varepsilon) > 0$ such that

$$|u_t - x_t|_{\mathscr{B}} < \varepsilon \quad \text{for all } t \ge t_0 + T(\varepsilon),$$

whenever $t_0 \ge 0$, $x \in \Omega_F(u)$ and $|u_{t_0} - x_{t_0}|_{\mathscr{B}} < \delta_0$.

Definition 1.5. The solution $u(t)$ is uniformly asymptotically stable in the large with respect to $\Omega_F(u)$ (in short, u.a.s.l. $\Omega_F(u)$), if it is u.s.$\Omega_F(u)$ and for any $\alpha > 0$ and $\varepsilon > 0$, there exists a $T(\alpha,\varepsilon) > 0$ such that

$$|u_t - x_t|_{\mathcal{B}} < \varepsilon \text{ for } t \ge t_0 + T(\alpha,\varepsilon),$$

whenever $t_0 \ge 0$, $x \in \Omega_F(u)$ and $|u_{t_0} - x_{t_0}|_{\mathcal{B}} < \alpha$.

The above definitions of stabilities are weaker than the usual ones (cf. Section 7.2), respectively, because $u(t)$ is not necessarily unique for initial value problem. For example, the solution $x(t) = 0$ of $\dot{x}(t) = x^{1/3}$ is not uniformly stable for $t \ge 0$, but u.a.s.$\Omega_{x^{1/3}}(x(t) = 0)$. Note that the above (u.a.s. or u.a.s.l.) stabilities can be realized, even if $\mathcal{B}$ is not a uniform fading memory space; cf. Theorem 7.2.6.

In the below, following Hino [4] we shall give some results on the existence of almost periodic solutions and periodic solutions for periodic systems.

Theorem 1.6. Suppose $F \in P_\omega$ and the solution $u(t)$ is u.s.$\Omega_F(u)$. Then $u(t)$ is an asymptotically almost periodic solution of (E). Consequently, System (E) has an almost periodic solution.

Proof. Let $\{\tau_k\}$ be a sequence such that $\tau_k \ge \omega$ and $\tau_k \to \infty$ as $k \to \infty$. For each τ_k, there exists a positive integer N_k such that $N_k\omega \le \tau_k < (N_k+1)\omega$. If we set $\tau_k = N_k\omega + \sigma_k$, then $0 \le \sigma_k < \omega$. By Proposition 1.1, we may assume that there exists a σ, $0 \le \sigma \le \omega$, and a function $v(t)$ such that $\sigma_k \to \sigma$ as $k \to \infty$ and $(u^{\tau_k}, F^{\tau_k}) \to (v, F^\sigma)$ compactly on $\mathbb{R}$. For a given $\varepsilon > 0$, select a positive integer $n_0(\varepsilon)$ so that

$$(1.5)\qquad |u_{\tau_k+\omega} - v_\omega|_{\mathcal{B}} < \delta(\varepsilon)/2$$

if $k \ge n_0(\varepsilon)$, where $\delta(\cdot)$ is the one for u.s.$\Omega_F(u)$ of $u(t)$. Since $v(t)$ is uniformly continuous and bounded on $\mathbb{R}$, Proposition 7.1.1 implies that the function $v_t : \mathbb{R} \to \mathcal{B}$ is uniformly continuous. Thus we may assume that

$$(1.6)\qquad |v_t - v_{\sigma_k - \sigma + t}|_{\mathcal{B}} < \delta(\varepsilon)/2$$

for all $k \ge n_0(\varepsilon)$ and $t \in \mathbb{R}$. Set $v^k(t) = v(t - N_k\omega - \sigma)$. Clearly, we have $v^k(t) \in \Omega_F(u)$. Note that, for $k \ge n_0(\varepsilon)$,

$$|u_{\tau_k+\omega} - (v^k)_{\tau_k+\omega}|_{\mathscr{B}} = |u_{\tau_k+\omega} - v_{\omega+\sigma_k-\sigma}|_{\mathscr{B}}$$

$$\leq |u_{\tau_k+\omega} - v_{\omega}|_{\mathscr{B}} + |v_{\omega} - v_{\omega+\sigma_k-\sigma}|_{\mathscr{B}} \leq \delta(\varepsilon)$$

by (1.5) and (1.6). Since $u(t)$ is u.s.$\Omega_F(u)$, it follows

$$(1.7) \qquad |u_{\tau_k+t} - (v^k)_{\tau_k+t}|_{\mathscr{B}} < \varepsilon$$

for all $t \geq \omega$ and all $k \geq n_0(\varepsilon)$. Furthermore, since

$$|u_{\tau_k+t} - u_{\tau_m+t}|_{\mathscr{B}} \leq |u_{\tau_k+t} - (v^k)_{\tau_k+t}|_{\mathscr{B}} + |(v^k)_{\tau_k+t} - (v^m)_{\tau_m+t}|_{\mathscr{B}}$$

$$+ |(v^m)_{\tau_m+t} - u_{\tau_m+t}|_{\mathscr{B}}$$

$$\leq |u_{\tau_k+t} - (v^k)_{\tau_k+t}|_{\mathscr{B}} + |v_{\sigma_k-\sigma+t} - u_{\sigma_m-\sigma+t}|_{\mathscr{B}}$$

$$+ |(v^m)_{\tau_m+t} - u_{\tau_m+t}|_{\mathscr{B}}$$

$$\leq |u_{\tau_k+t} - (v^k)_{\tau_k+t}|_{\mathscr{B}} + |v_{\sigma_k-\sigma+t} - v_t|_{\mathscr{B}}$$

$$+ |v_t - v_{\sigma_m-\sigma+t}|_{\mathscr{B}} + |(v^m)_{\tau_m+t} - u_{\tau_m+t}|_{\mathscr{B}},$$

(1.6) and (1.7) imply that

$$(1.8) \qquad |u_{\tau_k+t} - u_{\tau_m+t}|_{\mathscr{B}} < 3\varepsilon$$

for all $t \geq \omega$ and for $m \geq k \geq n_0(\varepsilon)$; hence

$$|u(\tau_m+t) - u(\tau_m+t)| < 3H\varepsilon \quad \text{for all} \quad t \geq \omega \quad \text{and for} \quad m \geq k_0 \geq n(\varepsilon).$$

Thus, $\{u(t+\tau_k)\}$ converges uniformly on $[\omega,\infty)$; consequently, $u(t)$ is an asymptotically almost periodic solution. The existence of an almost periodic solution follows immediately from Theorem 1.2.

<u>Theorem 1.7.</u> Let $F \in P_\omega$. If the solution $u(t)$ is u.a.s.$\Omega_F(u)$, then System (E) has a periodic solution of period $m\omega$ for some integer $m \geq 1$. In particular, if $u(t)$ is u.a.s.l.$\Omega_F(u)$, then System (E) has an ω-periodic solution.

Proof. By Proposition 1.1, there exist a sequence $\{k_j\}$, where k_j is a positive integer and $k_j \to \infty$ as $j \to \infty$, and a $v \in \Omega_F(u)$ such that $u_{t+k_j\omega} \to v_t$ as $j \to \infty$ compactly on $\mathbb{R}$, because $F(t,\varphi)$ is periodic with period ω. Then, there exist integers k_p and k_{p+1} such that $|u_{k_p\omega} - v_0|_{\mathscr{B}} < \delta_0$ and $|u_{k_{p+1}\omega} - v_0|_{\mathscr{B}} < \delta_0$, where δ_0 is the one for u.a.s.$\Omega_F(u)$ of $u(t)$. Set $m = k_{p+1} - k_p$, $u^m(t) = u(t+m\omega)$ and $v^{k_p}(t) = v(t-k_p\omega)$. Clearly, $v^{k_p}(t) \in \Omega_F(u)$ and $u^m(t)$ is the solution of (E). Thus, we have

$$|(u^m)_{k_p\omega} - (v^{k_p})_{k_p\omega}|_{\mathscr{B}} = |u_{k_{p+1}\omega} - v_0|_{\mathscr{B}} < \delta_0$$

and

$$|u_{k_p\omega} - (v^{k_p})_{k_p\omega}|_{\mathscr{B}} = |u_{k_p\omega} - v_0|_{\mathscr{B}} < \delta_0;$$

hence,

$$|(u^m)_{k_p\omega+t} - v_t|_{\mathscr{B}} \to 0 \quad \text{as} \quad t \to \infty$$

and

$$|u_{k_p\omega+t} - v_t|_{\mathscr{B}} \to 0 \quad \text{as} \quad t \to \infty.$$

Consequently,

$$(1.9) \qquad |(u^m)_t - u_t|_{\mathscr{B}} \to 0 \quad \text{as} \quad t \to \infty.$$

On the other hand, by Theorem 1.6, $u(t)$ is asymptotically almost periodic; hence $u(t)$ can be represented as

$$(1.10) \qquad u(t) = p(t) + q(t),$$

where $p(t)$ is almost periodic and $q(t) \to 0$ as $t \to \infty$. From (1.9) and (1.10) it follows that

$$|p(t) - p(t+m\omega)| = |u(t) - q(t) - u(t+m\omega) + q(t+m\omega)|$$

$$\leq |u(t) - u(t+m\omega)| + |q(t) - q(t+m\omega)|$$

$$\le H|u_t - (u^m)_t|_{\mathcal{B}} + |q(t) - q(t+m\omega)| \to 0$$

as $t \to \infty$. Therefore, we have $p(t) = p(t+m\omega)$ for all $t \in \mathbb{R}$, because $p(t)$ is almost periodic. If we consider a sequence $\{km\omega\}$, then

$$u(t+km\omega) = p(t) + q(t+km\omega),$$

and hence $p(t) \in \Omega_F(u)$. This shows that System (E) has a periodic solution $p(t)$ of period $m\omega$, because $p(t)$ also is a solution of System (E) by Proposition 1.1.

If u is u.a.s.l.$\Omega_F(u)$, then $|u_{t+\omega} - u_t|_{\mathcal{B}} \to 0$ as $t \to \infty$, from which we have $p(t) = p(t+\omega)$ on $\mathbb{R}$.

Next, we shall discuss the existence of almost periodic solutions for almost periodic systems. Let $K_\infty = \{\varphi^1, \varphi^2, \cdots\}$ be a countable dense set in $\mathcal{B}$, and set $K_m = \{\varphi^1, \cdots, \varphi^m\}$ for each $m = 1, 2, \cdots$. For any $G, H \in H(F)$, set

$$\rho(G, H) = \sum_{m=1}^{\infty} \frac{1}{2^m} \frac{\|G-H\|_m}{1+\|G-H\|_m},$$

where

$$\|G-H\|_m = \sup_{(t,\varphi)\in\mathbb{R}\times K_m} |G(t,\varphi) - H(t,\varphi)|.$$

By the same way as in the proof of Proposition 8.1.3, we see that for any $\{G^k\} \subset H(F)$ and $G \in H(F)$, $\rho(G^k, G) \to 0$ as $k \to \infty$ if and only if $G^k(t,\varphi) \to G(t,\varphi)$ as $k \to \infty$ uniformly on $\mathbb{R}\times W$ for any compact set W in $\mathcal{B}$.

<u>Definition 1.8.</u> The solution $u(t)$ is stable under disturbance from $\Omega(u,F)$ (in short, s.d.$\Omega(u,F)$), if for any $\varepsilon > 0$, there exists a $\delta(\varepsilon) > 0$ such that $|u_{t+\tau} - x_t|_{\mathcal{B}} < \varepsilon$ for $t \ge 0$, whenever $(x,G) \in \Omega(u,F)$, $|u_\tau - x_0|_{\mathcal{B}} < \delta(\varepsilon)$ and $\rho(F^\tau, G) < \delta(\varepsilon)$ for some $\tau \ge 0$.

<u>Theorem 1.9 (Hino [4]).</u> Let (H1) and (H2) hold. If the solution $u(t)$ is s.d.$\Omega(u,F)$, then it is an asymptotically almost periodic solution. Consequently, System (E) has an almost periodic solution.

<u>Proof.</u> Let $\{\tau_k\}$ be any sequence such that $\tau_k > 0$ and $\tau_k \to \infty$

as $k \to \infty$. By Proposition 1.1, we may assume that $(u^{\tau_k}, F^{\tau_k}) \to (x, G)$ compactly on $\mathbb{R}$ for some $(x(t), G(t,\varphi)) \in \Omega(u,F)$. For any $\varepsilon > 0$ there exists a $k_0(\varepsilon) > 0$ such that if $k \ge k_0(\varepsilon)$, then $|u_{\tau_k} - x_0|_{\mathcal{B}} < \delta(\varepsilon)$ and $\rho(F^{\tau_k}, G) < \delta(\varepsilon)$, where $\delta(\varepsilon)$ is the one for s.d.$\Omega(u,F)$ of $u(t)$, from which it follows that

$$|u_{\tau_k+t} - x_t|_{\mathcal{B}} < \varepsilon \quad \text{for} \quad t \ge 0.$$

Therefore, $|u_{\tau_k+t} - u_{\tau_m+t}|_{\mathcal{B}} < 2\varepsilon$ for all $t \ge 0$ if $m \ge k \ge k_0(\varepsilon)$. Thus, we see that $u(t)$ is asymptotically almost periodic. The existence of an almost periodic solution follows immediately from Theorem 1.2.

Corollary 1.10. Let (H1) and (H2) hold, and suppose the solution $u(t)$ is TS for System (E). Then $u(t)$ is s.d.$\Omega(u,F)$; consequently, System (E) has an almost periodic solution.

Proof. By the remark stated in the preceding, for any $\eta > 0$ we can choose a constant $\gamma(\eta) > 0$, $\gamma(\eta) < \eta$, so that

$$\sup_{(t,\varphi)\in\mathbb{R}\times\mathcal{O}^+(u)} |F(t,\varphi) - G(t,\varphi)| < \eta$$

whenever $G \in H(F)$ and $\rho(F,G) < \gamma(\eta)$. Now, suppose that the solution $u(t)$ is TS for System (E) but it is not s.d.$\Omega(u,F)$. Then there exist an $\varepsilon > 0$ and an $(x,G) \in \Omega(u,F)$ such that

$$(1.11) \quad |u_\tau - x_0|_{\mathcal{B}} < \gamma(\delta(\varepsilon)) \quad \text{and} \quad \rho(F^\tau, G) < \gamma(\delta(\varepsilon)) \quad \text{for some} \quad \tau \ge 0$$

and

$$(1.12) \quad |u_{t_1+\tau} - x_{t_1}|_{\mathcal{B}} = \varepsilon \quad \text{for some} \quad t_1 > 0,$$

where $\delta(\varepsilon)$ is the one for TS of $u(t)$. Set $p(t) = G(t-\tau, x_{t-\tau}) - F(t, x_{t-\tau})$ for $t \in \mathbb{R}$. Then $p(t)$ is a continuous function on $\mathbb{R}$, and it satisfies $\sup_{t\in\mathbb{R}}|p(t)| \le \delta(\varepsilon)$ by (1.11) and the definition of $\gamma(\delta(\varepsilon))$, because of $x_t \in \mathcal{O}^+(u)$ for $t \in \mathbb{R}$. Define $z(t) = x(t-\tau)$. Then $\dot{z}(t) = \dot{x}(t-\tau) = G(t-\tau, x_{t-\tau}) = F(t, x_{t-\tau}) + p(t) = F(t, z_t) + p(t)$ for $t \in \mathbb{R}$, and $|z_\tau - u_\tau|_{\mathcal{B}} = |x_0 - u_\tau|_{\mathcal{B}} < \gamma(\delta(\varepsilon)) < \delta(\varepsilon)$ by (1.11). Since $u(t)$ is TS for System (E), we have $|z_t - u_t|_{\mathcal{B}} < \varepsilon$ for all

$t \geq \tau$, and particularly $|u_{t_1+\tau} - x_{t_1}|_{\mathscr{B}} = |u_{t_1+\tau} - z_{t_1+\tau}|_{\mathscr{B}} < \varepsilon$. This contradicts (1.12). Therefore, $u(t)$ must be s.d.$\Omega(u,F)$. The latter conclusion of the corollary follows from Theorem 1.9.

Example 1.11. As an application of the preceding results, we consider an almost periodic system of integrodifferential equations

$$(1.13)\qquad (d/dt)x^i(t) = x^i(t)\{b_i(t)-a_i(t)x^i(t)$$

$$- \sum_{j=1}^{n} a_{ij}(t)\int_{-\infty}^{t} K_{ij}(t-\tau)x^j(\tau)d\tau\},$$

$$i = 1,2,\cdots,n;\ t \in \mathbb{R},\ x = (x^1,x^2,\cdots,x^n) \in \mathbb{R}^n;$$

where the functions a_i, b_i and a_{ij} are almost periodic. Furthermore, we assume that System (1.13) satisfies Conditions (H3), (H4) and (H5) in Section 8.5. Let α_i and β_i be the positive numbers defined in the paragraph foregoing Lemma 8.5.2. If $u(t) = (u^1(t),u^2(t),\cdots,u^n(t))$ be any solution of System (1.13) satisfying $u_0 \in BC$ and $\alpha_i \leq u^i(s) \leq \beta_i$ for all $s \leq 0$ and all $i = 1,2,\cdots,n$, then $\alpha_i \leq u^i(t) \leq \beta_i$ for all $t \geq 0$ and all $i = 1,2,\cdots,n$, and the solution u is TS for System (1.13) by Theorem 8.5.3. Therefore, Corollary 1.10 yields that System (1.13) has an almost periodic solution $v(t)$ which satisfies $\alpha_i \leq v^i(t) \leq \beta_i$ for all $t \geq 0$ and all $i = 1,2,\cdots,n$. Moreover, $v(t)$ is a unique almost periodic solution of System (1.13) whose components are positive functions. Indeed, if x is such a solution, then $|x(t)-v(t)| \to 0$ as $t \to \infty$ by the argument in the proof of Theorem 8.5.3, and hence $x(t) \equiv v(t)$, because the function $x(t)-v(t)$ is almost periodic. In particular, if the functions a_i, b_i and a_{ij} are ω-periodic, then v is an ω-periodic solution of System (1.13); cf. Gopalsamy [1], Murakami [5] Hamaya [1].

Next, we shall establish an existence theorem of almost periodic solutions under a separation condition.

Definition 1.12. $\Omega(u,F)$ is said to satisfy a separation condition if for any $G \in \Omega(F)$, $\Omega_G(u)$ is a finite set and if φ and ψ, $\varphi, \psi \in \Omega_G(u)$, are distinct solutions of (F_G), then there exists a constant $\lambda(G,\varphi,\psi) > 0$ such that

$$|\varphi_t - \psi_t|_{\mathscr{B}} \geq \lambda(G,\varphi,\psi) \quad \text{for all} \quad t \in \mathbb{R}.$$

Theorem 1.13 (Hino [4]). Let (H1) and (H2) hold, and suppose that $\Omega(u,F)$ satisfies the separation condition. Then $u(t)$ is an asymptotically almost periodic solution. Consequently, System (E) has an almost periodic solution.

If for any $G \in \Omega(F)$, $\Omega_G(u)$ consists of only one element, then $\Omega(u,F)$ clearly satisfies the separation condition. Thus, the following result is an immediate consequence of Theorem 1.13.

Corollary 1.14. If for any $G \in \Omega(F)$, $\Omega_G(u)$ consists of only one element, then (E) has an almost periodic solution.

Before proving Theorem 1.13, we shall introduce the following notations. For a sequence $\{\alpha_k\}$, we shall denote it by α and $\beta \subset \alpha$ means that β is a subsequence of α. For $\alpha = \{\alpha_k\}$ and $\beta = \{\beta_k\}$, $\alpha + \beta$ will denote the sequence $\{\alpha_k+\beta_k\}$. Moreover, $L_\alpha x$ will denote $\lim_{k\to\infty} x(t+\alpha_k)$, whenever $\alpha = \{\alpha_k\}$ and the limit exists for each t.

We shall need the following lemma.

Lemma 1.15. Suppose that $\Omega(u,F)$ satisfies the separation condition. Then one can choose a number λ_0 independent of G, φ and ψ for which $|\varphi_t - \psi_t|_{\mathscr{B}} \geq \lambda_0$ for all $t \in \mathbb{R}$. The number λ_0 is called the separation constant for $\Omega(u,F)$.

Proof. Obviously, we can assume that the number $\lambda(G,\varphi,\psi)$ is independent of φ and ψ. Let G_1 and G_2 be in $\Omega(F)$. Then there exists a sequence $r' = \{r_k'\}$ such that

$$G_2(t,\varphi) = \lim_{k\to\infty} G_1(t+r_k',\varphi)$$

uniformly on $\mathbb{R} \times W$ for any compact set W in $\mathscr{B}$, that is, $L_{r'}G_1 = G_2$ uniformly on $\mathbb{R} \times W$ for any compact set W in $\mathscr{B}$. Let $x^1(t)$ and $x^2(t)$ be solutions in $\Omega_{G_1}(u)$. In virtue of Theorem 8.1.5, there exist a subsequence $r \subset r'$, $(y^1,G_2) \in H(x^1,G_1)$ and $(y^2,G_2) \in H(x^2,G_1)$ such that $L_r x^1{}_t = y^1{}_t$ and $L_r x^2{}_t = y^2{}_t$ in $\mathscr{B}$ compactly on $\mathbb{R}$, where $y^1(t)$ and $y^2(t)$ are solutions of

$$(G_2) \qquad \dot{x}(t) = G_2(t,x_t)$$

defined on $\mathbb{R}$. Since $H(x^i,G_1) \subset \Omega(u,F)$, $i = 1,2$, $y^1(t)$ and $y^2(t)$ also are in $\Omega_{G_2}(u)$. Let $x^1(t)$ and $x^2(t)$ be distinct solutions.

Then

$$\inf_{t\in\mathbb{R}} |x^1{}_{t+r_k} - x^2{}_{t+r_k}|_{\mathscr{B}} = \inf_{t\in\mathbb{R}} |x^1{}_t - x^2{}_t|_{\mathscr{B}} = \alpha_{12} > 0,$$

and hence

$$\text{(1.14)} \qquad \inf_{t\in\mathbb{R}} |y^1{}_t - y^2{}_t|_{\mathscr{B}} = \beta_{12} \geq \alpha_{12} > 0,$$

which means that $y^1(t)$ and $y^2(t)$ are distinct solutions of System (G_2). Let $p_1 \geq 1$ and $p_2 \geq 1$ be the numbers of distinct solutions of $\dot{x}(t) = G_1(t,x_t)$ and System (G_2), respectively. Clearly, $p_1 \leq p_2$. In the same way, we have $p_2 \leq p_1$. Therefore, $p_1 = p_2 =: p$.

Now, let $\alpha = \min\{\alpha_{ik} : i,k = 1,2,\cdots,p,\ i \neq k\}$ and $\beta = \min\{\beta_{jm} : j,m = 1,2,\cdots,p,\ j \neq m\}$. By (1.14), we have $\alpha \leq \beta$. In the same way, we have $\alpha \geq \beta$. Therefore $\alpha = \beta$, and we may set $\lambda_0 = \alpha = \beta$.

<u>Proof of Theorem 1.13.</u> For any sequence $\{\tau_k'\}$ such that $\tau_k' \to \infty$ as $k \to \infty$, there is a subsequence $\{\tau_k\}$ of $\{\tau_k'\}$ and a $(v,G) \in \Omega(u,F)$ such that $(u^{\tau_k},F^{\tau_k}) \to (v,G)$ compactly on $\mathbb{R}$.

Suppose that $u(t+\tau_k)$ is not convergent uniformly on $\mathbb{R}^+ := [0,\infty)$. Then for some $\varepsilon > 0$ such that $\varepsilon < \lambda_0/2$, where λ_0 is the separation constant, there are sequences $\{t_j'\}$, $\{k_j\}$ and $\{m_j\}$ such that $k_j \to \infty$, $m_j \to \infty$ as $j \to \infty$ and

$$|u(t_j'+\tau_{k_j}) - u(t_j'+\tau_{m_j})| \geq H\varepsilon,$$

that is,

$$|u_{t_j'+\tau_{k_j}} - u_{t_j'+\tau_{m_j}}|_{\mathscr{B}} \geq \varepsilon.$$

Since $u_{\tau_k} \in \mathcal{O}^+(u)$, we may assume that $\{u_{\tau_k}\}$ is convergent in $\mathscr{B}$, and hence $|u_{\tau_{k_j}} - u_{\tau_{m_j}}|_{\mathscr{B}} < \lambda_0/2$ if j is sufficiently large. Set $\psi^j(t) = u(t+\tau_{k_j})-u(t+\tau_{m_j})$. Then $|\psi^j{}_0|_{\mathscr{B}} < \lambda_0/2$ and $|\psi^j{}_{t_j'}|_{\mathscr{B}} \geq \varepsilon$ for all large j. Since $\varepsilon < \lambda_0/2$, there exists a t_j such that $\varepsilon \leq |\psi^j{}_{t_j}|_{\mathscr{B}} < \lambda_0/2$. Thus we have sequences $\{t_j\}$, $\{\tau_{k_j}\}$ and $\{\tau_{m_j}\}$ for which

(1.15) $$\varepsilon \le | u_{t_j+\tau_{k_j}} - u_{t_j+\tau_{m_j}} |_{\mathcal{B}} < \lambda_0/2.$$

Now, we shall denote by r the sequence $\{\tau_k\}$. Then $r' = \{\tau_{k_j}\} \subset r$ and $r'' = \{\tau_{m_j}\} \subset r$. Set $\alpha = \{t_j\}$. Since $F \in AP$, there exists $\alpha' \subset \alpha$, $\beta \subset r'$ and $\beta' \subset r''$ such that $L_{\alpha'+\beta}F = L_{\alpha'}L_{\beta}F$, $L_{\alpha'+\beta'}F = L_{\alpha'}L_{\beta'}F$ exist uniformly on $\mathbb{R} \times W$ for any compact set W in $\mathcal{B}$.

Moreover we may assume that

$$L_{\alpha'+\beta}u = x,\ L_{\alpha'+\beta'}u = y \quad \text{exist compactly on } \mathbb{R}.$$

Since $L_{\beta}F = L_{\beta'}F = G$, we have $L_{\alpha'+\beta}F = L_{\alpha'+\beta'}F = L_{\alpha'}G =: \tilde{G}$. Thus, $x(t)$ and $y(t)$ are in $\Omega_{\tilde{G}}(u)$ by Proposition 1.1.

On the other hand, we have, by (1.15),

$$\varepsilon \le |x_0 - y_0|_{\mathcal{B}} \le \lambda_0/2,$$

which shows that $x(t)$ and $y(t)$ are distinct solutions of $\dot{x}(t) = \tilde{G}(t,x_t)$, and hence

$$|x_0 - y_0|_{\mathcal{B}} \ge \lambda_0.$$

This is a contradiction. Therefore, $u(t+\tau_k)$ converges uniformly on $\mathbb{R}^+$; consequently, $u(t)$ is an asymptotically almost periodic solution. Hence, System (E) has an almost periodic solution by Theorem 1.2.

Before concluding this section, we shall derive some relationships between the separation condition and the stability properties in almost periodic systems.

The solution $u(t)$ is said to be unique for initial value problem with respect to $\Omega_F(u)$, if $u_t = x_t$ for all $t \ge t_0$ whenever $x \in \Omega_F^+(u)$ and $u_{t_0} = x_{t_0}$ for some $t_0 \ge 0$.

<u>Proposition 1.16.</u> Suppose that $u(t)$ is unique for initial value problem with respect to $\Omega_F(u)$ and that $\Omega(u,F)$ satisfies the separation condition. Then $u(t)$ is s.d.$\Omega(u,F)$.

<u>Proof.</u> Suppose $u(t)$ is not s.d.$\Omega(u,F)$. Then there exist an $\varepsilon > 0$ and sequences $(x^k, G_k) \in \Omega(u,F)$, $\tau_k \ge 0$ and $t_k > 0$ such that

$$(1.16)\qquad |u_{t_k+\tau_k} - (x^k)_{t_k}|_{\mathcal{B}} = \varepsilon\ (< \lambda_0/2).$$

$$(1.17)\qquad |u_{\tau_k} - (x^k)_0|_{\mathcal{B}} < 1/k$$

and

$$(1.18)\qquad \rho(F^{\tau_k}, G_k) < 1/k.$$

First we shall show that $t_k+\tau_k \to \infty$ as $k \to \infty$. Suppose not. Then there exists a subsequence of $\{\tau_k\}$, which we shall denote by $\{\tau_k\}$ again, and a constant $\tau \geq 0$ such that $\tau_k \to \tau$ as $k \to \infty$, because $0 \leq \tau_k < \tau_k+t_k < \infty$. Since

$$\rho(F^\tau, G_k) \leq \rho(F^\tau, F^{\tau_k}) + \rho(F^{\tau_k}, G_k),$$

(1.18) implies that

$$(1.19)\qquad \rho(F^\tau, G_k) \to 0 \quad \text{as} \quad k \to \infty.$$

Moreover, by Theorem 8.1.5 we can assume that

$$(x^k, G_k) \to (y, G) \quad \text{compactly on} \quad \mathbb{R}$$

for some $(y,G) \in \Omega(u,F)$. From (1.19) it follows that $G(t,\varphi) = F(t+\tau,\varphi)$ on $\mathbb{R} \times \mathcal{B}$. Hence $y(t)$ is a solution on $\mathbb{R}$ of $\dot{x}(t) = F(t+\tau, x_t)$. Note that $|u_\tau - y_0|_{\mathcal{B}} = 0$. Indeed,

$$|u_\tau - y_0|_{\mathcal{B}} \leq |u_\tau - u_{\tau_k}|_{\mathcal{B}} + |u_{\tau_k} - (x^k)_0|_{\mathcal{B}} + |(x^k)_0 - y_0|_{\mathcal{B}} \to 0$$

as $k \to \infty$ by (1.17). However, we have $|u_{t_k+\tau_k} - y_{t_k}|_{\mathcal{B}} \geq \varepsilon/2$ for a sufficiently large k by (1.16), because

$$|u_{t_k+\tau_k} - y_{t_k}|_{\mathcal{B}} \geq |u_{t_k+\tau_k} - (x^k)_{t_k}|_{\mathcal{B}} - |(x^k)_{t_k} - y_{t_k}|_{\mathcal{B}}.$$

Since t_k is bounded, this is a contradiction to the uniqueness of $u(t)$ with respect to $\Omega_F(u)$. Thus we must have $t_k+\tau_k \to \infty$ as $k \to \infty$.

Now, set $q_k = t_k+\tau_k$ and $v^k(t) = x^k(t_k+t)$. Then $u(t+q_k)$ and $v^k(t)$ are solutions of $\dot{x}(t) = F(t+q_k, x_t)$ and $\dot{x}(t) = G_k(t+t_k, x_t)$, respectively. By Proposition 1.1, we may assume that

$(u^{q_k}, F^{q_k}) \to (\eta, P)$ compactly on $\mathbb{R}$ for some $(\eta, P) \in \Omega(u,F)$, because $q_k \to \infty$ as $k \to \infty$. Since

$$\rho(P,(G_k)^{t_k}) \le \rho(P,F^{q_k}) + \rho(F^{q_k},(G_k)^{t_k})$$

$$\le \rho(P,F^{q_k}) + \rho(F^{\tau_k},G_k),$$

we see that $\rho(P,(G_k)^{t_k}) \to 0$ as $k \to \infty$ by (1.18). Hence, by Theorem 8.1.5 we can choose a subsequence $\{v^{k_j}(t)\}$ of $\{v^k(t)\}$ and a $z(t) \in \Omega_P(u)$ such that

$$(v^{k_j},(G_{k_j})^{t_{k_j}}) \to (z,P) \quad \text{compactly on } \mathbb{R}.$$

Since

$$\lim_{j\to\infty}\{|u_{t_{k_j}+\tau_{k_j}} - (x^{k_j})_{t_{k_j}}|_{\mathscr{B}} - |(v^{k_j})_0 - z_0|_{\mathscr{B}} - |\eta_0 - u_{q_{k_j}}|_{\mathscr{B}}\}$$

$$\le |\eta_0 - z_0|_{\mathscr{B}}$$

$$\le \lim_{j\to\infty}\{|u_{t_{k_j}+\tau_{k_j}} - (x^{k_j})_{t_{k_j}}|_{\mathscr{B}} + |(v^{k_j})_0 - z_0|_{\mathscr{B}} + |\eta_0 - u_{q_{k_j}}|_{\mathscr{B}}\},$$

it follows from (1.16) that $|\eta_0 - z_0|_{\mathscr{B}} = \varepsilon$, which contradicts the separation condition of $\Omega(u,F)$.

<u>Proposition 1.17.</u> The following statements are equivalent:

(i) $\Omega(u,F)$ satisfies the separation condition;

(ii) there exists a number $\delta_0 > 0$ with the property that for any $\varepsilon > 0$, any $t_0 \in \mathbb{R}$ and $G \in \Omega(F)$, $|x_{t_0} - y_{t_0}|_{\mathscr{B}} < \delta_0$ implies $|x_t - y_t|_{\mathscr{B}} < \varepsilon$ for $t \ge t_0+T(\varepsilon)$, whenever $x(t), y(t) \in \Omega_G(u)$.

<u>Proof.</u> [(i) implies (ii)]. If we set $\delta_0 = \lambda_0$, then (ii) clearly holds.

[(ii) implies (i)]. First of all, we shall verify that any distinct solutions $x(t), y(t)$ in $\Omega_G(u)$ satisfy

$$\text{(1.20)} \qquad \underline{\lim}_{t\to-\infty} |x_t - y_t|_{\mathscr{B}} \ge \delta_0.$$

Suppose not. Then for some $G \in \Omega(F)$, there exist two distinct solutions $x(t)$ and $y(t)$ in $\Omega_G(u)$ which satisfy

(1.21) $$\varliminf_{t\to-\infty} |x_t - y_t|_{\mathscr{B}} < \delta_0.$$

Since $x(t)$ and $y(t)$ are distinct solutions, we have $|x_{t_0} - y_{t_0}|_{\mathscr{B}} = \varepsilon$ at some t_0 and for some $\varepsilon > 0$. Then there is a t_1 such that $t_1 < t_0 - T(\varepsilon/2)$ and $|x_{t_1} - y_{t_1}|_{\mathscr{B}} < \delta_0$ by (1.21). Then $|x_{t_0} - y_{t_0}|_{\mathscr{B}} < \varepsilon/2$, which contradicts $|x_{t_0} - y_{t_0}|_{\mathscr{B}} = \varepsilon$. Thus we have (1.20). Since $\mathcal{O}^+(u)$ is a compact set, there are a finite number of coverings which consists of m_0 balls with diameter $\delta_0/4$. We shall show that the number of solutions in $\Omega_G(u)$ is at most m_0. Suppose not. Then there are m_0+1 solutions in $\Omega_G(u)$, $x^j(t)$, $j = 1,2,\cdots,m_0+1$, and a t_2 such that

(1.22) $$|x^j_{t_2} - x^i_{t_2}|_{\mathscr{B}} \geq \delta_0/2 \quad \text{for } i \neq j,$$

by (1.20). Since $x^j_{t_2}$, $j = 1,2,\cdots,m_0+1$, are in $\mathcal{O}^+(u)$, some of these solutions, say $x^i(t)$, $x^j(t)$ $(i\neq j)$, are in one ball at time t_2, and hence $|x^j_{t_2} - x^i_{t_2}|_{\mathscr{B}} < \delta_0/4$, which contradicts (1.22). Therefore the number of solutions in $\Omega_G(u)$ is $m \leq m_0$. Thus

(1.23) $$\Omega_G(u) = \{x^1(t), x^2(t), \cdots, x^m(t)\}$$

and

(1.24) $$\varliminf_{t\to-\infty} |x^i_t - x^j_t|_{\mathscr{B}} \geq \delta_0, \quad i \neq j.$$

Consider a sequence $\{\tau_k\}$ such that $\tau_k \to -\infty$ as $k \to \infty$ and $\rho(G^{\tau_k}, G) \to 0$ as $k \to \infty$. For each $j = 1,2,\cdots,m$, set $v^{j,k}(t) = x^j(t+\tau_k)$. Since $(v^{j,k}, G^{\tau_k}) \in H(x^j, G)$, by Theorem 8.1.5, we can assume that

$$(v^{j,k}, G^{\tau_k}) \to (y^j, G) \quad \text{compactly on } \mathbb{R}$$

for some $(y^j, G) \in H(x^j, G) \subset \Omega(u,F)$. Then it follows from (1.24) that

(1.25) $$|y^j_t - y^i_t|_{\mathscr{B}} \geq \delta_0 \quad \text{for all } t \in \mathbb{R} \text{ and } i \neq j.$$

Since the number of solutions in $\Omega_G(u)$ is m, $\Omega_G(u)$ consists of $y^1(t), y^2(t), \cdots, y^m(t)$ and we have (1.25); this shows that $\Omega(u,F)$ satisfies the separation condition.

9.2. Existence theorems for linear systems. In this section, we shall consider a linear system of FDEs

$$\text{(L)} \qquad \dot{x}(t) = L(t, x_t) + f(t), \quad t \in \mathbb{R}.$$

We assume the following conditions on (L):

(H3) $L(t,\varphi)$, $f(t) \in AP$ and $L(t,\varphi)$ is linear in φ;

(H4) System (L) has a solution u on $[0,\infty)$ such that $\sup_{t\geq 0}|u(t)| = H < \infty$.

If we set $F(t,\varphi) = L(t,\varphi) + f(t)$, then $F(t,\varphi) \in AP$. Moreover, we easily see that $\sup\{|F(t,\varphi)| : t \in \mathbb{R}, |\varphi|_{\mathcal{B}} \leq 1\} < \infty$. Thus System (L) satisfies Conditions (H1) and (H2) in the previous section. Hence, the following result is a direct consequence of Proposition 1.1.

Proposition 2.1. Assume (H3) and (H4). Then System (L) has an $\mathbb{R}$-bounded solution x which satisfies $|x|_\infty \leq H$.

Let $h(t)$ be a positive and continuous function defined on $\mathbb{R}$ such that $\int_{-\infty}^{\infty} h(s)ds < \infty$. For each $x \in BU(\mathbb{R},\mathbb{E})$ and $t \in \mathbb{R}$, define $|T_t x|_*$, $\mu(x)$ and $\lambda(x)$ by

$$|T_t x|_* = \int_{-\infty}^{\infty} \|x(t+s)\|^2 h(s)ds,$$

$$\mu(x) = \inf_{t\in\mathbb{R}} |T_t x|_*,$$

and

$$\lambda(x) = \sup_{t\in\mathbb{R}} |T_t x|_*,$$

respectively, here $\|x\|$ denotes the Euclidean norm of $x \in \mathbb{E}$. We set

$$\Lambda(L+f) = \inf\{\lambda(x) : x(t) \text{ is an } \mathbb{R}\text{-bounded solution of System (L) which satisfies } |x|_\infty \leq H\}.$$

Since the norms $|\cdot|$ and $\|\cdot\|$ in $\mathbb{E}$ are equivalent, we have $\Lambda(L+f) < \infty$ by Proposition 2.1.

Lemma 2.2. There exists an $\mathbb{R}$-bounded solution $p(t)$ of System (L) which satisfies $\lambda(p) = \Lambda(L+f)$. Such a solution $p(t)$ is called minimal.

Proof. By the definition of $\Lambda(L+f)$, there exists a sequence $\{x^k\}$ of $\mathbb{R}$-bounded solutions of (L) which satisfy $\lambda(x^k) \le \Lambda(L+f) + 1/k$ and $|x^k|_\infty \le H$. Since $\{x^k\}$ is equicontinuous and uniformly bounded on $\mathbb{R}$, $\{x^k\}$ has a subsequence which converges to an $\mathbb{R}$-bounded solution $p(t)$ of (L). Since $|T_t x^k|_* \le \lambda(x^k) \le \Lambda(L+f) + 1/k$, we have $|T_t p|_* \le \Lambda(L+f)$ by Lebesgue's convergence theorem, and hence $\lambda(p) \le \Lambda(L+f)$. That is, $\lambda(p) = \Lambda(L+f)$, because $\lambda(p) \ge \Lambda(L+f)$ by the definition.

Lemma 2.3. If $p^1(t)$ and $p^2(t)$ are minimal solutions of (L). Then $\mu(p^1-p^2) = 0$ and there exists a sequence $\{t_k\}$ such that $p^1(t_k+s) - p^2(t_k+s) \to 0$ as $k \to \infty$ compactly on $\mathbb{R}$; consequently, $\inf_{t\in\mathbb{R}} |p^1{}_t - p^2{}_t|_{\mathcal{B}} = 0$.

Proof. Put

$$y(t) = \frac{p^1(t) + p^2(t)}{2} \quad \text{and} \quad z(t) = \frac{p^1(t) - p^2(t)}{2}.$$

By the parallelogram theorem, we have

$$|T_t z|_* + |T_t y|_* = \{|T_t p^1|_* + |T_t p^2|_*\}/2 = \Lambda(L+f),$$

and hence

$$|T_t y|_* \le \Lambda(L+f) - |T_t z|_* \le \Lambda(L+f) - \mu(z).$$

This implies $\mu(z) = 0$, that is, $\mu(p^1-p^2) = 0$. There exists a sequence $\{t_k\}$ and a continuous function $r(t)$ such that

$$\lim_{k\to\infty} |T_{t_k} p^1 - T_{t_k} p^2|_* = 0$$

and

$$p^1(t_k+s) - p^2(t_k+s) \to r(s) \quad \text{as} \quad k \to \infty \quad \text{compactly on } \mathbb{R}.$$

Then, Lebesgue's convergence theorem yields that

$$\int_{-\infty}^{\infty} \|r(s)\|^2 h(s)ds = 0;$$

that is, $r(t) = 0$ on $\mathbb{R}$, because $h(t)$ is positive on $\mathbb{R}$. Hence, $\lim_{k\to\infty}|p^1_{t_k} - p^2_{t_k}|_{\mathscr{B}} = 0$ by Axiom (C2), and consequently $\inf_{t\in\mathbb{R}}|p^1_t - p^2_t|_{\mathscr{B}} = 0$.

Lemma 2.4. Suppose $p(t)$ is a minimal solution of (L). Then, for every $(q,M+g) \in H(p,L+f)$, we have

$$\lambda(p) = \Lambda(L+f) = \Lambda(M+g) = \lambda(q),$$

where $H(p,L+f)$ is the hull of $(p(t),F(t,\varphi)+f(t))$.

Proof. Let $p(t)$ be a minimal solution of (L), and let $(q,M+g) \in H(p,L+f)$. There exists a sequence $\{\tau_m\}$ such that

$$(p^{\tau_m}, L^{\tau_m}+f^{\tau_m}) \to (q,M+g) \quad \text{compactly on } \mathbb{R}.$$

Letting $m \to \infty$ in the inequality

$$\Lambda(L+f) = \lambda(p) \geq \int_{-\infty}^{\infty} \|p(\tau_m+t+s)\|^2 h(s)ds, \quad t \in \mathbb{R},$$

we have

$$\Lambda(L+f) = \lambda(p) \geq \int_{-\infty}^{\infty} \|q(t+s)\|^2 h(s)ds, \quad t \in \mathbb{R},$$

and hence

$$\Lambda(L+f) = \lambda(p) \geq \lambda(q) \geq \Lambda(M+g).$$

On the other hand, $L+f \in H(M+g)$, and hence $\Lambda(M+g) \geq \Lambda(L+f)$. Thus,

$$\lambda(p) = \Lambda(L+f) = \Lambda(M+g) = \lambda(q).$$

Now, we shall give a result on the existence of periodic solutions for periodic systems. The following result was proved in Section 7.5 in a different manner, where $\mathscr{B}$ was a uniform fading memory space.

<u>Theorem 2.5 (Hino & Murakami [1])</u>. Let (H3) and (H4) hold, and suppose $L(t,\varphi)$, $f(t) \in P_\omega$. Then System (L) has a solution in P_ω.

<u>Proof.</u> By Lemma 2.2, System (L) has a minimal solution $p(t)$. Since System (L) is ω-periodic, $p(t+\omega)$ also is a minimal solution of System (L), and hence there exist a sequence $\{t_k\}$ and a function $q(t)$ such that $p(t_k+t) - p(t_k+\omega+t) \to 0$ and $p(t_k+t) \to q(t)$ as $k \to \infty$ compactly on $\mathbb{R}$ by Lemma 2.3. Clearly, $q(t) = q(t+\omega)$. Let $t_k = N_k\omega+\sigma_k$, where N_k is an integer and $0 \le \sigma_k < \omega$. We may assume $\sigma_k \to \sigma$, $0 \le \sigma \le \omega$, as $k \to \infty$. Since $p(t+N_k\omega)$ is a solution of System (L) and $p(t+N_k\omega) \to q(t-\sigma)$ as $k \to \infty$ compactly on $\mathbb{R}$, we easily see that $q(t-\sigma)$ is a solution of System (L). Then, $q(t-\sigma)$ is the desired solution.

Next, we shall discuss the existence of almost periodic solutions for almost periodic systems. In the below, Theorem 2.6 corresponds to Favard's theorem (Favard [1]) for ordinary differential equations. On the one hand, Theorem 2.7 implies that the existence of almost periodic solutions follows from the existence of a bounded solution which is uniformly stable (for ordinary differential equations, see Nakajima [1]).

<u>Theorem 2.6.</u> Suppose (H3) and (H4). If for every $M \in \Omega(L)$, every nontrivial $\mathbb{R}$-bounded solution of the system

$$\dot{x}(t) = M(t, x_t) \tag{2.1}$$

satisfies the condition

$$\inf_{t\in\mathbb{R}} |x_t|_{\mathcal{B}} > 0, \tag{2.2}$$

then System (L) has a solution in AP.

<u>Proof.</u> By Lemma 2.2, System (L) has a minimal solution $p(t)$. Lemma 2.3 and Condition (2.2) imply that $p(t)$ is a unique minimal solution. Furthermore, combining Lemma 2.4 with Condition (2.2), we see that if $M+g \in \Omega(L+f)$, then $\Omega_{M+g}(p)$ consists of only one element. Hence the conclusion of the theorem follows from Corollary 1.14.

Theorem 2.7. Let (H3) and (H4) hold, and suppose the zero solution of

$$\dot{x}(t) = L(t,x_t)$$

is uniformly stable. Then System (L) has a solution in AP.

Proof. Let $p(t)$ be a minimal solution of System (L). We shall show that for every $M+g \in \Omega(L+f)$, $\Omega_{M+g}(p)$ consists of only one element. Then the conclusion follows from Corollary 1.14.

Let $(q^1,M+g)$, $(q^2,M+g) \in \Omega(p,L+f)$, and assume that $|q^1{}_s - q^2{}_s|_{\mathscr{B}} > 0$ for some $s \in \mathbb{R}$. Then there exists an $\varepsilon > 0$ such that $|q^1{}_t - q^2{}_t|_{\mathscr{B}} \geq \varepsilon$ for all $t \leq s$, because the zero solution of (2.1) is uniformly stable (cf. Proposition 8.2.2) and $q^1(t) - q^2(t)$ is a solution of (2.1). Furthermore, there exists a sequence $\{t_k\}$, $t_k \to -\infty$ as $k \to \infty$, and $(r^i,\hat{M}+\hat{g}) \in H(q^i,M+g) \subset \Omega(p,L+f)$, $i = 1,2$, such that

$$((q^i)^{t_k}, M^{t_k}+g^{t_k}) \to (r^i, \hat{M}+\hat{g}) \quad \text{compactly on } \mathbb{R}$$

by Theorem 8.1.5. Since $|q^1{}_{t_k+t} - q^2{}_{t_k+t}|_{\mathscr{B}} \geq \varepsilon$ for all $t \leq -t_k+s$, we have

$$|r^1{}_t - r^2{}_t|_{\mathscr{B}} \geq \varepsilon \quad \text{for all} \quad t \in \mathbb{R}.$$

However, this is a contradiction by Lemma 2.3, because $r^1(t)$ and $r^2(t)$ are minimal solutions of $\dot{x}(t) = \hat{M}(t,x_t) + \hat{g}(t)$ by Lemma 2.4.

Example 2.8. Consider a system of linear Volterra equations

$$(2.3) \qquad \dot{x}(t) = D(t)x(t) + \int_{-\infty}^{t} E(t,s)x(s)ds + f(t), \qquad t \in \mathbb{R},$$

where x is an n-vector, D is an $n \times n$ matrix of functions continuous on $\mathbb{R}$, E is an $n \times n$ matrix of functions continuous for $-\infty < s \leq t < \infty$ and $f : \mathbb{R} \to \mathbb{R}^n$ is continuous. In addition to the above conditions, we assume that

(2.4) for each $\delta > 0$ there is a $T > 0$ such that $\int_{-\infty}^{-T} |E(t,t+s)|ds \leq \delta$ for all $t \in \mathbb{R}$;

(2.5) D, $f \in P_\omega$ and $E(t+\omega, s+\omega) = E(t,s)$ for $-\infty < s \le t < \infty$;

(2.6) System (2.3) has a bounded solution u on $[0,\infty)$ such that $u_0 \in BC$.

Under the above conditions, we shall prove that System (2.3) has a solution in P_ω; cf. Hino & Murakami [2]. To do this, by Theorem 2.5 it suffices to show that the function

$$F(t,\varphi) = \int_{-\infty}^{0} E(t,t+s)\varphi(s)ds$$

can be considered as an element in $P_\omega(\mathbb{R}\times\mathcal{B},\mathbb{R}^n)$ for an appropriate fading memory space $\mathcal{B}$. We can establish this assertion by the same argument as in Section 8.5. Indeed, by (2.5), we can choose a sequence $\{T_m\}$, $T_{m+1} \ge T_m+1$, $T_1 \ge 1$, so that

$$\int_{-\infty}^{-T_m} |E(t,t+s)|ds < 1/[m^2(m+2)], \quad m = 1,2,\cdots.$$

Define a continuous function g by

$$g(s) = \begin{cases} 1 & \text{if } s = 0 \\ m+1 & \text{if } s = -T_m, \\ \text{linear} & \text{if } -T_{m+1} \le s \le -T_m, \end{cases}$$

$m = 1,2,\cdots$, for $s \le 0$. $\mathcal{B} = C_g^0$ is a separable fading memory space. For $\varphi \in C_g^0$, we have

$$\begin{aligned} |F(t,\varphi)| &\le \int_{-T_1}^{0} |E(t,t+s)|ds\cdot g(-T_1)|\varphi|_g \\ &\quad + \sum_{m=1}^{\infty}\int_{-\infty}^{-T_m} |E(t,t+s)|ds\cdot g(-T_{m+1})|\varphi|_g \\ &\le \int_{-T_1}^{0} |E(t,t+s)|ds\cdot 2|\varphi|_g + \sum_{m=1}^{\infty}(1/m^2)|\varphi|_g. \end{aligned}$$

Hence F is well-defined on $\mathbb{R}\times C_g^0$ and it is ω-periodic in t by (2.5). Moreover, by (2.4) and (2.5) it is not difficult to see that $\int_{-\infty}^{0} |E(t,t+s)|ds$ is bounded and continuous on $\mathbb{R}$. Thus $F(t,\varphi) \in P_\omega(\mathbb{R}\times C_g^0,\mathbb{R}^n)$.

In the above, we can conclude the existence of a solution in P_ω of (2.3), even if Condition (2.5) is replaced by the conditions

(2.5′) if $x(t)$ is a solution of (2.3) on $[a,\infty)$, then $x(t+\omega)$ is also a solution of (2.3) on $[a-\omega,\infty)$

and

(2.5″) $\int_{-\infty}^{t} |E(t,s)|ds$ is bounded on $\mathbb{R}$

(cf. Burton [1], Wu, Li and Wang [1]). Indeed, for the function constructed in the above, System (2.3) can be considered as a system of FDEs on $\mathbb{R}\times C_g^0$, and moreover, by carefully reading the proof of Theorem 2.5, we see that the ω-periodicity of the system can be replaced by Condition (2.5′).

APPENDIX

A.1. Some recent results on fundamental theorems. In what follows, we shall summarize some recent results presented by Kato [5], which are applicable to wider classes of functional differential equations than those treated throughout this note.

Let $\mathcal{F}(J)$ be the space of E-valued functions defined on a domain $J \subset \mathbb{R}$. Then, the fact that the system

$$\text{(D)} \qquad \dot{x}(t) = f(t,x(\cdot))$$

is of retarded type requires that the function $f(t,\cdot)$ is defined for $x \in \mathcal{F}(I_t)$ for an $I_t \subset (-\infty,t]$. As will be mentioned later we are concerned with the case where the domain X_t of $f(t,\cdot)$ is contained in $\mathcal{F}(I_t)$ for an $I_t \subset (-\infty,t]$.

The equation (D) covers many classes of the equations but does not allow to introduce the concepts of an autonomous system or a periodic system under this formulation. To clear this hazard, we need to accept the translation operator

$$(1.1) \qquad \pi : x(\cdot) \to x_t(\cdot), \; x_t(s) := x(t+s),$$

which transforms $\mathcal{F}(I_t)$ into $\mathcal{F}(I_t - t)$, where clearly $I_t - t \subset (-\infty,0]$. Putting

$$(1.2) \qquad F(t,x_t) := f(t,x(\cdot)) \quad \text{or} \quad F(t,\varphi) := f(t,[\pi_t^{-1}\varphi](\cdot)),$$

we can represent the equation (D) in the form

$$\text{(H)} \qquad \dot{x}(t) = F(t,x_t),$$

which is of the form discussed throughout this note. The function $F(t,\varphi)$ is considered to be defined on $[0,\infty) \times X$ for an $X \subset \mathcal{F}(I)$, where $I := \bigcup_t \{I_t - t\} \subset (-\infty,0]$, and if $F(t,\varphi)$ is independent of t, then an autonomous system will arise. Moreover, if the transformation (1.1) is allowed, then we are able to make use of concept of the limiting equation.

On the contrary, by accepting the transformation (1.1) there arise some defects. For example, the set I may not be bounded even though I_t is bounded for each $t \geq 0$, whence we may loose the advantages of the boundedness of I_t, and furthermore some particular restriction will be required for the phase space.

The initial value problem for the equation (D) is to find a function $x(t)$, which is absolutely continuous in $t \geq \tau$, satisfying

$$(1.3) \qquad \begin{cases} x(t) = \xi(\tau) + \int_\tau^t f(s, x(\cdot))ds & t \geq \tau \\ x(t) = \xi(t) & t \leq \tau \end{cases}$$

for given $\tau \in [0,\infty)$ and $\xi \in X_t$, where ξ is called the initial function.

Let $C_0([\tau,a],E)$ be the space of continuous E-valued functions u defined on the interval $[\tau,a]$ satisfying $u(\tau) = 0$, which is a Banach space with the norm

$$\|u\|_{[\tau,a]} := \max_{\tau \leq s \leq a} |u(s)|_E .$$

A general scheme to get a solution of the initial value problem (1.3) is reduced to finding a fixed point of the operator T_ξ defined on $C_0([\tau,a],E)$ by

$$(1.4) \qquad [T_\xi u](t) := \int_\tau^t f(s, x(\xi,u)(\cdot))ds$$

for a suitable $a > \tau$, where

$$x(\xi,u)(s) := \begin{cases} \xi(\tau) + u(s) & s \geq \tau \\ \xi(s) & s \in I_\tau . \end{cases}$$

Namely, $x(\xi,u)$ is a solution of (1.3) if and only if u is a fixed point of the operator T_ξ given by (1.4).

In order that the operator T_ξ is well-defined on $C_0([\tau,a],E)$ for a given (τ,ξ), $\xi \in X_\tau$, we must verify the following

(i) $I_t \subset I_\tau \cup [\tau,t]$;

(ii) $\xi(\tau)$ is determined for $\xi \in X_\tau$;

(iii) $x(\xi,u)|_{I_t} \in X_t$ for any $u \in C_0([\tau,a],E)$;

(iv) X_t is equipped with a topology; and

(v) $f(t,x(\xi,u)(\cdot))$ is integrable in t, or more desirably, continuous in t for any $u \in C_0([\tau,a],\mathbb{F})$.

Moreover, the continuous dependence on the initial function requires that

(vi) $\xi(\tau)$ and $f(t,x(\xi,u)(\cdot))$ are continuous in ξ.

In order to see the existence of a fixed point of the operator T_ξ defined by (1.4), an eligible fixed point theorem is the Schauder-Tikhonov fixed point theorem. To apply the theorem the following requirement are natural:

(vii) $f(t,x(\xi,u)(\cdot))$ is continuous in (t,u),

while the global existence demands that

(viii) $f(t,x(\xi,u)(\cdot))$ is completely continuous in (t,u).

Putting together these requirements, we are proper to expect the following axioms on the phase space for the functional differential equation (D) and the hypotheses, which will be given in the later, to the right hand side of the equation: Here and henceforth X_tC denotes the set of functions $\varphi \in \mathcal{F}(\mathbb{R})$ such that $\varphi|_{I_t} \in X_t$ and $\varphi|_{[t,\infty)}$ is continuous.

(a_1) X_t is a linear pseudometric space with an invariant pseudometric $\|\cdot\|_{X_t}$;

(a_2) $x|_{I_t} \in X_t$ for any $x \in X_\tau C$ and $t \ge \tau$;

(a_3) for any $\varepsilon > 0$ and $t \ge 0$ there is a positive number $\nu = \nu(\varepsilon,t)$, continuous in (ε,t), such that $\|\varphi\|_{X_t} \le \nu$ implies $|\varphi(t)| \le \varepsilon$ for any $\varphi \in X_t$ and that $\nu(\varepsilon,t) \to \infty$ as $\varepsilon \to \infty$; and

(a_4) for any $\varepsilon > 0$ and $t \ge \tau \ge 0$ there is a positive number $\mu = \mu(\varepsilon,t,\tau)$, continuous in (ε,t,τ), such that $\|x\|_{X_t} \le \varepsilon$ if $x \in X_\tau C$, $\|x\|_{X_\tau} \le \mu$ and $\|x\|_{[\tau,t]} \le \mu$

and that $\mu(\varepsilon,t,\tau) \to \infty$ as $\varepsilon \to \infty$.

In the above and hereafter, $\|x\|_{X_t}$, $\|x\|_{[\tau,t]}$ and $|x(t)|$ stand for $\|x|_{I_t}\|_{X_t}$, $\|x|_{[\tau,t]}\|_{[\tau,t]} = \sup\{|x(s)|_{\mathbb{E}} : s \in [\tau,t]\}$ and $|x(t)|_{\mathbb{E}}$, respectively. The axiom (a_1) is made so as to simplify the situation, which requires $X_t \subset \mathcal{F}(I_t)$ for an $I_t \subset (-\infty,t]$ as an inevitable consequence. The axioms (a_2), (a_3) and (a_4) are a matter of course from verifying the operator T_ξ defined by (1.4), which naturally postulate the (completely) continuity of the mappings

$$\varphi \to \varphi(t) : X_t \to \mathbb{E}$$

and

$$x \to x|_{I_t} : X_\tau C \to X_t$$

when $t \geq \tau$, where we shall note that $x \in X_\tau C$ if and only if $x|_{I_\tau} \in X_\tau$ and $(x-x(\tau))|_{[\tau,\infty)} \in C_0([\tau,\infty),\mathbb{E})$.

In the most of the cases we may expect that (a_1) is strengthened to

(b_1) X_t is a semi-norm space with the semi-norm $\|\cdot\|_{X_t}$.

Under the axiom (b_1) each axiom (a_k) is endowed with a special feature, which will be denoted by (b_k).

Furthermore, in order to see the t-uniformity of the properties of solutions, we need the axiom:

(c) $\|\pi_t\varphi\|_X = \|\varphi\|_{X_t}$ for any $\varphi \in X_t$ and $t \geq 0$,

where $X := X_0$. The axiom (c) also requires that $I := I_t - t \subset (-\infty,0]$ must be independent of t and that the space X_t is isometric to X_0 along π_t for any $t \geq 0$. Similarly to (b_k), the axiom (c_k) denotes the axiom corresponding to (a_k) under the axiom (c), and furthermore, the notion (bc_k) will be used under the both axioms (b_1) and (c). Of course, the above axiom is a generalization of Axiom (A) in the following sense; if $\mathcal{B}$ satisfies Axiom (A), then $X_t := \mathcal{F}((-\infty,t])$ satisfies the axioms (bc_k), $k = 1,\cdots,4$, where X_t is equipped with the seminorm $\|\cdot\|_{X_t}$ defined by $\|\varphi\|_{X_t} = |\pi_t\varphi|_{\mathcal{B}}$ for $\varphi \in X_t$.

A typical example of the phase space X_t is the space considered with $\varphi \in \mathcal{F}(I_t)$ such that $\varphi(s)$ is continuous on J^1_t, measurable on J^2_t and satisfies

$$\sup_{s\in J^1_t} g^1(t,s)|\varphi(s)| < \infty, \qquad \int_{J^2_t} g^2(t,s)|\varphi(s)|^p ds < \infty,$$

where J^i_t are measurable sets satisfying $J^1_t \cup J^2_t \subset I_t$, $g^i(t,s)$ are positive functions continuous in $(t,s) \in [0,\infty) \times J^i_t$, $p \geq 1$ is a constant and i stands for 1 or 2.

By setting

$$(1.5) \qquad \|\varphi\|_{X_t} := \sup_{s\in J^1_t} g^1(t,s)|\varphi(s)| + \left(\int_{J^2_t} g^2(t,s)|\varphi(s)|^p ds\right)^{\frac{1}{p}}$$

we can verify the axiom (b_1), and (b_3) is satisfied if

$$(1.6) \qquad t \in J^1_t.$$

Moreover, $(b_2) = (a_2)$ and $(b_4) = (a_4)$ hold if

$$(1.7) \qquad J^i_t \cap (-\infty,\tau] \subset J^i_\tau \quad \text{for any} \quad t \geq \tau,$$

and if

$$\sup_{s\in J^i_t\cap(-\infty,\tau]} \frac{g^i(t,s)}{g^i(\tau,s)} < \infty.$$

We can verify that the axiom (c) holds if and only if $J^i := J^i_t - t$ and $g^i(s) := g^i(t,t+s)$ are independent of t. Consider the case where (c) holds. Then from these and (1.7) we have

$$J^i \cap (-\infty,-t] \subset J^i - t \quad \text{for any} \quad t \geq 0,$$

and hence J^i must be an empty set or an interval containing 0, that is, the semi-norm defined by (1.5) becomes

$$\|\varphi\|_X = \sup_{-r_1<s\leq 0} g^1(s)\|\varphi(s)\| + \left(\int_{-r_2}^0 g^2(s)\|\varphi(s)\|^p ds\right)^{\frac{1}{p}}$$

for constants r_1 and $r_2 \in [0,\infty]$ (namely, r_1 and r_2 may be zero or infinity, and the second term of the right hand side will be annihilated if $r_2 = 0$), where we should note that J^1 cannot be empty under the assumption (1.6).

Another interesting example is the space $C((\alpha,t],E)$ of continuous functions for an $\alpha \geq -\infty$ with the compact open topology. Define the invariant metric $\|\cdot\|_{X_t}$ by

$$(1.8) \qquad \|x\|_{X_t} := |x(t)| + \sup\Big(\min\{\tfrac{1}{k}, \sup_{\alpha_k(t)\leq s\leq t} |x(s)|\};\ k = 1,2,\cdots\Big),$$

where $\alpha_k(t) = -k$ when $\alpha = -\infty$ and $\alpha_k(t) = \min\{t,\alpha+\frac{1}{k}\}$ otherwise. In this case, the axiom (b_1) is not satisfied, but there are no difficulties to verify the axiom (a_1) through (a_4).

On the space BC_t of bounded continuous functions on $(-\infty,t]$ we can define two metric spaces BC_t^u and BC_t^c introducing the uniform norm:

$$\|\varphi\|_{BC_t^u} := \sup_{s\leq t} |\varphi(s)|$$

and the compact open topology:

$$\|\varphi\|_{BC_t^c} := \sup_{s\leq t} |\varphi(s)|\cdot\sup\Big(\min\{\tfrac{1}{k}, \sup_{t-k\leq s\leq t} \|\varphi(s)\|\};\ k = 1,2,\cdots\Big)$$

which endows the topology equivalent to the one given by (1.8).

Now, consider the functional differential equation of the form

$$(D) \qquad \dot{x}(t) = f(t,x(\cdot))$$

defined on $\{(t,x) : x \in X_t,\ t \geq 0\}$, or

$$(H) \qquad \dot{x}(t) = F(t,x_t)$$

defined on $[0,\infty) \times X$, and give the following definitions:

Definition 1.1. The function $f(t,x(\cdot))$ is said to be :

(1^0) continuous in (t,x) if for any $\varepsilon > 0$, any compact interval J and a y such that $y|_{I_s} \in X_s$ for all $s \in J$, there exists a $\delta = \delta(\varepsilon,J,y) > 0$ such that

$$|f(t,x(\cdot)) - f(s,y(\cdot))| < \varepsilon$$

whenever $s \le t \le s+\delta$, $s \in J$, $t \in J$, $x \in X_t$ and $\|x - y\|_{X_t} < \delta$;

(2^0) completely continuous in (t,x) if it is continuous in (t,x) and if it is compact, that is, for given $a, b > 0$ there is a $B = B(a,b) > 0$ such that

$$\|x\|_{X_t} \le b \quad \text{and} \quad t \in [0,a] \quad \text{imply} \quad |f(t,x(\cdot))| \le B.$$

Definition 1.2. $X := \{X_t\}_{t\ge0}$ is said to be an admissible phase space for the equation (D) or the equation (D) be admissible on the phase space X or the pair $\{(D),X\}$ is admissible, if X_t satisfies the axioms (a_1) through (a_4) and if $f(t,x(\cdot))$ is completely continuous.

The following lemma is obvious:

Lemma 1.3. (1^0) Suppose that $f(t,x(\cdot))$ is continuous in $t \ge \tau$ for each fixed $\tau \ge 0$ and $x \in X_\tau C$ with X satisfying (a_1) through (a_4). If $f(t,x(\cdot))$ satisfies a Lipschitz condition, that is, for any $a, b > 0$ there is an $L = L(a,b) > 0$ such that $|f(t,x(\cdot)) - f(t,y(\cdot))| \le L\|x - y\|_{X_t}$ for $t \in [0,a]$ and $x, y \in X_t$, $\|x\|_{X_t}$, $\|y\|_{X_t} \le b$, then it is completely continuous in (t,x).

(2^0) Suppose that in addition to the axioms (c_1) through (c_4) the axiom

(c_0) for any $x \in X_\tau C$, the function $x_t := \pi_t(x|_{I_t})$ is continuous in $t \ge \tau$.

is fulfilled. Then, $f(t,x(\cdot))$ is (completely) continuous in (t,x) if, and only if, $F(t,\varphi)$ defined by (1.2) is (completely) continuous in (t,φ) in the usual sense.

In the above, the same letter X will be used in double meaning. However, since $X_t \simeq X_0$ for all $t \ge 0$ under the axiom (c), there will arise no confusions.

By Lemma 1.3(1^0) above we can see that the following theorem covers the existence theorem given by Driver [1] on the space BC^u_t of bounded continuous functions, while Lemma 1.3(2^0) asserts that the same conclusions of the theorem holds for the equation (H) if $F(t,\varphi)$ is completely continuous in (t,φ) under the axioms (c) and (c_0); cf. Sections 2.1 and 2.2.

Theorem 1.4. If $X = \{X_t\}_{t\geq 0}$ is admissible for the equation (D), then the equation (D) has a solution for the initial value problem and the solution is continuable as long as it remains bounded. Furthermore, if the solution is unique for each initial function, then it is continuous as a function of the initial function.

The proof of the existence is clear by applying the Schauder-Tikhonov fixed point theorem for the operator T_ξ defined by (1.4). The rest of the proof will be given in the standard arguments.

Now back to the spaces discussed before. For the space with the norm (1.8) the axiom (c_0) always holds good. On the other hand, for the space with the norm (1.5) the axiom (c_0) holds if and only if $g^i(s+u)/g^i(s)$ converges to 1 as $u \to 0$ uniformly in $s \in J^i$,

$$(1.9) \qquad g^1(s)\varphi(s) \text{ is uniformly continuous on } J^1$$

and

$$\int_{J^2} g^2(s)\left|\left(\frac{g^2(s+u)}{g^2(s)}\right)^{1/p}\varphi(s+u) - \varphi(s)\right|^p ds \to 0 \quad \text{as} \quad u \to 0$$

for every $\varphi \in X$. Clearly, the condition (1.9) is satisfied if

$$\lim_{s\to r_1+0} g^1(s)\varphi(s) \quad \text{exists}$$

when $J^1 = (-r,0]$.

Some results on the stability problem, corresponding to those in Section 7.2, can also be developed in the context of the above axioms. For the detail, we refer the reader to Kato [5].

A.2. Browder's theorem.

Theorem 2.1 (Browder [1]). Let T be a closed linear operator densely defined in the Banach space X with finite dimensional generalized eigenspace for the complex number ξ_0. Then the point ξ_0 of the spectrum of T does not lie in the essential spectrum of T if and only if the resolvent $R(\xi:T) := (\xi I-T)^{-1}$ is analytic in the neighborhood of ξ_0 and has a pole at ξ_0.

Proof. Suppose $\xi_0 \in \sigma(T)\setminus ess(T)$, and let C be a closed rectifiable Jordan curve which contains ξ_0 in its interior but no other points of $\sigma(T)$. From the well-known result in the operator calculus for closed linear operators (e.g., Taylor [1, chapter 5]), it follows that the bounded linear operator P_0 on X defined by

$$(2.1) \qquad 2\pi i P_0 = \int_C R(\xi;T)d\xi$$

is an idempotent linear mapping of X into itself such that $P_0(X) \subset \mathcal{D}(T)$, $X_0 = P_0(X)$ is invariant under T and closed in X, and the spectrum of T_0, the restriction of T to X_0, consists of the single point ξ_0. Moreover, the range $\mathcal{R}(T_0-\xi_0 I)$ of the operator $T_0-\xi_0 I$ in X_0 equals $\mathcal{R}(T-\xi_0 I) \cap X_0$, since if u lies in $\mathcal{D}(T)$ while $(T-\xi_0 I)u = v \in X_0$, then $(T_0-\xi_0 I)P_0 u = P_0(T-\xi_0 I)u = v$ implies that v lies in $\mathcal{R}(T_0-\xi_0 I)$. In particular, since ξ_0 does not lie in ess(T), $\mathcal{R}(T-\xi_0 I)$ is closed and therefore so is $\mathcal{R}(T_0-\xi_0 I)$.

Set $\mathcal{N}_0 = \bigcup_{r\geq 0} \mathcal{N}((T-\xi_0 I)^r)$. By the assumption that $\xi_0 \bar{\in}$ ess(T), $\mathcal{N}_0$ is of finite dimension and hence closed in X, and in particular $\mathcal{N}_0 = \mathcal{N}((T-\xi_0 I)^k)$ for some fixed sufficiently large positive integer k. We assert first that $\mathcal{N}_0 \subset X_0$. Indeed, suppose that we have shown that $\mathcal{N}((T-\xi_0 I)^{r-1}) \subset X_0$ for some $r \geq 1$. Then for $u \in \mathcal{N}((T-\xi_0 I)^r)$, $u_0 = (T-\xi_0 I)u$ lies in X_0. Let $\omega_\xi = R(\xi;T)u$ for $\xi \in C$. Then

$$u_0 = Tu - \xi_0 u = (Tu - \xi u) + (\xi-\xi_0)u,$$

which implies that

$$u = (\xi-\xi_0)^{-1}u_0 + (\xi-\xi_0)^{-1}(\xi I-T)u$$

and

$$(2.2) \qquad \omega_\xi = R(\xi;T)u = (\xi-\xi_0)^{-1}R(\xi;T)u_0 + (\xi-\xi_0)^{-1}u.$$

But for u_0 in X_0, $R(\xi;T)u_0 = v_0$ also lies in X_0, since the equation $(\xi I-T)v_0 = u_0$ immediately implies that $(\xi I-T)(P_0 v_0) = u_0$, and by the uniqueness of the inverse, $v_0 = P_0 v_0$. Integrating (2.2) with respect to ξ around the closed curve C, we obtain

$$(2.3) \qquad P_0 u = u + (2\pi i)^{-1}\int_C (\xi-\xi_0)^{-1}R(\xi;T)u_0 d\xi$$

and, since the second term on the right in (2.3) lies in X_0, u lies in X_0. Proceeding by recursion, it follows that $\mathcal{N}_0 \subset X_0$.

We now assert that $\mathcal{N}_0 = X_0$, under our present hypothesis. To establish this fact, we remark first that the transformation $T_0' = T_0 - \xi_0 I$ of X_0 has only zero in its spectrum, i.e. T_0' is quasi-nilpotent. In addition, $T_0'(\mathcal{N}_0) \subset \mathcal{N}_0$, while if $T_0'u \in \mathcal{N}_0$, then u must lie in $\mathcal{N}((T-\xi_0 I)^{k+1}) = \mathcal{N}_0$. Let X_1 be the quotient space $X_0/\mathcal{N}_0$ with the usual Banach space structure, and let T_1 be the linear transformation of X_1 induced by T_0', i.e.,

$$(2.4) \qquad T_1(u+\mathcal{N}_0) = T_0'u + \mathcal{N}_0.$$

Since for every $r \geq 1$, $T_1{}^r$ is the transformation of X_1 induced by $(T_0')^r$ and since the norm of the induced transformation is not more than the norm of the inducing transformation, it follows that $|T_1{}^r| \leq |(T_0')^r|$, so that the spectral radius of the transformation T_1 must be zero since T_0' is quasi-nilpotent. On the other hand, $\mathcal{R}(T_1)$, the range of T_1 in X_1, is a closed subspace of X_1, since its inverse in X_0 under the quotient mapping is $\mathcal{R}(T_0')+\mathcal{N}_0$. $\mathcal{R}(T_0')$ is a closed subspace of X_0 and the sum of a closed subspace and a finite dimensional subspace of a Banach space is always closed. Finally, T_1 has a trivial null-space, since $T_1(u+\mathcal{N}_0) = \mathcal{N}_0$ implies that $T_0'u$ lies in $\mathcal{N}_0$, and by our preceding remark, it follows that u must lie in $\mathcal{N}_0$. Hence T_1 is a one-to-one mapping of X_1 into the closed subspace $\mathcal{R}(T_1)$ of X_1. Suppose X_1 is not trivial. By the open mapping theorem, it follows that there exists a constant $c > 0$ such that $|T_1 u| \geq c|u|$ for all u in X_1. But then by recursion, it follows that $|T_1{}^r u| \geq c^r|u|$ for all u in X_1, i.e. $|T_1{}^r| \geq c^r$ for all r, and hence $|T_1{}^r|^{1/r} \geq c$ for all r. This contradicts the fact that T_1 is quasi-nilpotent, thus proving that X_1 must be trivial, hence $X_0 = \mathcal{N}_0$.

Let us now consider the Laurent expansion of the analytic operator-valued function $R(\xi;T)$ at its isolated singularity ξ_0. We have

$$R(\xi;T) = \sum_{j=-\infty}^{\infty} A_j(\xi-\xi_0)^j,$$

where the operator A_j is given by

$$A_j = (2\pi i)^{-1}\int_C (\xi-\xi_0)^{-j-1}R(\xi;T)d\xi.$$

In particular, for $j = -k-1$,

$$A_{-k-1} = (2\pi i)^{-1}\int_C (\xi-\xi_0)^k R(\xi;T)d\xi$$

while for u in $\mathscr{D}(T^k)$,

$$(\xi-\xi_0)^k R(\xi;T)u = (T-\xi_0 I)^k R(\xi;T)u + g(\xi)u,$$

where $g(\xi)$ is analytic within C. Hence, for u in $\mathscr{D}(T^k)$,

$$A_{-k-1}u = (2\pi i)^{-1}(T-\xi_0 I)^k \int_C R(\xi,T)u d\xi = (T-\xi_0 I)^k P_0 u = 0,$$

because of $P_0 u \in X_0 = \mathscr{N}_0$. Since $\mathscr{D}(T^k) = \mathscr{D}((T-\xi I)^k)$ for any ξ, the set $\mathscr{D}(T^k)$ is dense in X (for indeed $(T^*-\xi I)^k$ for any given ξ on C has a bounded inverse). Hence the bounded operator A_{-k-1} is zero on all X. Moreover, for $r \geq 1$ we have

$$\begin{aligned}(2\pi i)A_{-r-1} &= \int_C (\xi-\xi_0)^r R(\xi;T)d\xi \\ &= \int_C (\xi-\xi_0)^r R(\xi;T)d\xi - \int_C (\xi-\xi_0)^{r-1}(\xi I-T)R(\xi;T)d\xi \\ &= \int_C (\xi-\xi_0)^{r-1}(T-\xi_0 I)R(\xi;T)d\xi \\ &= (2\pi i)(T-\xi_0 I)A_{-r}.\end{aligned}$$

Hence, all the A_j for $j \leq -k-1$ must vanish. Thus the function $R(\xi;T)$ has at most a pole at its isolated singularity ξ_0. This completes the proof of the "only if" part of the theorem.

Suppose conversely that $R(\xi;T)$ has a pole of order k at ξ_0. Then for a closed rectifiable Jordan curve C containing ξ_0 and no other points of the spectrum of T in its interior, we have

$$(2\pi i)^{-1}\int_C (\xi-\xi_0)^k R(\xi;T)d\xi = 0.$$

As before, for u in $\mathscr{D}(T^k)$,

$$(2\pi i)^{-1}\int_C (\xi-\xi_0)^k R(\xi;T)u d\xi = (T-\xi_0)^k P_0 u.$$

But P_0 maps all of X into $\mathscr{D}(T^k)$, so that $(T-\xi_0 I)^k P_0 = 0$ by the density of $\mathscr{D}(T^k)$ in X. In particular, P_0 is an idempotent mapping of X whose range, being contained in the generalized eigenspace of with respect to ξ_0, must be finite-dimensional since that generalized eigenspace is finite-dimensional. The range of the complementary idempotent $(I-P_0)$ will be denoted by X'. By the spectral mapping theorem for closed linear operators (e.g., Taylor [1, Chapter 5]), the spectrum of the mapping T' on X' obtained by restricting T to $\mathcal{D}(T) \cap X'$ is the set $\sigma(T)\setminus\{\zeta_0\}$. Hence $\mathcal{R}(T-\zeta_0 I) = (T-\zeta_0 I)P_0 X + (T-\zeta_0 I)(\mathcal{D}(T)\cap X') = (T-\zeta_0 I)P_0 X + X'$. Thus, $R(T-\zeta_0 I)$ is closed, because it is the sum of closed subspace X' and finite-dimensional subspace $(T-\zeta_0 I)P_0 X$. This completes the proof of the "if" part of theorem.

Corollary 2.2. Let λ be a normal eigenvalue of a densely defined, closed linear operator T on a Banach space X. Then λ is in $P_\sigma(T)$, for which the generalized eigenspace $\mathcal{M}_\lambda$ of T is of the form $\mathcal{N}((T-\lambda I)^k)$ for some integer $k > 0$. In this case, $\mathcal{R}_\lambda = \mathcal{R}((T-\lambda I)^k)$ is a closed subspace of X, and $\mathcal{M}_\lambda \subset \mathcal{D}(T)$, $T\mathcal{M}_\lambda \subset \mathcal{M}_\lambda$, $T(\mathcal{R}_\lambda \cap \mathcal{D}(T)) \subset \mathcal{R}_\lambda$, $\sigma(T|_{\mathcal{M}_\lambda}) = \{\lambda\}$ and $\sigma(T|_{\mathcal{R}_\lambda\cap\mathcal{D}(T)}) = \sigma(T)\setminus\{\lambda\}$. Moreover, X can be represented as

$$X = \mathcal{M}_\lambda \oplus \mathcal{R}_\lambda .$$

Proof. Let X_0, X', $\mathcal{N}_0$ and T' be the ones arising in the proof of Theorem 2.1. Then $\mathcal{M}_\lambda = \mathcal{N}_0 = X_0$. Since $X = X_0 \oplus X'$, we obtain

$$\begin{aligned}\mathcal{R}((T-\lambda I)^k) &= (T-\lambda I)^k[\mathcal{N}_0 \oplus (X'\cap\mathcal{D}(T^k))] \\ &= (T-\lambda I)^k\mathcal{N}_0 \oplus (T'-\lambda I)^k[\mathcal{D}(T'^k)] \\ &= \{0\} \oplus X' \\ &= X'\end{aligned}$$

since $\sigma(T') = \sigma(T)\setminus\{\lambda\}$. Hence, the assertions of Corollary 2.2 follows from this fact and the proof of Theorem 2.1.

A.3. Levinger's theorem.

Theorem 3.1 (Levinger [1]). Let $H(\lambda) = \sum_{i=0}^{\infty} (\lambda-\lambda_0)^i H_i$ be an $n \times n$ matrix whose elements are analytic functions of λ. If

$$\text{(3.1)} \qquad \det H(\lambda) = (\lambda - \lambda_0)^m p(\lambda), \qquad p(\lambda_0) \neq 0,$$

then for each $N \geq m$, the $n(N+1) \times n(N+1)$ matrix $\mathcal{H}_N$,

$$(3.2)\qquad \mathcal{H}_N = \begin{bmatrix} H_0 & H_1 & \cdots & H_N \\ 0 & H_0 & \cdots & H_{N-1} \\ \vdots & & \ddots & \vdots \\ 0 & \cdots & 0 & H_0 \end{bmatrix}$$

has rank $n(N+1) - m$.

Obviously, it suffices to prove the theorem in the case $\lambda_0 = 0$. Now, to establish the theorem, we need a series of lemmas.

Lemma 3.2. Let $A(\lambda) = \sum_{i=0}^{\infty} \lambda^i A_i$ be a $\nu \times \nu$ matrix whose elements are analytic functions of λ. Let $A_0 \neq 0$ and let ρ denote the rank of A_0. If $\nu_1 = \nu - \rho > 0$, then there exist $\nu \times \nu$ matrices

$$X,\quad Y,\quad C(\lambda;\beta) = I - \sum_{i=1}^{\beta-1} \lambda^i C_i(\lambda),\qquad \beta = 1,2,\cdots,$$

of determinant 1 such that, for $\beta = 1,2,\cdots,$

$$(3.3)\qquad \bar{A}(\lambda;\beta) := C(\lambda;\beta)XA(\lambda)Y = \begin{bmatrix} \bar{A}_{1,1}(\lambda) & \lambda\bar{A}_{1,2}(\lambda) \\ \lambda^{\beta}\bar{A}_{2,1}(\lambda;\beta) & \lambda\bar{A}_{2,2}(\lambda;\beta) \end{bmatrix}$$

where $\bar{A}_{1,1}(0) \equiv \bar{A}_{1,1}$ is a nonsingular $\rho \times \rho$ matrix and $\bar{A}_{2,2}(0;\beta)$ is a $\nu_1 \times \nu_1$ matrix whose coefficients are independent of β. Furthermore, if $\det A(\lambda) = \lambda^{\alpha} f(\lambda)$ with $f(0) \neq 0$, then

$$(3.4)\qquad \nu_1 \leq \alpha$$

and, for all $\beta \geq \alpha$,

$$(3.5)\qquad \det \bar{A}_{2,2}(\lambda;\beta) = \lambda^{\alpha-\nu_1} f_1(\lambda;\beta),$$

where $f_1(0;\beta)$ is independent of β and $f_1(0;\beta) \neq 0$.

Proof. Since $A_0 \neq 0$, $0 < \rho \leq \nu$, we can find matrices X, Y of determinant 1, such that

$$(3.6) \qquad XA_0Y = \begin{bmatrix} \bar{A}_{1,1} & 0 \\ 0 & 0 \end{bmatrix},$$

where $\bar{A}_{1,1}$ is a nonsingular $\rho \times \rho$ matrix. If $\rho < \nu$, we obtain

$$(3.7) \qquad \bar{A}(\lambda;1) := XA(\lambda)Y = \begin{bmatrix} \bar{A}_{1,1}(\lambda) & \lambda\bar{A}_{1,2}(\lambda;1) \\ \lambda\bar{A}_{2,1}(\lambda;1) & \lambda\bar{A}_{2,2}(\lambda;1) \end{bmatrix},$$

which is of the form (3.3) with $\beta = 1$.

Now, if $\bar{A}(\lambda;\beta)$ is given by (3.3) for some $\beta \geq 1$, let

$$(3.8) \qquad \Gamma_{\beta+1} := I - \lambda^{\beta}C_{\beta}(\lambda)C(\lambda;\beta)^{-1} = \begin{bmatrix} I & 0 \\ -\lambda^{\beta}\bar{A}_{2,1}(0;\beta)\bar{A}_{1,1}{}^{-1} & I \end{bmatrix}.$$

Then $C(\lambda;\beta+1) = \Gamma_{\beta+1}(\lambda)C(\lambda;\beta)$ and hence

$$(3.9) \qquad \bar{A}(\lambda;\beta+1) = \Gamma_{\beta+1}(\lambda)\bar{A}(\lambda;\beta) = \begin{bmatrix} \bar{A}_{1,1}(\lambda) & \lambda\bar{A}_{1,2}(\lambda) \\ \lambda^{\beta+1}\bar{A}_{2,1}(\lambda;\beta+1) & \lambda\bar{A}_{2,2}(\lambda,\beta+1) \end{bmatrix},$$

where

$$\bar{A}_{2,1}(\lambda;\beta+1) = \lambda^{-1}\left(\bar{A}_{2,1}(\lambda;\beta) - \bar{A}_{2,1}(0;\beta)\bar{A}_{1,1}{}^{-1}\bar{A}_{1,1}(\lambda)\right)$$

and

$$\bar{A}_{2,2}(\lambda;\beta+1) = \bar{A}_{2,2}(\lambda;\beta) - \lambda^{\beta}\bar{A}_{2,1}(0;\beta)\bar{A}_{1,1}{}^{-1}\bar{A}_{1,2}(\lambda).$$

Thus $\bar{A}_{2,2}(0;\beta+1) = \bar{A}_{2,2}(0;\beta) = \bar{A}_{2,2}(0;1)$ for $\beta \geq 1$ and $\bar{A}(\lambda;\beta+1)$ also satisfies (3.3).

Compute $\det A(\lambda) = \det \bar{A}(\lambda;\beta)$ by expanding (3.3) by minors from the first ρ columns. This gives

$$\det \bar{A}(\lambda;\beta) = \lambda^{\nu_1} \det \bar{A}_{2,2}(\lambda;\beta)\cdot \det \bar{A}_{1,1}(\lambda) + \lambda^{\beta+\nu_1}(\cdot\cdot\cdot\cdot\cdot)$$

$$= \lambda^{\alpha} f(\lambda) \tag{3.10}$$

or

$$\det \bar{A}_{2,2}(\lambda;\beta) = \left(\det \bar{A}_{1,1}(\lambda)\right)^{-1}\left(\lambda^{\alpha-\nu_1} f(\lambda) - \lambda^{\beta}(\cdot\cdot\cdot\cdot)\right),$$

and consequently (3.4) and (3.5) follow, because the coefficients of $\bar{A}_{2,2}(\lambda;\beta)$ are analytic functions of λ.

Lemma 3.3. Let $H(\lambda) = \sum_{i=0}^{\infty} \lambda^i H_i$ be an $n \times n$ matrix whose elements are analytic functions of λ. Suppose

$$\det H(\lambda) = \lambda^m p(\lambda), \quad p(0) \neq 0. \tag{3.11}$$

Then there exist integers $0 \leq d_1 < d_2 < \cdots < d_k$ and a partition of $n = \rho_1 + \rho_2 + \cdots + \rho_k$ such that

$$m = \rho_1 d_1 + \cdots + \rho_k d_k. \tag{3.12}$$

Further, there exist $n \times n$ matrices X_i, Y_i,

$$C_i = I - \sum_{j=1}^{\beta_i} \lambda^j C_{i,j}, \qquad i = 1,\cdots,k,$$

such that

$$K(\lambda) = C(\lambda)H(\lambda)Y$$

$$:= \begin{bmatrix} \lambda^{d_1} K_{1,1}(\lambda) & \lambda^{(d_1+1)} K_{1,2}(\lambda) & \cdots & \lambda^{(d_1+1)} K_{1,k}(\lambda) \\ \lambda^m K_{2,1}(\lambda) & \lambda^{d_2} K_{2,2}(\lambda) & \cdots & \lambda^{(d_2+1)} K_{2,k}(\lambda) \\ \vdots & & & \vdots \\ \lambda^m K_{k,1}(\lambda) & \lambda^m K_{k,2}(\lambda) & \cdots & \lambda^{d_k} K_{k,k}(\lambda) \end{bmatrix} \tag{3.13}$$

where $C(\lambda) = C_k(\lambda)X_k \cdots C_2(\lambda)X_2C_1(\lambda)X_1$ and $Y = Y_1Y_2\cdots Y_k$. The (i,j)-th block of $K(\lambda)$ is a $\rho_i \times \rho_j$ matrix $\lambda^{\alpha_{i,j}} K_{i,j}(\lambda)$, where

$$(3.14) \qquad \alpha_{i,j} = \begin{cases} m, & 1 \le j < i \le k \\ d_i, & i = j \\ d_i+1, & i < j \end{cases}$$

and $K_{i,j}(0)$ is nonsingular for $i = 1, 2, \cdots, k$.

Proof. To prove the lemma we apply Lemma 3.2 successively to matrices $H(\lambda) = H_1(\lambda), H_2(\lambda), \cdots, H_k(\lambda)$ of decreasing order and for which the determinants satisfy $\det H_i(\lambda) = \lambda^{m_i} p_i(\lambda)$ with $m_1 > m_2 > \cdots > m_k$.

Let d_1 be the smallest index for which $H_{d_1} \neq 0$ and $\rho_1 = \operatorname{rank} H_{d_1}$. Then $A_1(\lambda) = \lambda^{-d_1} H(\lambda)$ satisfies the hypothesis of Lemma 3.2. If $\rho_1 = n$, then $H(\lambda)$ is already in the form (3.13) with $K_{1,1}(\lambda) = A_1(\lambda)$. Otherwise $\nu_1 = n - \rho_1 > 0$, and by Lemma 3.2 we can find $n \times n$ matrices

$$X_1, \; Y_1, \; C_1(\lambda) = I - \sum_{j=1}^{m-d_1-1} \lambda^j C_{1,j}(\lambda),$$

of determinant 1 such that

$$(3.15) \qquad C_1(\lambda) X_1 A_1(\lambda) Y_1 = \begin{bmatrix} K_{1,1}(\lambda) & \lambda K_{1,12}(\lambda) \\ \lambda^{(m-d_1)} K_{1,21}(\lambda) & \lambda H_2(\lambda) \end{bmatrix},$$

where $K_{1,1}(0)$ is a $\rho_1 \times \rho_1$ nonsingular matrix, $\nu_1 = n-\rho_1 \le m-nd_1$ and $\det H_2(\lambda) = \lambda^{(m-nd_1-\nu_1)} p_2(\lambda)$ with $p_2(0) \neq 0$. Thus, the proof of the lemma for $H(\lambda)$ is reduced to the proof of a slightly different statement for $H_2(\lambda)$. Let $H_2(\lambda) = \sum_{i=0}^{\infty} \lambda^i H_{2,i}$ and let δ_2 be the smallest index for which $H_{2,\delta_2} \neq 0$. If ρ_2 is the rank of H_{2,δ_2}, then, as before, we can find $\nu_1 \times \nu_1$ matrices

$$X_2', \; Y_2', \; C_2'(\lambda) = I - \sum_{j=1}^{m-\delta_2-d_1-2} \lambda^j C_{2,j}(\lambda),$$

of determinant 1 such that

$$(3.16)\qquad C_2{}'X_2{}'H_2(\lambda)Y_2{}' = \begin{bmatrix} \lambda^{\delta_2}K_{2,2}(\lambda) & \lambda^{(\delta_2+1)}K_{2,23}(\lambda) \\ \lambda^{(m-d_1-1)}K_{2,32}(\lambda) & \lambda^{(\delta_2+1)}H_3(\lambda) \end{bmatrix},$$

where $K_{2,2}(0)$ is a $\rho_2 \times \rho_2$ nonsingular matrix. We set $d_2 = d_1 + 1 + \delta_2$,

$$X_2 = \begin{bmatrix} I & 0 \\ 0 & X_2{}' \end{bmatrix},\quad Y = \begin{bmatrix} I & 0 \\ 0 & Y_2{}' \end{bmatrix},\quad C_2(\lambda) = \begin{bmatrix} I & 0 \\ 0 & C_2{}'(\lambda) \end{bmatrix}$$

and obtain

$$C_2(\lambda)X_2C_1(\lambda)X_1H(\lambda)Y_1Y_2$$

$$(3.17)\qquad = \left[\begin{array}{c|c|c} \lambda^{d_1}K_{1,1}(\lambda) & \lambda^{(d_1+1)}K_{1,2}(\lambda) & \lambda^{(d_1+1)}K_{1,13}(\lambda) \\ \lambda^{m}K_{2,1}(\lambda) & \lambda^{d_2}K_{2,2}(\lambda) & \lambda^{(d_2+1)}K_{2,23}(\lambda) \\ \lambda^{m}K_{1,31}(\lambda) & \lambda^{m}K_{2,32}(\lambda) & \lambda^{(d_2+1)}H_3(\lambda) \end{array}\right].$$

We can proceed similarly with $H_3(\lambda)$ until, after $k \le n$ steps, the process terminates. This proves that $H(\lambda)$ is equivalent to a canonical matrix $K(\lambda)$ given by (3.13). Relation (3.12) follows from (3.11) and (3.13) if one expands $\det K(\lambda)$ by minors.

<u>Lemma 3.4.</u> Let $K(\lambda) = \sum \lambda^i K_i$, $H(\lambda) = \sum \lambda^i H_i$, $C(\lambda) = \sum \lambda^i C_i$, Y, be $n \times n$ matrices. For $N = 0,1,2,\cdots$, define the $(N+1)n \times (N+1)n$ matrix

$$(3.2)\qquad \mathscr{H}_N = \begin{bmatrix} H_0 & H_1 & \cdots\cdots & H_N \\ 0 & H_0 & \cdots\cdots & H_{N-1} \\ \vdots & & \ddots & \vdots \\ 0 & \cdots\cdots & 0 & H_0 \end{bmatrix}$$

and analogously $\mathscr{K}_N$, $\mathscr{C}_N$, and $\mathscr{Y}_N$ (which is block diagonal). If $K(\lambda) = C(\lambda)H(\lambda)Y$, then

$$(3.18)\qquad \mathscr{K}_N = \mathscr{C}_N\mathscr{H}_N\mathscr{Y}_N,\qquad N = 0,1,2,\cdots.$$

Hence, if C_0 and Y are nonsingular, $\mathscr{K}_N$ is equivalent to $\mathscr{H}_N$ for each N.

The proof is a simple computation with power series.

Lemma 3.5. Let $K(\lambda) = \sum \lambda^i K_i$ be an $n \times n$ matrix given by (3.13) where $d_0 = 0 \le d_1 < d_2 < \cdots < d_k$ and, for each $i = 1, \cdots, k$, $K_{i,i}(0)$ is a nonsingular $\rho_i \times \rho_i$ matrix. Then, for each $N = 0, 1, 2, \cdots$,

$$\text{rank } \mathcal{K}_N = (N+1) \sum_{i=1}^{j} \rho_i - \left(\sum_{i=1}^{j} \rho_i d_i \right), \tag{3.19}$$

where j is the largest index such that $d_j \le N$.

Proof. We compute the rank of

$$\mathcal{K}_N = \begin{bmatrix} K_0 & K_1 & \cdots\cdots & K_N \\ 0 & K_0 & \cdots\cdots & K_{N-1} \\ \vdots & & \ddots & \vdots \\ 0 & \cdots\cdots & 0 & K_0 \end{bmatrix}$$

by columns. From (3.13), the $\rho_i \times \rho_i$ block $K_{i,i}(0)$ is a nonsingular block in K_{d_i} and the (i,j)-block is 0 in each K_h with $h < d_i$ and $j = 1, \cdots, k$. Thus any column of $\mathcal{K}_N$ which contains a column of such a block is linearly independent of all columns to the left. Further, a column of $\mathcal{K}_N$ can intersect at most one of the nonsingular blocks $K_{i,i}(0)$. Thus the rank of $\mathcal{K}_N$ is just the number of these blocks occurring in $\mathcal{K}_N$.

This number is easily counted. If j is the largest index with $d_j \le N$, then

$$\text{rank } \mathcal{K}_N = \sum_{i=1}^{j} \rho_i + \text{rank } \mathcal{K}_{N-1}. \tag{3.20}$$

For $N = 0$,

$$\text{rank } \mathcal{K}_0 = \begin{cases} 0 & \text{if } d_1 > 0, \\ \rho_1 & \text{if } d_1 = 0, \end{cases}$$

which agrees with (3.19). Suppose (3.19) holds for $N-1$ and

$$d_{j'} \le N - 1 < d_{j'+1}.$$

If $j = j'$, then (3.20) yields

$$\operatorname{rank} \mathcal{X}_N = \sum_{i=1}^{j} \rho_i + N \sum_{i=1}^{j'} \rho_i - \sum_{i=1}^{j'} \rho_i d_i$$

$$= (N+1) \sum_{i=1}^{j} \rho_i - \sum_{i=1}^{j} \rho_i d_i .$$

If $j \ne j'$, then $d_j = N$ and $j = j'+1$. Hence

$$\operatorname{rank} \mathcal{X}_N = \sum_{i=1}^{j} \rho_i + N \sum_{i=1}^{j'} \rho_i - \sum_{i=1}^{j'} \rho_i d_i$$

$$= (N+1) \sum_{i=1}^{j} \rho_i - \sum_{i=1}^{j} \rho_i d_i - N\rho_j + \rho_j d_j$$

$$= (N+1) \sum_{i=1}^{j} \rho_i - \sum_{i=1}^{j} \rho_i d_i .$$

Thus, Relation (3.19) holds in either case.

Now, we shall prove Theorem 3.1 with $\lambda_0 = 0$. Since $n = \sum_{i=1}^{k} \rho_i$, we obtain $\operatorname{rank} \mathcal{X}_N = (N+1)n - \sum_{i=1}^{k} \rho_i d_i$ for each $N \ge d_k$. Since $\det C(\lambda) = \det Y = 1$ in (3.13), it follows from (3.18) that $\mathcal{K}_N$ is equivalent to $\mathcal{X}_N$ for each N. Consequently, for any $N \ge m$ Relation (3.18) yields that $N \ge d_k$ and $\operatorname{rank} \mathcal{X}_N = (N+1)n - m$, which completes the proof of the theorem.

A.4. The proof of Relation (8.3.9). It suffices to show this result for $t = 1$. Suppose it does not hold. Then, we can assume that there exists an $\varepsilon_0 > 0$ such that

$$\varepsilon_0 \le \left| \int_{\tau_k + r_k}^{1+\tau_k+r_k} g(s,(x^k)_s)ds \right|, \qquad k = 1,2,\cdots.$$

Fix a natural number N. Dividing the interval $I_k = [\tau_k+r_k, \tau_k+r_k+1]$ into N subintervals of the length $1/N$, one can find at least one subinterval $J_k = [\sigma_k, \sigma_k + 1/N]$ of I_k such that

$$\varepsilon_0/N \le \left| \int_{J_k} g(s,(x^k)_s)ds \right|, \qquad k = 1,2,\cdots.$$

Consider the inequality

$$|\int_{J_k} g(s,(x^k)_s)ds| \le |\int_{J_k} \{g(s,(x^k)_s) - g(s,(x^k)_{\sigma_k})\}ds|$$

$$+ |\int_{J_k} g(s,(x^k)_{\sigma_k})ds|.$$

Since we can assume that $(x^k)_{\sigma_k} \to y_\nu$ as $k \to \infty$ for some $\nu \in [0,1]$, from (E3) it follows that

$$|\int_{J_k} g(s,(x^k)_{\sigma_k})ds| \to 0 \quad \text{as} \quad k \to \infty.$$

On the other hand, if we set

$$\delta(N) = \sup\{|(x^k)_t - (x^k)_s|_{\mathscr{B}} : t,\ s \in [s_k,\ t_k+a],\ |t-s| \le \frac{1}{N},$$
$$k = 1,2,\cdots\},$$

we have

$$|\int_{J_k} \{g(s,(x^k)_s) - g(s,(x^k)_{\sigma_k})\}ds \le \frac{1}{N} b(\delta(N)) + \int_{J_k} h^r(s)ds.$$

Combining these results, we have

$$\frac{\varepsilon_0}{N} \le \frac{1}{N} b(\delta(N)) + \overline{\lim_{k\to\infty}} \int_{J_k} h^r(s)ds + \overline{\lim_{k\to\infty}} |\int_{J_k} g(s,(x^k)_{\sigma_k})ds|$$

$$= \frac{1}{N} b(\delta(N)),$$

or $\varepsilon_0 \le b(\delta(N))$, which is a contradiction since $\delta(N) \to 0$ as $N \to \infty$.

REFERENCES

O.A. Arino, T.A. Burton and J.R. Haddock,

1. Periodic solutions to functional differential equations, Proc. Royal Soc. Edinburgh, 101 A (1985), 253-271.

Z. Artstein,

1. Uniform asymptotic stability via the limiting equations, J. Differential Equations, 27 (1978), 172-189.

F.V. Atkinson and J.R. Haddock,

1. On determining phase spaces for functional differential equations, Funkcial. Ekvac., 31 (1988), 331-347.

V. Barbu and S.I. Grossman,

1. Asymptotic behavior of linear integrodifferential systems, Trans. Amer. Math. Soc., 173 (1972), 277-288.

P.L. Butzer and H. Berens,

1. Semi-groups of Operators and Approximations, Springer, Berlin, 1967.

P. Bondi, V. Moauro and F. Visentin,

1. Limiting equations in the stability problem, Nonlinear Analysis, T.M.A., 1 (1977), 123-128.

F.E. Browder,

1. On the spectral theory of elliptic differential operators, I. Math. Ann., 142 (1961), 22-130.

T.A. Burton,

1. Periodic solutions of linear Volterra equations, Funkcial. Ekvac., 27 (1984), 229-253.
2. Volterra Integral and Differential Equations, Academic Press, New York, 1983.
3. Stability and Periodic Solutions of Ordinary and Functional Differential Equations, Academic Press, New York, 1985.
4. Periodic solutions of nonlinear Volterra equations, Funkcial.

Ekvac., 27 (1985), 301-317.

R.H. Cameron and W.T. Martin,

1. An unsymmetric Fubini theorem, Bull. Amer. Math. Soc., 47 (1941), 121-125.

S.N. Chow and J.K. Hale,

1. Strongly limit-compact maps, Funkcial. Ekvac., 17 (1974), 31-38.

P.H. Clément, O. Diekmann, M. Gyllenberg, H.J.A.M. Heijmans and H.R. Thieme,

1. Perturbation theory for dual semigroups, I. The sun-reflexive case. Report AM-R8605, Center for Mathematics and Computer Sciences, (1986).

B.D. Coleman and V.J. Mizel,

1. Norms and semi-groups in the theory of fading memory, Arch. Rational Mech. Anal., 23 (1966), 87-123.
2. On the general theory of fading memory, Arch. Rational Mech. Anal., 30 (1968), 18-31.
3. On the stability of solutions of functional differential equations, Arch. Rational Mech. Anal., 30 (1968), 173-196.

B.D. Coleman and D.R. Owen,

1. On the initial value problem for a class of functional differential equations, Arch. Rational Mech. Anal., 55 (1974), 275-299.

W.A. Coppel,

1. Almost periodic properties of ordinary differential equations, Ann. Mat. Pura. Appl., 76 (1967), 27-49.

C. Corduneanu,

1. Integral Equations and Stability of Feedback Systems, Academic Press, New York and London, 1973.
2. Integral Equations and Applications, a preprint.

C. Corduneanu and V. Lakshmikantham,

1. Equations with unbounded delay: survey, Nonlinear Anal., 4 (1980) 831-877.

G. Darbo,

1. Punti uniti in transformazioni a condiminio non compatto, Rend. Sem. Mat. Univ. Padova, 24 (1955), 84-92.

O. Diekmann,

1. Perturbed dual semigroups and delay equations, Report AM-R8604, Center for Mathematics and Computer Sciences, 1986.

R.D. Driver,

1. Existence and stability of solutions of a delay differential system, Arch. Rational Mech. Anal., 10 (1962), 401-426.

N. Dunford and J.T. Schwartz,

1. <u>Linear Operators, Part I,</u> Wiley-Interscience, New York, 1966.

D.E. Edmunds and W.D. Evans,

1. <u>Spectral Theory and Differential Operators,</u> Oxford Science Publications, Clarendon Press, Oxford 1987.

R.E. Edwards,

1. <u>Functional Analysis, Theory and Applications,</u> Holt, Rinehart and Winston, 1965.

J. Favard,

1. <u>Leçons sur les Fonctions Presque périodiques,</u> Gauthier-Villars, Paris, 1933.

T. Furumochi,

1. On the converse theorem for integral stability in functional differential equations, Tohoku Math. J., 27 (1975), 461-477.
2. Stability and boundedness in functional differential equations, J. Math. Anal. Appl., 113 (1986), 473-489.

G. Gripenberg, S-O. Londen and O. Staffans,

1. <u>Volterra Integral and Functional Equations,</u> Encyclopedia of Math. and its Appl., 34, Cambridge Univ. Press, Cambridge, New York, Melbourne, 1990.

K. Gopalsamy,

1. Global asymptotic stability in a periodic integrodifferential system, Tohoku Math. J., 37 (1985), 323-332.

J.R. Haddock,

1. Friendly spaces for functional differential equations with infinite delay, in "Trends in the Theory and Practice of Non-linear Analysis," Elsevier Science Publishers B.V., North-Holland, (1985), 173-182.

J.R. Haddock and W.E. Hornor,

1. Precompactness and convergence in norm of positive orbits in a certain fading memory space, Funkcial. Ekvac., 31 (1988), 349-361.

J.R. Haddock, T. Krisztin and J. Terjéki,

1. Invariance principles for autonomous functional differential equations, J. Integral Equations 10 (1985), 123-136.
2. Comparison theorems and convergence properties for functional differential equations with infinite delay, Acta. Sci. Math., 52 (1988), 399-414.

J.R. Haddock and J. Terjéki,

1. On the location of positive limit sets for autonomous functional differential equations with infinite delay, J. Differential Equations, 86(1990), 1-32.

T. Hagemann,

1. Adjungierte gleichungen und randwertprobleme bei linearen differentialgleichungen mit unendlicher verzögerung, Doktors Dissertation, Ruhr-Universität, Bochum, 1986.

T. Hagemann and T. Naito,

1. Functional differential equations with infinite delay on the space C_γ, Proc. EQUADIFF 82, Lecture Notes in Math. 1017, Springer-Verlag, Berlin, Heidelberg, New York, Tokyo, (1983), 207-214.

A. Halanay,

1. <u>Differential Equations; Stability, Oscillations, Timelags,</u> Academic Press, New York, 1966.

J.K. Hale,

1. <u>Ordinary Differential Equations,</u> Wiley, New York, 1969.
2. Dynamical systems and stability, J. Math. Anal. Appl., 26 (1969), 35-59.

3. Functional differential equations with infinite delays, J. Math. Anal. Appl., 48 (1974), 276-283.
4. The solution operator with infinite delays, International Conference on Differential Equations, Academic Press, New York, (1975), 330-336.
5. Theory of Functional Differential Equations, Applied Math. Sciences, vol.3, Springer-Verlag, New York, Heidelberg, Berlin, 1977.

J.K. Hale and J. Kato,
1. Phase space for retarded equations with infinite delay, Funkcial. Ekvac., 21 (1978), 11-41.

Y. Hamaya,
1. Periodic solutions of nonlinear integrodifferential equations, Tohoku Math. J., 41 (1989), 105-116.

P. Hartman,
1. Ordinary Differential Equations, Birkhäuser, Boston, Basel, Stuttgart, 1982.

T.H. Hildebrandt,
1. Introduction to the Theory of Integration, Academic Press, New York and London, 1963.

E. Hille and R.S. Phillips,
1. Functional Analysis and Semigroups, Amer. Math. Soc., Providence, R. I., 1957.

Y. Hino,
1. Asymptotic behavior of solutions of some functional differential equations. Tohoku Math. J., 22 (1970), 98-108.
2. Continuous dependence for functional differential equations, Tohoku Math. J., 23 (1971), 565-571.
3. On the stability of the solution of some functional differential equations, Funkcial. Ekvac., 14 (1971), 47-60.
4. Stability and existence of almost periodic solutions of some functional differential equations, Tohoku Math. J., 28 (1976), 389-409.
5. Favard's separation theorem in functional differential equations with infinite retardations, Tohoku Math. J., 30 (1978), 1-12.

6. Almost periodic solutions of functional differential equations with infinite retardation, Funkcial. Ekvac., 21 (1978), 139-150.
7. Almost periodic solutions of functional differential equations with infinite retardations, II, Tohoku Math. J., 32 (1980), 525-530.
8. Total stability and uniformly asymptotic stability for linear functional differential equations with infinite delay, Funkcial. Ekvac., 24 (1981), 345-349.
9. Stability properties for functional differential equations with infinite delay, Tohoku Math. J., 35 (1983), 597-605.

Y. Hino and S. Murakami,
1. Favard's property for linear retarded equations with infinite delay, Funkcial. Ekvac., 29 (1986), 11-17.
2. Periodic solutions of a linear Volterra system, in "Differential Equations : Proc. of the EQUADIFF Conference," Lecture Notes in Pure and Appl. Math., vol. 118, Marcel Dekker, New York and Basel, (1989), 319-326.

Y. Hino and T. Yoshizawa,
1. Total stability property in limiting equations for a functional differential equation with infinite delay, Casopis pro Pestrovani Matematiky, 111 (1986), 62-69.

T. Kaminogo,
1. Kneser's property and boundary value problems for some retarded functional differential equations, Tohoku Math. J., 30 (1978), 471-486.
2. Continuous dependence of solutions for integrodifferential equations with infinite delay, J. Math. Anal. Appl., 129 (1988), 307-314.

F. Kappel and W. Schappacher,
1. Some considerations to the fundamental theory of infinite delay equations, J. Differential Equations, 37 (1980), 141-183.

J. Kato,
1. Uniformly asymptotic stability and total stability, Tohoku Math. J., 22 (1970), 254-269.
2. Razumikhin type theorem for differential equations with infinite delay, Dynamical Systems, Academic Press, (1977),

441-443.

3. Stability problem in functional differential equations with infinite delay, Funkcial. Ekvac., 21 (1978), 63-80.
4. Asymptotic behavior in functional differential equations with infinite delay, Proc. EQUADIFF 82, Lecture Notes in Math., vol. 1017. Springer-Verlag, Berlin, Heidelberg, New York, Tokyo, (1983), 300-312.
5. Phase space for the functional differential equations, in "Qualitative Theory of Differential Equations," Colloq. Math. Soc. Janós Bolyai, 53, North-Holland, Amsterdam, Oxford, New York, (1990), 307-325.

J. Kato and T. Yoshizawa,

1. Stability under the perturbation by a class of functions, in "Dynamical Systems, Internat. Symp.," Vol. 2, (1976), 217-2.

C. Kuratowski,

1. Sur les espaces completes, Fund. Math., 15 (1930), 301-309.

B.W. Levinger,

1. A folk theorem in functional differential equations, J. Differential Equations 4 (1968), 612-619.

P.F. Lima,

1. Hopf Bifurcation in Equations with Infinite Delay, Ph. D. Thesis, Brown University, Providence, R. I., 1977.

J.L. Massera and J.J. Schäffer,

1. Linear Differential Equations and Function Spaces, Academic Press, New York, 1966.

R.K. Miller,

1. Asymptotic stability properties of linear Volterra integrodifferential equations, J. Differential Equations, 10 (1971), 485-506.
2. Linear Volterra integrodifferential equations as semigroups, Funkcial. Ekvac., 17 (1974), 39-55.

S. Murakami,

1. Perturbation theorems for functional differential equations with infinite delay via limiting equations, J. Differential Equations, 59 (1985), 314-335.

2. Asymptotic behavior of solutions of some differential equations, J. Math. Anal. Appl., 109 (1985), 534-545.
3. Stability in functional differential equations with infinite delay, Tohoku Math. J., 37 (1985), 561-570.
4. Linear periodic functional differential equations with infinite delay, Funkcial. Ekvac., 29 (1986), 335-361.
5. Almost periodic solutions of a system of integrodifferential equations, Tohoku Math. J., 39 (1987), 71-79.

S. Murakami and T. Naito,
1. Fading memory spaces and stability properties for functional differential equations with infinite delay, Funkcial. Ekvac., 32, (1989), 91-105.
2. Some properties of phase spaces for functional differential equations with infinite delay, in "Differential Equations : Proc. of the EQUADIFF Conference," Lecture Notes in Pure and Appl. Math., vol. 118, Marcel Dekker, New York and Basel, (1989), 507-514.

T. Naito,
1. Integral manifolds for linear functional differential equations on some Banach space, Funkcial. Ekvac., 13 (1970), 199-213.
2. Adjoint equations of autonomous linear functional differential equations with infinite retardations, Tohoku Math. J., 28 (1976), 135-143.
3. On autonomous linear functional differential equations with infinite retardations, J. Differential Equations, 21 (1976), 297-315.
4. On linear autonomous retarded equations with an abstract phase space for infinite delay, J. Differential Equations, 33 (1979), 74-91.
5. Fundamental matrices of linear autonomous retarded equations with infinite delay, Tohoku Math. J., 32 (1980), 539-556.
6. A modified form of the variation-of-constants formula for equations with infinite delay, Tohoku Math. J., 36 (1984), 33-40.
7. Adjoint semigroups associated with linear functional differential equations with infinite delays, to appear in the Proc. of US-Japan Seminar on Dynamical Systems, Kyoto, 1989.

8. Asymptotic stability of linear functional differential equations with the fading memory space, to appear in the Proc. of International Symposium on Functional Differential Equations and Related Topics, Kyoto, 1990.

F. Nakajima,

1. Separation conditions and stability properties in almost periodic systems, Tohoku Math. J., 26 (1974), 305-314.

R.D. Nussbaum,

1. The radius of essential spectrum, Duke Math. J., 37 (1970), 473-478.

W. Rudin,

1. Real and Complex Analysis, 3rd ed. McGraw-Hill, New York, 1987.
2. Functional Analysis, McGraw-Hill, New York, 1973.

K.P. Rybakowski,

1. Wazewski's principle for retarded functional differential equations, J. Differential Equations, 36 (1980), 117- 138.
2. A topological principle for retarded functional differential equations of Caratheodory type, J. Differential Equations, 39 (1981), 131-150.

B.N. Sadovskii,

1. Limit-compact and condensing operator, Russian Math. Surveys, 27 (1972), 85-155.

S. Saks,

1. Theory of The Integral, 2nd ed., "Monografje Matematyczne," Vol. 7, Warsaw, 1937.

K. Sawano,

1. Exponential asymptotic stability for functional differential equations with infinite retardations, Tohoku Math. J., 31 (1979), 363-382.
2. Positively invariant sets for functional differential equations with infinite delay, Tohoku Math. J., 32 (1980), 557-566.
3. Some considerations on the fundamental theorems for functional differential equations with infinite delay,

Funkcial. Ekvac., 25 (1982), 97-104.

K. Schumacher,

1. Existence and continuous dependence for differential equations with unbounded delay, Arch. Rational Mech. Anal., 67 (1978), 315-335.
2. Dynamical systems with memory on history-spaces with monotonic seminorms, J. Differential Equations, 34 (1979), 440-463.

G. Seifert,

1. Almost periodic solutions and asymptotic stability, J. Math. Anal. Appl. 21 (1968), 136-149.
2. Liapunov-Razumikhin conditions for stability and boundedness of functional differential equations of Volterra type, J. Differential Equations, 14 (1973), 424-430.
3. Positively invariant closed sets for systems of delay differential equations, J. Differential Equations, 22 (1976), 292-304.
4. Uniform stability for delay differential equations with infinite delays, Funkcial. Ekvac., 25 (1982), 347-356.
5. On Caratheodory conditions for functional differential equations with infinite delays, Rocky Mountain J. Math., 12 (1982), 615-619.

G.R. Sell,

1. Topological Dynamics and Ordinary Differential Equations, Van Nostrand Reinhold Company, London, 1971.

J.S. Shin,

1. Existence of periodic solutions in linear periodic functional differential equations with infinite delay (in Korean), Bull. Association of Korean Scientists in Japan (1982), 63-68.
2. Uniqueness and global convergence of successive approximations for solutions of functional integral equations with infinite delay, J. Math. Anal. Appl., 120 (1986), 71-88.
3. An Existence theorem of functional differential equations with infinite delay in a Banach space, Funkcial. Ekvac., 30 (1987), 19-29.
4. On the uniqueness of solutions for functional differential equations with infinite delay, Funkcial. Ekvac., 30 (1987),

225-236.

5. Global convergence of successive approximation for functional differential equations with infinite delay, Tohoku Math. J., 39 (1987), 557-574.
6. Existence of solutions and Kamke's theorem for functional differential equations in Banach spaces, J. Differential Equations, 81 (1989), 294-312.

O.L. Staffans

1. A neutral FDE with stable D-operator is retarded, J. Differential Equations, 49 (1983), 208-217.
2. On a neutral functional differential equation in a fading memory space, J. Differential Equations, 50 (1983), 183-217.
3. The null space and the range of a convolution operator in a fading memory space, Trans. Amer. Math. Soc., 281 (1984), 361-388.
4. Extended initial and forcing function semigroups generated by a functional equations, SIAM J. Math. Anal., 16 (1985), 1034-1048.
5. Semigroups generated by a neutral functional differential equations, SIAM J. Math. Anal., 17 (1986), 46-57.
6. Convolution operators in a fading memory space: the critical case, SIAM J. Math. Anal., 18 (1987), 366-386.
7. Subspaces of a stable and unstable solutions of a functional differential equations in a fading memory space: the critical case, SIAM J. Math. Anal., 18 (1987), 1323-1340.
8. On the stable and unstable subspaces of a critical functional differential equation, J. Integral Equations and Appl., 2 (1990), 225-236.

H.W. Stech,

1. <u>Contributions to the Theory of Functional Differential Equations with Infinite Delay,</u> Ph. D. Thesis, Michigan State University, East Lansing, Mich., 1976.
2. On the adjoint theory for autonomous linear functional differential equations with unbounded delays, J. Differential Equations, 27 (1978), 421-443.

A. Strauss and J.A. Yorke,

1. Perturbation theorems for ordinary differential equations, J. Differential Equations, 3 (1967), 15-30.

2. Perturbing uniform asymptotically stable nonlinear systems, J. Differential Equations, 6 (1969), 452-483.

A.E. Taylor,

1. Introduction to Functional Analysis, New York, John Wiley and Sons, Inc. 1958.

M.V. Volterra,

1. Sur la théorie mathématique des phénomènes héréditaires, J. Math. Pures Appl., 7 (1928), 249-298.

J. Wu, Z. Li and Z. Wang,

1. Remarks on "Periodic solutions of linear Volterra equations", Funkcial. Ekvac., 30 (1987), 105-109.

K. Yosida,

1. Functional Analysis, Springer, Berlin, 1965.

T. Yoshizawa,

1. Stability Theory by Liapunov's Second Method, Math. Soc. Japan, 1966.
2. Stability Theory and the Existence of Periodic Solutions and Almost Periodic Solutions, , Applied Math. Sciences, vol. 14, Springer-Verlag, New York, Heidelberg, Berlin, 1975.
3. Asymptotic behaviors of solutions of differential equations, in "Differential Equations: Qualitative Theory," Colloq. Math. Soc. János Bolyai, 47, North-Holland, Amsterdam, Oxford, New York, (1987), 1141-1164.

LIST OF AXIOMS

(A) If x is a function mapping $(-\infty, \sigma+a)$ into $\mathbb{E}$, $a > 0$, such that x is in $\mathcal{B}$ and x is continuous on $[\sigma, \sigma+a)$, then for every t in $[\sigma, \sigma+a)$ the following conditions hold;

(i) x_t is in $\mathcal{B}$,

(ii) $|x(t)|_{\mathbb{E}} \le H|x_t|_{\mathcal{B}}$,

(iii) $|x_t|_{\mathcal{B}} \le K(t-\sigma)\sup\{|x(s)|_{\mathbb{E}} : \sigma \le s \le t\} + M(t-\sigma)|x_\sigma|_{\mathcal{B}}$,

where H is a constant, $K, M : [0,\infty) \to [0,\infty)$, K is continuous, M is locally bounded, and they are independent of x.

(A1) For the function x in (A), x_t is a $\mathcal{B}$-valued continuous function for t in $[\sigma, \sigma+a)$.

(A) (ii)' $|\varphi(0)|_{\mathbb{E}} \le H|\varphi|_{\mathcal{B}}$ for $\varphi \in \mathcal{B}$.

(B) The space $\mathcal{B}$ is complete.

(C1) If $\{\varphi^n\}$ is a Cauchy sequence in $\mathcal{B}$ with respect to the seminorm, and if $\{\varphi^n(\theta)\}$ converges to a function $\varphi(\theta)$ compactly on $(-\infty, 0]$, then φ is in $\mathcal{B}$ and $|\varphi^n - \varphi|_{\mathcal{B}} \to 0$ as $n \to \infty$.

(C2) If a uniformly bounded sequence $\{\varphi^n(\theta)\}$ in C_{00} converges to a function $\varphi(\theta)$ compactly on $(-\infty, 0]$, then φ is in $\mathcal{B}$ and $|\varphi^n - \varphi|_{\mathcal{B}} \to 0$ as $n \to \infty$.

(D) $\mathcal{B}$ is a separable space.

INDEX